Bioelectrochemistry:
Principles and
Practice

Volume 1

Bioelectrochemistry:
General Introduction

Edited by
S. R. Caplan
I. R. Miller
G. Milazzo[†]

Coordinated by D. Walz

Birkhäuser Verlag
Basel · Boston · Berlin

Editors:

S. Roy Caplan, Ph.D.
Professor Emeritus
Department of Membrane Research
   and Biophysics
The Weizmann Institute of Science
Rehovot 76100
Israel

Israel R. Miller, Ph.D.
Professor Emeritus
Department of Membrane Research
   and Biophysics
The Weizmann Institute of Science
Rehovot 76100
Israel

Giulio Milazzo†
formerly Professor
Istituto Superiore di Sanità
Rome
Italy

Coordinator:
PD Dr. Dieter Walz
Biozentrum
University of Basel
Klingelbergstrasse 70
CH-4056 Basel
Switzerland

Library of Congress Cataloging-in-Publication Data

Bioelectrochemistry: general introduction / edited by S. R. Caplan.
   I. R. Miller, G. Milazzo.
      p. cm. – (Bioelectrochemistry ; v. 1)
   Includes bibliographical references and index.

1. Bioelectrochemistry. I. Caplan, S. Roy. II. Miller, I. R.
(Israel R.), 1922–   . III. Milazzo, Giulio. IV. Series:
Bioelectrochemistry (Basel, Switzerland) ; v. 1.
QP517. B53B544   1995-30062
574. 19' 283 – dc20

Die Deutsche Bibliothek - CIP - Einheitsaufnahme

Bioelectrochemistry : general introduction / ed. by S. R.
Caplan... – Basel ; Boston ; Berlin : Birkhäuser, 1995
      (Bioelectrochemistry ; Vol. 1)

NE: Caplan, S. R. [Hrsg.]; GT

ISBN-13: 978-3-0348-7320-8   e-ISBN-13: 978-3-0348-7318-5
DOI: 10.1007/978-3-0348-7318-5

© 1995 Birkhäuser Verlag,
P.O. Box 133
CH-4010 Basel, Switzerland
Softcover reprint of the hardcover 1st edition 1995

9 8 7 6 5 4 3 2 1

GIULIO MILAZZO (1912–1993)

Giulio Milazzo, the father of Bioelectrochemistry, died on January 6, 1993 in Rome. He often pointed out that the roots of the subject go back two hundred years to Galvani and Volta, and that he had only resurrected the science. But the Bioelectrochemistry he started was certainly different from the Natural Philosophy of the past, and in many ways quite different from parallel modern developments.

At a time when science is becoming more narrowly focused and scientists more specialized, Giulio Milazzo catalyzed the formation of an interdisciplinary grouping that was broad in scope and inclusive in its organization. He envisaged Bioelectrochemistry as a discipline including all aspects of the overlap of biology and electrochemistry. He believed that science is international and that one should use all scientific means possible to foster cooperation across national barriers. His ideas catalyzed the founding of the *Bioelectrochemical Society* which to this day attempts to follow the high standards set by him.

Giulio Milazzo was convinced that communication between scientists from various fields would be greatly facilitated if a comprehensive textbook written in a common language existed. He therefore initiated the preparation of what he called a *Treatise on Bioelectrochemistry*, but his untimely death prevented him from finishing this enormous task. The Bioelectrochemical Society, under whose auspices the work had been commenced, considers it both an honor and a duty to pursue the project to completion. But with the guiding spirit no longer with us, the project of a comprehensive textbook seemed too ambitious, and the Treatise was therefore converted to the present Series of Texts. May it nevertheless not only serve the purpose envisaged by Giulio Milazzo, but also be a living memory to a great scientist and a dear friend.

Martin Blank   Roy Caplan   Dieter Walz

Contents

Contributors

Marie-Claire Bellissent-Funel, Laboratoire Léon Brillouin,
C.E.A.-C.N.R.S., C.E. de Saclay, 91191 Gif-sur-Yvette, France
Patrick Calmettes, Laboratoire Léon Brillouin, C.E.A.-C.N.R.S.,
C.E. de Saclay, 91191 Gif-sur-Yvette, France
S. Roy Caplan, Department of Membrane Research and Biophysics,
The Weizmann Institute of Science, Rehovot 76100, Israel
Werner Kunz, Université de Technologie de Compiègne,
Département Génie Chimique, B.P. 649, 60206 Compiègne, France
Donald C. Mikulecky, Department of Physiology and Biophysics,
Medical College of Virginia, Virginia Commonwealth University,
Richmond, Virginia 23298, USA
Israel R. Miller, Department of Membrane Research and Biophysics,
The Weizmann Institute of Science, Rehovot 76100, Israel
Shinpei Ohki, Department of Biophysical Sciences, State University
of New York at Buffalo, Buffalo, New York 14214-3005, USA
Hiroyuki Ohshima, Faculty of Pharmaceutical Sciences, Science
University of Tokyo, Tokyo, Japan
David R. L. Scriven, Department of Physiology, University of
the Witwatersrand Medical School, Parktown, Johannesburg 2193,
South Africa
Dieter Walz, Biozentrum, University of Basel, CH-4056 Basel,
Switzerland

Introduction

Volume 1 of this Series is intended to give the reader a fundamental understanding of the key areas deemed essential to the study of bioelectrochemistry. A thorough grasp of the theory and methodology of these basic topics is vital to cope successfully with the complex phenomena that currently face investigators in most bioelectrochemical laboratories.

Chapter 1 outlines the nonequilibrium thermodynamics and kinetics of the processes involved, stressing the connection between the two approaches. Particular emphasis is placed on the enzymes catalyzing cytosolic reactions and membrane transport. The techniques discussed are sufficient for the study of systems in the steady state, but systems that are evolving towards the steady state, or show some other time-dependent behavior, require in addition the techniques of mathematical modelling. These are dealt with in some detail in Chapter 2, where network representation of the system is treated at length as the method of choice in carrying out appropriate simulations. In Chapter 3 attention is directed to the twin problems of water structure and ionic hydration. Since water is ubiquitous in bioelectrochemical systems, and is characterized by a high degree of complexity even when pure, it is important to be aware of the experimental and computational methods used to study both pure water and aqueous solutions, and of the models that have emerged. Chapter 4 examines double layer phenomena at biocolloidal surfaces and interfaces. The structure of the double layer is reviewed, and its effects on membrane properties, including electrical potentials, ion binding, and adsorption are subjected to careful theoretical analysis. In contrast, Chapter 5 treats adsorption and surface reaction from a more experimental point of view. The adsorption of macromolecules is a primary concern here, and a comprehensive account is given of the practical methods available for the determination of surface concentration and surface structure.

In a volume of this nature some degree of overlap between the chapters is inevitable. In some ways this is a good thing, as the reader is forced to consider different approaches to a given problem, reflecting the current state of the art. Nevertheless we have tried to keep such overlap to a minimum, while at the same time providing a broad introduction to a difficult but fascinating area of research.

S. Roy Caplan
Israel Miller

Bioelectrochemistry: General Introduction
ed. by S. R. Caplan, I. R. Miller and G. Milazzo†
© 1995 Birkhäuser Verlag Basel/Switzerland

CHAPTER 1
Nonequilibrium thermodynamics and kinetics

Dieter Walz[1] and S. Roy Caplan[2]

[1]*Biozentrum, University of Basel, Basel, Switzerland*
[2]*Department of Membrane Research and Biophysics, The Weizmann Institute of Science, Rehovot, Israel*

1. Introduction

This chapter is intended to furnish the reader with the fundamentals of the nonequilibrium thermodynamics and kinetics of bioelectrochemical processes. Particular emphasis is given to cytosolic and membrane-bound enzymes catalyzing reaction and transport phenomena. Equilibrium thermodynamics is treated as a special steady state within the framework of nonequilibrium thermodynamics. The connection between thermodynamics and kinetics is stressed and applied to several illustrative examples. In particular the energetics of enzymes is analyzed in these terms. Certain frequently-invoked concepts such as the notion of rate-limiting steps are critically examined. A more comprehensive discussion of many of the points dealt with here can be found in Ref. 1.

2. Thermodynamic systems

A thermodynamic system is composed of a number of conceptual elements, each of which is well defined in terms of exchange of matter, heat, and work with its surroundings. Both heat and work are, of course, forms of energy; what distinguishes them is that in principle work is fully convertible into the potential energy of a weight, while heat is not. The exchange occurs via the boundaries or walls of the elements, which may be classified according to the scheme given in Table 1. From this table it is seen that no wall exists which permits the exchange of matter but prohibits the exchange of heat. The classification given in Table 1 is also applied to the thermodynamic system itself. In this case the outermost wall of the system, i.e. that which separates it from the rest of the universe, determines the classification. The most appropriate system for thermodynamic consideration is an isolated system, since it is not subject to uncontrolled disturbances from the

Table 1. Classification of elements in a thermodynamic system

	Permitted* (+) or prohibited (−) exchange [through a wall] of		
Element	matter	heat	work[†]
open	+	+	+
closed	−	+	+
adiabatic	−	−	+
isolated[‡]	−	−	−

*Permitted means allowed but not necessarily required.
[†]Work includes the action of electric and/or magnetic fields.
[‡]Isolation here includes electrical isolation in the most general sense. An isolated element is also said to be surrounded by rigid adiabatic walls since the rigidity of the wall prevents the exchange of work.

outside world. Indeed this is exactly what one attempts to do in properly designing an experiment.

In bioelectrochemistry the elements will, for the most part, consist of phases, compartments, and reservoirs. Compartments usually contain a single phase, separated from other compartments by a boundary (e.g. another phase such as a membrane). A reservoir is an element whose capacitance for heat or matter or electric charge (or whatever else applies) is very much greater than that of any other element. Clearly such a reservoir is bounded by appropriate walls and acts as source or sink for heat or matter etc. In constructing a thermodynamic system, the elements of the experimental system are substituted by equivalent elements of the thermodynamic system. The actual topology of the experimental system need not be retained. In fact it is often advantageous to combine physically disconnected spaces in the experimental system, which are expected to have identical composition due to the same processes occurring in and around them, into one space in the thermodynamic system.

Suppose we have a suspension of membrane vesicles (e.g. isolated mitochondria, thylakoids or vesicles obtained from a reconstitution of a purified protein with lipids) in a suspending medium with known composition. In order to eliminate effects of a varying ambient temperature we place the vessel containing the suspension in a thermostat which should guarantee a constant temperature in the system. The vessel is shielded from stray electric or electromagnetic fields by using a Faraday cage. Moreover, an effective stirring of the suspension prevents the formation of gradients in the suspending medium. Since the reaction vessel is open the atmospheric pressure acts as a barostat, i.e. the suspension is under constant pressure (see Fig. 1A). In the case of photochemical processes we illuminate the sample by means of a light source.

When translating such an experimental system into a thermodynamic system, we first define a heat reservoir whose heat capacity is assumed to be so large that it can act as a heat sink or heat source, i.e. heat supplied to or absorbed from the rest of the system does not change the temperature (see Fig. 1B). Similarly, we define a "reservoir for mechanical work" which provides the energy for the stirring device. Next we recognize that the system contains two *phases*, a phase being defined by the chemical nature of its major constituent. One phase is constituted by the membranes of the vesicles, while the other is the aqueous phase in the suspending medium and in the internal space of the vesicles. If we assume that the vesicles are identical in composition, we can combine all membranes into one extended and e.g. plane membrane. In doing so we have implicitly assumed that the aqueous phases in the internal spaces of the vesicles can also be collected and placed into one *compartment*. This compartment is separated by the membrane from another

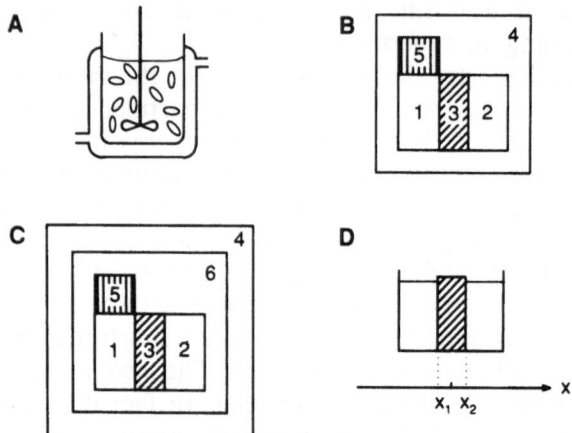

Fig. 1. Translation of an experimental system (A) into a thermodynamic system (B or C). In B and C two compartments with an aqueous phase (1, 2) are separated by a membrane (3) and surrounded by two reservoirs, one for heat (4) and one for mechanical work (5). In C an extra compartment (6) has been added which includes facilities for imposing an electric or magnetic field on the experimental system. The two-compartment system in D is the part of B or C in which the processes of interest take place. The x-axis is used as a space coordinate. For further explanation see text.

compartment which contains the aqueous phase constituted by the suspending medium (see Fig. 1B). Note that the aqueous phases in the two compartments in general have different physico-chemical properties. Hence these aqueous solutions are considered in physical chemistry as two *different phases*. In biology, however, they are considered as representatives of the same phase but being present in *different compartments*. The reader should be aware of this conceptual difference between phases and compartments.

In the case of photochemical processes the thermodynamic system would in addition include a reservoir for electrical energy (e.g. a battery) which supplies the light source. Heat produced by the conversion of electrical into radiant energy is absorbed by the heat reservoir. Similarly, heat produced by the absorption of photons not leading to a photochemical reaction is also taken up by the heat reservoir.

Circumstances may arise where one wishes to subject an experimental system to an external electric or magnetic field. This could be achieved by placing the vessel between the metal plates of a condenser or in the core of a solenoid. In such a case the thermodynamic system would contain an additional element, which can be thought of as an extra reservoir for the form of energy appropriate to maintain the electric or magnetic field (see Fig. 1C).

The thermodynamic system thus constructed is totally isolated from its surroundings and comprises all elements of the experimental system. Neither energy nor matter need to be supplied from the outside world during the experiment. The walls between the reservoirs and the compartments are selectively permeable to heat and mechanical energy. Note that the heat formed by the stirring of the sample is absorbed by the heat reservoir. We assume a constant pressure throughout the system due to the barostatic effect of the atmospheric pressure. The aqueous phases are homogeneous, i.e. temperature, composition and electrical potential are the same throughout a given compartment. In the suspending medium this is achieved by stirring. The internal spaces of the vesicles are obviously not stirred. However, due to the small dimension of these spaces, diffusion of matter and heat are usually much faster than the processes which cause a change in composition and give rise to a production of heat. Hence we can treat the compartment representing the combined internal spaces as if it were stirred. It should be kept in mind, however, that in a thin layer adjacent to a charged surface homogeneity is not necessarily found. The thermodynamics and kinetics of such "interfacial domains", while in principle not different from the rest of the topics discussed in this chapter, constitute an important special case and hence are dealt with in detail in chapters 4 and 5.

When constructing the thermodynamic system, we have neglected the analytical devices needed to follow the processes in the experimental system. This is legitimate since monitoring the parameters of the system should not disturb the processes. Although no information can be obtained without some interaction of an analytical device with the component to be assessed, the effect of this interaction should be negligibly small. Thus, the light intensity of a spectroscopic device should not give rise to photochemical processes, or the current through a pair of electrodes should be so small that it does not change the composition.

3. Nonequilibrium thermodynamics

3.1. Thermodynamic functions and the first law

The state of a system is uniquely defined by its internal energy, U, which is the sum of all sorts of energies present such as mechanical energy, heat, or chemical energy. In particular, U includes the electrical energy arising from charges and electrical potentials [2]. The internal energy $U(S, V, l, n_i)$ is a unique function of the state parameters volume V, elongation l (a length with respect to an appropriate reference point which is taken as a measure for performing mechanical work), chemical composition as indicated by the mole number n_i of the ith species, and a quantity S called entropy (see, for example ref. 3). It is important to note that U and S are thermodynamic potentials, i.e. their value is uniquely defined for a given state of the system independent of the path on which this state was reached. The parameters V, l, and n_i which define the state are implicitly independent of such paths. In a thermodynamic system like that depicted in Fig. 1B, the internal energy U_k can be defined separately for the kth element of the system (the elements being the reservoirs, the compartments and the membrane). The internal energy of the whole system is then

$$U = \sum_k U_k(S_k, V_k, l_k, n_{i,k}) \tag{1}$$

where the sum has to be taken over all elements.

The processes occurring in a system cause a transition of the system from one state to another. In the following discussion we dissect this transition into small increments which allows us to apply differentials for the variations of the parameters of the system. The change in internal energy of a given element of the system due to such an incremental transition amounts to

$$dU_k = T \, dS_k - p \, dV_k + X_k \, dl_k + \sum_i \tilde{\mu}_{i,k} \, dn_{i,k} \tag{2}$$

where p and T denote, respectively, the pressure and the absolute temperature in the system which are both constant as discussed in the preceding section. When writing Eqn. 2 the *electrochemical potential*, $\tilde{\mu}_{i,k}$, of the ith species in the kth element was introduced which is defined as the partial derivative of U_k with respect to $n_{i,k}$ at constant S, V, l, n_j in all elements (see, for example, refs. 2 or 3),

$$\tilde{\mu}_{i,k} = [\partial U_k / \partial n_{i,k}]_{S,V,l,n_{j,k}} \tag{3}$$

and the mechanical force X_k (which arises from a potential for mechanical work) similarly defined as $\partial U_k / \partial l_k$. Again, $\tilde{\mu}_{i,k}$ and X_k are thermodynamic potentials like U and S.

The first law of thermodynamics states that energy cannot be created or annihilated but only converted from one form into another. Since the total system is isolated (see preceding section) it does not exchange energy or matter with its surroundings, therefore, U = constant or dU = 0 for an incremental transition. It then follows from Eqns. 1 and 2 that

$$\sum_k \left[T \, dS_k - p \, dV_k + X_k \, dl_k + \sum_i \tilde{\mu}_{i,k} \, dn_{i,k} \right] = 0 \tag{4}$$

Since the total volume of the system is constant, $\sum_k p \, dV_k = 0$. No change in chemical composition occurs in the reservoirs shown in Fig. 1B, i.e. $dn_{i,4} = dn_{i,5} = 0$. Moreover, $dl_k \neq 0$ only for k = 5. Hence, for the system in Fig. 1B, Eqn. 4 reads

$$T \sum_k dS_k + \sum_{k=1}^{3} \sum_i \tilde{\mu}_{i,k} \, dn_{i,k} + X_5 \, dl_5 = 0 \qquad (4a)$$

A new thermodynamic potential, G, called *free enthalpy* (or Gibbs free energy) can be defined as

$$G = U - TS + pV \qquad (5)$$

This quantity measures the system's capability to perform useful work either from a mechanical potential or from the electrochemical potentials $\tilde{\mu}_i$. If temperature and pressure are constant, its change in the kth element of the system becomes (cf. Eqn. 2)

$$dG_k = dU_k - T \, dS_k + p \, dV_k = X_k \, dl_k + \sum_i \tilde{\mu}_{i,k} \, dn_{i,k} \qquad (6)$$

It is important to realize that dG_k in Eqn. 6 merely indicates how much the free enthalpy changes in a transition. The actual fate of dG_k is considered in the following section.

3.2. The uniqueness of entropy and the second law

Among the thermodynamic potentials entropy has a unique feature expressed by the second law of thermodynamics. This law states that the change in entropy of a system has to be positive (or zero in a special case, see below) when the system changes from one state to another. Rearranging Eqn. 4a and introducing Eqn. 6 yields

$$T \, dS_{tot} = T \sum_k dS_k = -X_5 \, dl_5 - \sum_{k=1}^{3} \sum_i \tilde{\mu}_{i,k} \, dn_{i,k} = -dG_{tot} \qquad (7)$$

where dS_{tot} and dG_{tot} are the change in entropy and free enthalpy, respectively, for the total system. Thus, the mandatory increase in entropy is covered by a decrease in free enthalpy.

It is possible to further analyze the entropy changes dS_k for each element of the system [4]. It is then recognized that $dS_k = \delta_{ex}S_k + \delta_{in}S_k$, where $\delta_{ex}S_k$ is called the *exchange* contribution and $\delta_{in}S_k$ the *internal* contribution. Note that $\delta_{ex}S_k$ and $\delta_{in}S_k$ are not total differentials, in contrast to their sum, dS_k. The exchange contribution is mandatory and is present even if the transition of the system is conceptually made to occur from one equilibrium state (or very close to it) to another, i.e. if all changes in the system associated with the transition are brought about by so-called *reversible processes*. In contrast, the internal contribution arises only if *irreversible processes* are involved in the transition. It is found that $\Sigma_k \, \delta_{ex}S_k = 0$ which means that $dS_{tot} = 0$ for a reversible transition[1]. Thus, no free enthalpy has to be spent for an entropy change (cf. Eqn. 7) in a reversible transition and all can in principle be used in energy conversion by a suitable device. However, *reversible transitions are hypothetical limiting cases virtually never realized in practice because they imply that the processes associated with them occur at an infinitely small rate.* Processes of interest proceed at finite rates and therefore always give rise to an expenditure of free enthalpy and an increase of entropy which amounts to $dS_{tot} = \Sigma_k \delta_{in} S_k$. As a consequence, the free enthalpy available for the energy conversion is always less in real processes than expected from the corresponding hypothetical reversible transitions.

3.3. The dissipation function: flows and forces

The change in entropy dS_{tot} in Eqn. 7 includes a contribution arising from the irreversible process of stirring of the sample which is driven by the term $X_5 \, dl_5$. This contribution is of no relevance to the processes we are interested in because a partial conversion of mechanical energy invested in stirring into the energy involved in these processes can be excluded. Therefore, in what follows, we consider only the entropy change dS due to the processes of interest, which is equivalent to singling out the two-compartment system shown in Fig. 1D from the total thermodynamic system.

3.3.1. The dissipation function: When taking the time derivative of dS, we obtain from Eqn. 7

$$\Phi = T \, dS/dt = - \sum_{k=1}^{3} \sum_i \tilde{\mu}_{i,k} \, dn_{i,k}/dt \geq 0 \qquad (8)$$

The quantity Φ is called *dissipation function* and indicates the rate of loss (or dissipation) of free enthalpy due to irreversible processes other than stirring. The dissipation function is of central importance because it is the key relation which fully describes a system in terms of nonequilibrium thermodynamics.

In order to determine the dissipation function in Eqn. 8 at any state of the system in the course of an experiment, it suffices to estimate the electrochemical potentials $\tilde{\mu}_{i,k}$ and the pertinent time derivatives $dn_{i,k}/dt$

---

[1]It should be clearly understood that thermodynamic reversibility is not to be confused with "kinetic reversibility" in biochemical usage. A reversible reaction is said to be one whose backward rate is commensurable with its forward rate while, a so-called irreversible reaction is one with a very high equilibrium constant, i.e. a very low backward rate. In this case "irreversibility" is the hypothetical limiting case, frequently resorted to as a means of simplifying the kinetics (see also the "rate-limiting" step in section 5.3.2). However, real chemical reactions are always reversible in the kinetic sense.

for the constituents of the system. No knowledge of the actual processes which take place in the system and cause $dn_{i,k}/dt \neq 0$ is required. This allows us to define these processes in any convenient form with the only limitation that a given set of definitions has to comply with the dissipation function. In what follows, we assume that the processes in the membrane phase (index $k = 3$ in Eqn. 8) have reached a steady state (see section 3.4) so that $dn_{i,3}/dt = 0$. We then are only concerned with the processes occurring in the two compartments of Fig. 1D.

3.3.2. Flows and forces for chemical reactions and transport processes: We first consider a chemical reaction which occurs in compartment k. It is given the index j, k (in order to distinguish it from other reactions) and is represented by

$$\sum_s \nu_{Ss(j,k)} S_{s(j,k)} \rightleftarrows \sum_p \nu_{Pp(j,k)} P_{p(j,k)} \tag{9}$$

The reaction converts the initial reactants[2] $S_{s(j,k)}$ with stoichiometric coefficients $\nu_{Ss(j,k)}$ into the final reactants[2] $P_{p(j,k)}$ with stoichiometric coefficients $\nu_{Pp(j,k)}$ and vice versa. Mass balance imposes a strict relation on the mole numbers of substrates and products of a reaction. It is most conveniently expressed by means of a quantity called degree of advancement, $\xi_{j,k}$, defined as [5]

$$\xi_{j,k} = [n_{Ss(j,k)}(0) - n_{Ss(j,k)}(t)]/\nu_{Ss(j,k)} = [n_{Pp(j,k)}(t) - n_{Pp(j,k)}(0)]/\nu_{Pp(j,k)} \tag{10}$$

where the arguments t and 0 indicate the mole numbers, respectively, at time t and at the beginning $(t = 0)$ of the experiment. Note that the expression for substrates in Eqn. 10 becomes identical with that for products if we adopt the convention that the *stoichiometric coefficients for initial reactants have a negative sign and those for final reactants a positive sign.* Moreover, as already implied with the notation initial and final reactants, we have introduced a *positive direction for the reaction* when going from initial to final reactants in the sense that $d\xi_{j,k} \geq 0$. Taking the time derivative of $\xi_{j,k}$ in Eqn. 10 yields a measure for the rate of the chemical reaction which we denote by the *flow of the reaction*

$$J_{j,k} = d\xi_{j,k}/dt = (dn_{Rr(j,k)}/dt)/\nu_{Rr(j,k)} \tag{11}$$

where reactant $R_{r(j,k)}$ now stands for both initial and final reactants which need no longer be distinguished if the above sign convention is used. When introducing the time derivatives for the mole numbers in Eqn. 11 into Eqn. 8, the corresponding $\tilde{\mu}_{Rr(j,k)}$ can be collected and used to define a quantity which we call the *affinity*, $\mathscr{A}_{j,k}$, of the *j*th reaction in

---

[2]Initial and final reactants are also called substrates and products of the reaction, respectively.

compartment k [5]

$$\mathscr{A}_{j,k} = -\sum_r \nu_{Rr(j,k)} \tilde{\mu}_{Rr(j,k)} \tag{12}$$

The quantity $\mathscr{A}_{j,k}$ is an example of a thermodynamic force.

A redox reaction just consists of an electron transfer between initial reactants whereby the final reactants are formed. Such a reaction is more appropriately formulated in terms of redox couples [6]. A redox couple in the system is represented by

$$R_{ox} + \nu_{e(R)} \, e^- \rightleftarrows R_{rd} \tag{13}$$

where R_{ox} and R_{rd} denote the oxidized and the reduced species, respectively, while $\nu_{e(R)}$ is the number of electrons which the couple can exchange. Despite the fact that $\nu_{e(R)}$ is clearly a stoichiometric coefficient the symbol n is frequently used in the literature which is misleading since n usually denotes a mole number. Let us arbitrarily select one of the redox couples in the jth redox reaction occurring in compartment k as being the electron donating couple and denote its oxidized and reduced species by $D_{ox(j,k)}$ and $D_{rd(j,k)}$, respectively; the other redox couple is then the electron accepting couple with species $A_{ox(j,k)}$ and $A_{rd(j,k)}$. This notation also defines the positive direction of the redox reaction which reads

$$\nu_{e(A,j)} D_{rd(j,k)} + \nu_{e(D,j)} A_{ox(j,k)} \rightleftarrows \nu_{e(A,j)} D_{ox(j,k)} + \nu_{e(D,j)} A_{rd(j,k)} \tag{14}$$

where $\nu_{e(D,j)}$ and $\nu_{e(A,j)}$ are the $\nu_{e(R)}$ (see Eqn. 13) of the donating and accepting redox couple, respectively. The flow in Eqn. 11 when multiplied by $\nu_{e(D,j)} \nu_{e(A,j)}$ is then equivalent to a flow of electrons, $J_{ej,k}$, associated with the redox reaction, while the force in Eqn. 12, divided by $\nu_{e(D,j)} \nu_{e(A,j)}$, is equivalent to the affinity of the electron [6]

$$\mathscr{A}_{ej,k} = (\tilde{\mu}_{Drd(j,k)} - \tilde{\mu}_{Dox(j,k)})/\nu_{e(D,j)} - (\tilde{\mu}_{Ard(j,k)} - \tilde{\mu}_{Aox(j,k)})/\nu_{e(A,j)} \tag{15}$$

When dealing with transport of species across the membrane, it is necessary to arbitrarily choose a *positive direction of transport which has to be the same for all transport processes (sign convention)*. Here we choose the direction from compartment 1 to compartment 2 (Fig. 1D) as positive. We then define the *flow for the transport* of the ith species as

$$J_i = -dn_{i,1}/dt = dn_{i,2}/dt \tag{16}$$

where the second part of Eqn. 16 arises from mass balance. Introducing the time derivatives into Eqn. 8 again shows that the $\tilde{\mu}_{i,k}$'s can be collected and used to define the *thermodynamic force for transport*

$$\Delta\tilde{\mu}_i = \tilde{\mu}_{i,1} - \tilde{\mu}_{i,2} \tag{17}$$

i.e. the difference in electrochemical potential between the compartments.

The thermodynamic equivalence of the forces is evident on examination of Eqns. 12, 15, and 17. It is seen that thermodynamic forces always consist of algebraic sums of electrochemical potentials. In this respect the affinities of chemical reactions and electrochemical potential differences between compartments do not differ in any fundamental aspect.

3.3.3. Flows and forces in the dissipation function:

3.3.3. Flows and forces in the dissipation function: Chemical reactions and transport processes are connected by the effect of their respective flows on the mole number of the species involved. Thus, for the ith species which is transported and takes part as reactant $R_{r(j,1)}$ or $R_{r(j',2)}$ in the jth chemical reaction in compartment 1 or the j'th reaction in compartment 2, respectively,

$$dn_{i,1}/dt = -J_i + \sum_j \nu_{Rr(j,1)} J_{j,1} \qquad (18a)$$

and

$$dn_{i,2}/dt = J_i + \sum_{j'} \nu_{Rr(j',2)} J_{j',2} \qquad (18b)$$

The changes in mole number given by Eqns. 18 have to be identical with those in the dissipation function, and the chosen set of definitions of transport processes and reactions is then appropriate. In terms of flows and forces, the dissipation function (see Eqn. 8) reads

$$\Phi = \sum_i J_i \, \Delta\tilde{\mu}_i + \sum_{k=1}^{2} \sum_j J_{j,k} \mathscr{A}_{j,k} = \sum_p J_p X_p \geq 0 \qquad (19)$$

where J_p and X_p denote, respectively, the flow and the force of the pth process in general notation. The sums in Eqn. 19 have to be taken over all transported species and all chemical reactions occurring in both compartments. We recognize that the dissipation function is a sum of products of flows J_p and conjugated thermodynamic forces X_p. The flow of a process has the same sign as its force when it runs "downhill", i.e. when it occurs spontaneously, and the product of flow and conjugate force is positive. If flow and conjugate force have opposite signs the flow-force product is negative and the process is driven "uphill" by another process (or processes). This is the thermodynamic expression of the energy conversion due to some form of coupling between processes. It is possible because only the sum of all flow-force products has to be positive according to Eqn. 19 and not each product by itself.

We have arrived at the above conclusion in most general terms without considering any mechanisms by which transport or chemical reactions are brought about. This is the great advantage of thermodynamics but, at the same time, its severe limitation. Thermodynamics cannot tell us how flows are related to their conjugate forces and, in the case of energy conversion, to the forces of other processes. Answers to this question can in general only be obtained on the basis of molecular or kinetic schemes (see section 4).

In practice it is frequently found that a set of flows and forces arrived at by considering the most elementary processes in the system is experimentally inconvenient. In such cases it often proves possible to transform the dissipation function so that it becomes a function of forces and flows which can be readily fixed or measured. For example, consider a two-compartment system containing an aqueous solution of uni-univalent salt. In the most general case water and the ionic species are transported independently across the membrane. Hence the dissipation function consists of a sum of the flows pertinent to each species, each multiplied by its conjugate force, i.e. the appropriate electrochemical potential difference. From the experimental point of view, the measurement of the electrochemical potential differences and the ion flows by means of reversible electrodes (see section 3.5.2) poses no problem. However, for water the measurement of flow and its conjugate force is impossible by direct means. This difficulty is overcome by transforming the flows in the dissipation function into volume flow and salt flow. This is achieved by making use of relations pertinent to osmotic and electric phenomena, and leads to the well-known Kedem-Katchalsky equations [7].

3.4. Steady states

Steady states are states in which the system has become stationary in the sense that its parameters are no longer time-dependent. Under certain circumstances the time-dependence does not vanish but it becomes negligibly small, in which case we refer to pseudo-steady states. In general the parameters of a system are constrained by "clamping". For example, epithelial tissues are frequently studied in a two-compartment system where both compartments are accessible to manipulation (see Fig. 1D). The electrical potential difference across the membrane is controlled by a voltage clamp, and the concentrations of ionic species in each compartment may be kept constant by suitable devices.

We turn now to a consideration of certain important and characteristic properties of stationary states. If all forces are fixed, i.e. the maximum number of constraints is applied, the steady state is fully defined, since no more degrees of freedom are left. If some of the forces are fixed, the remainder will reach values in the steady state such that their conjugate flows become zero. If no constraints at all are applied, the forces will all tend to decrease until the system eventually reaches equilibrium.

If for technical reasons the forces cannot be clamped, it may nevertheless be possible to keep them approximately constant, i.e. to allow the system to reach a pseudo-steady state, by choosing appropriate chemical capacities. The chemical capacity of a particular compartment for a given species relates the change in mole number to the change in concentration of that species in the particular compartment. From this definition it can be shown that the chemical capacity is just the volume of the compartment which is formally increased when a buffering system for the species is added [1] (see also section 5.3.2 in chapter 2).

A vanishingly small change in the parameters of a compartment can also come about if the sum of the flows given on the right-hand side of Eqn. 18 is close to zero. This happens if the given compartment has considerably smaller capacities for the species present than the other compartment(s) in the system. A particular element which always satisfies this condition is a membrane which reaches a steady state (or, strictly speaking, a pseudo-steady state) long before the other elements of the system. Then the flows of charged species with charge number z_i are restricted by

$$\sum_i z_i J_i = 0 \qquad (20)$$

which arises from the condition of electroneutrality in the compartments. Eqn. 20 can serve to estimate the membrane potential which is the difference of the electrical potentials in the two compartments separated by the membrane (see section 4.3.1).

3.5. The equilibrium state

This state is a special steady state characterized by $\Phi = 0$ with the corollary that all processes have ceased. It also implies that all flows vanish because *all forces of unconstrained processes are zero*. The latter condition allows us to introduce relations pertinent to the equilibrium state. Before doing so, we expand the electrochemical potential of the ith species in the kth compartment or phase into three terms,

$$\tilde{\mu}_{i,k} = \mu^0_{i,k} + RT \ln\{a_{i,k}/a^0\} + z_i \mathscr{F} \phi_k \qquad (21)$$

Here R and $\mathscr{F}$ denote the gas constant and the Faraday constant, respectively. The standard chemical potential, $\mu^0_{i,k}$, is a substance-specific constant which also depends on the type of phase (hence index k is retained) and, in general, on pressure and temperature[3]. The next term in Eqn. 21 comprises the activity, $a_{i,k}$; it is normalized by the standard activity a^0 which, by definition, amounts to 1 M. The activity of the species is related to its concentration by

$$a_{i,k} = y_{i,k} c_{i,k} \qquad (22)$$

---

[3]Standard conditions are $p = 10^5$ Pa (1 bar) and T = 298.15 K (25°C) [2].

where the factor $y_{i,k}$, called the activity coefficient, expresses the interaction of the ith species with all other constituents of the phase. The sum of the standard chemical potential and the activity dependent term is usually abbreviated by

$$\mu_{i,k} = \mu_{i,k}^0 + RT \ln\{a_{i,k}/a^0\} \tag{23}$$

and called the chemical potential of the species. The last term in Eqn. 21 is frequently referred to as the electrical part of $\tilde{\mu}$. It comprises the charge number z_i of the species and ϕ_k which is the electrical potential of the phase with respect to an arbitrarily chosen reference point [2]. It has sometimes been questioned if the splitting of $\tilde{\mu}$ into a chemical and an electrical part is meaningful or legitimate. However, when following consistently the notion presented in the context of reference electrodes (see section 3.5.2) Eqn. 21 does not pose any problems.

An electrochemical potential, $\tilde{\mu}_e$, can be formally assigned to the electron although this species does not exist as a free entity [6]. The assignment is based on the presence of redox couples (see Eqn. 13), and the electrochemical potential of the electron associated with the redox couple R_{ox}/R_{rd} in the kth compartment is defined as

$$\tilde{\mu}_{e(R,k)} = \mu_{e(R,k)} - \mathscr{F}\phi_k = \mu_{e(R,k)}^0 + [RT/\nu_{e(R)}] \ln\{\rho_{R,k}\} - \mathscr{F}\phi_k \tag{24a}$$

where $\mu_{e(R,k)}$ is the chemical potential of the electron, and

$$\rho_{R,k} = a_{Rrd,k}/a_{Rox,k} \tag{24b}$$

is called the reduction state of the couple. The standard chemical potential of the electron is

$$\mu_{e(R,k)}^0 = (\mu_{Rrd,k}^0 - \mu_{Rox,k}^o)/\nu_{e(R)} = -\mathscr{F} E_{R,k}^0 \tag{24c}$$

where the quantity $E_{R,k}^0$ is known as the standard redox potential of the couple. Each couple in the system thus determines its own $\tilde{\mu}_e$ which, however, causes no difficulties if consistently dealt with. This and other aspects, including the case of ligand binding (particularly of H^+ ions) to the redox species, are discussed in Ref. 6.

3.5.1. Equilibrium constants: The vanishing force for transport at equilibrium means $\Delta\tilde{\mu}_i = 0$ or $\tilde{\mu}_{i,1} = \tilde{\mu}_{i,2}$ (see Eqn. 17). We can generalize this condition and state that *at equilibrium $\tilde{\mu}_i$ has the same value in all phases and compartments into which the ith species can move.* The qualification of movement in this statement is important, although at equilibrium nothing moves. As discussed by Walz and Caplan [8] an element in the system which sets a restriction to the movement of the species (e.g. an impermeable membrane) may cause different electrochemical potentials at equilibrium in the phases or compartments separated by the restrictive element. It is therefore important to specify which possible constraints exist in a given equilibrium state.

The condition of equal electrochemical potential applied to the transport of the ith species between two phases (or compartments) with index k and k′ yields, by virtue of Eqn. 21,

$$K_{pi,kk'} = \exp\{(\mu_{i,k}^0 - \mu_{i,k'}^0)/RT\}$$

$$= [a_{i,k'}/a_{i,k}]_{eq} \exp\{z_i \mathscr{F}(\phi_{k'} - \phi_k)/(RT)\} \qquad (25)$$

The quantity $K_{pi,kk'}$ is the equilibrium constant for partitioning of a species called *the partition coefficient*. It relates the activities of the species in the two phases at equilibrium. Note that these activities are affected by a possible difference in electrical potential between the phases. Obviously, $K_{pi,kk'} = 1$ for partitioning of a species between two equal phases (e.g. between the aqueous phases in two compartments).

The condition $X_j = 0$ for the jth chemical reaction at equilibrium is transformed by means of Eqn. 21 into

$$K_{rj,k} = \exp\{-\Delta G_{j,k}^0/(RT)\} = \prod_r [a_{Rr(j,k)}/a^0]_{eq}^{\nu_{Rr(j,k)}} \qquad (26a)$$

where the standard free enthalpy of the reaction (usually called the standard Gibbs free energy), $\Delta G_{j,k}^0$, is defined as

$$\Delta G_{j,k}^0 = \sum_r \nu_{Rr(j,k)} \mu_{Rr(j,k)}^0 = \Delta H_{j,k}^0 - T \Delta S_{j,k}^0 \qquad (26b)$$

It can be related to the standard enthalpy and the standard entropy of the reaction, $\Delta H_{j,k}^0$ and $\Delta S_{j,k}^0$, respectively. $K_{rj,k}$ is the *equilibrium constant* for the reaction which determines the ratio of the reactant's activities at equilibrium, as indicated by the mass action ratio[4] on the right hand side of Eqn. 26a. Similarly for a redox reaction, by virtue of Eqns. 24,

$$K_{ej,k} = \exp\{\mathscr{F}(E_{Aj,k}^0 - E_{Dj,k}^0)/(RT)\} = [\rho_{Aj,k}^{1/\nu_{e(Aj,k)}}/\rho_{Dj,k}^{1/\nu_{e(Dj,k)}}]_{eq} \qquad (26c)$$

It has to be stressed that equilibrium constants which emerge from standard chemical potentials are always valid irrespective of the state of the system. The term equilibrium in their name merely refers to the fact that they relate the activities of reactants at equilibrium.

3.5.2. Reversible and reference electrodes: For certain ionic species there exist what are termed *reversible electrodes* whose characteristic feature can be expressed by the relation

$$\tilde{\mu}_{i,k} = z_i \mathscr{F}(\phi_{el}' + \phi_{el,k}) \qquad (27)$$

---

[4]Note that the equilibrium constants are dimensionless. Nevertheless, equilibrium constants with a dimension of concentration units can be found in the literature, due to the fact that the mass action ratio is usually written without normalization by a^0.

Here, $\tilde{\mu}_{i,k}$ is the electrochemical potential of the species to which the electrode is sensitive and which is in the kth compartment contacted by the electrode. The quantity $\phi_{el,k}$ is the electrical potential in the metal terminal of the selective electrode, while ϕ'_{el} denotes an electrode specific parameter depending on temperature and pressure only. Well-known examples of such electrodes are the silver/silver chloride electrode (a silver wire covered with solid silver chloride) which responds to chloride ions, or the glass electrode which responds to protons. Similarly, an inert metal wire in a phase containing a redox couple constitutes a reversible electrode for electrons. In every case, Eqn. 27 is obtained by applying the equilibrium condition to all processes involved. Thus, for the silver/silver chloride electrode,

$$\tilde{\mu}_{Cl} + \tilde{\mu}_{Ag} = \mu^0_{AgCl} \tag{28a}$$

and

$$\tilde{\mu}_{Ag} + \tilde{\mu}_{e,el} = \mu^0_{Ag,el} \tag{28b}$$

Note that the electrochemical potential of substances in the solid state comprises, by definition, only the standard chemical potential. The quantity $\tilde{\mu}_{e,el}$ is the electrochemical potential of the electron in the metal terminal which amounts to

$$\tilde{\mu}_{e,el} = \mu^0_{e,el} = \mathscr{F}\phi_{el} \tag{29}$$

In Eqn. 29 the activity dependent term is included in the standard potential since the changes in electron activity due to different electrical potentials are negligibly small. Substitution of $\tilde{\mu}_{Ag}$ and $\tilde{\mu}_{e,el}$ in Eqn. 28b from Eqns. 28a and 29, respectively, yields Eqn. 27 with $\mathscr{F}\phi'_{el} = \mu^0_{Ag,el} - \mu^0_{AgCl} - \mu^0_{e,el}$. In the case of an inert metal electrode, which is reversible for "electrons in solution", $\phi'_{el} = 0$. In a two-compartment system like that shown in Fig. 2 the difference in electrical potential, $\Delta\phi_{el,12}$, between the metal terminals of the reversible electrodes is

$$\Delta\phi_{el,12} = \phi_{el,1} - \phi_{el,2} = (\tilde{\mu}_{i,1} - \tilde{\mu}_{i,2})/(z_i\mathscr{F}) \tag{30}$$

and thus indicates the difference in electrochemical potential of the ith species between the compartments.

A *reference electrode* consists of a reversible electrode for chloride ions (the silver/silver chloride or the mercury/calomel electrode) in a solution of a chloride salt with a constant chemical potential (e.g. a saturated aqueous solution of KCl). Hence, the electrical potential in the metal terminal is clamped to a constant value. The reference electrode contacts a given phase via this salt solution ("salt bridge") and the assumption made is that, at least for an aqueous phase, the potential difference between salt bridge and phase is independent of the composition of the phase. Then, from Eqn. 27

$$\phi_{ref,k} = \phi_k + \phi'_{ref} \tag{31}$$

where ϕ_k and $\phi_{ref,k}$ denote the electrical potential in the kth phase and in the metal terminal of the reference electrode contacting this phase, respectively. The quantity ϕ'_{ref} is an electrode-specific parameter which comprises ϕ'_{el} of the chloride sensitive electrode as well as the chemical potential of chloride in the salt bridge and the unknown but constant electrical potential difference between salt bridge and phase (which in some cases may be zero). The electrical potential difference $\Delta\phi_{ref,12}$, between the metal terminals of the reference electrodes in the two-compartment system shown in Fig. 2 is equal to the difference in electrical potential between the (equal) phases in the compartments

$$\Delta\phi_{ref,12} = \phi_{ref,1} - \phi_{ref,2} = \phi_1 - \phi_2 = \Delta\phi_m \qquad (32)$$

where $\Delta\phi_m$ is usually called the "membrane potential". This is the rationale for measuring the membrane potential of a cell by means of a microelectrode since the latter represents a reference electrode whose salt bridge is a very fine needle.

The difference in electrical potential, $\Delta\phi_{er,k}$, between the metal terminals of a reversible electrode and a reference electrode which both

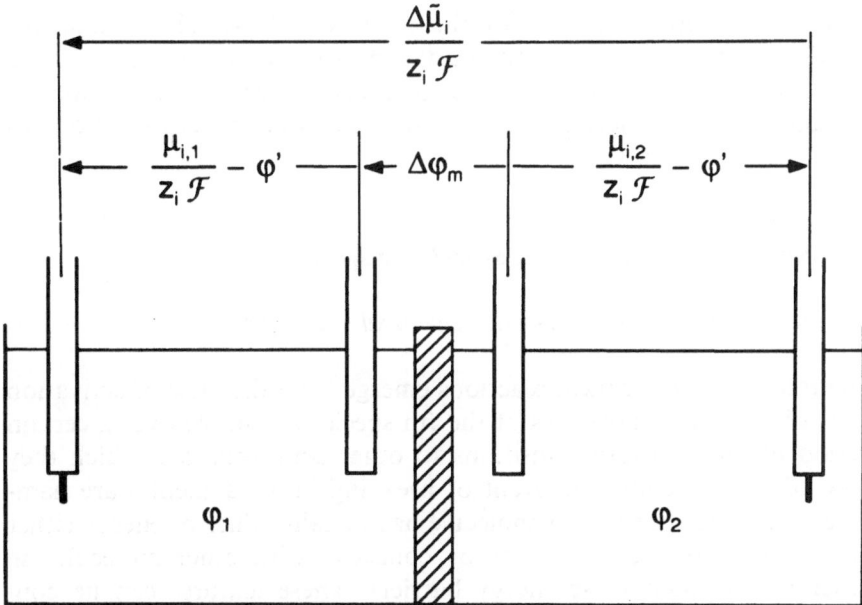

Fig. 2. Two compartment system with electrodes. Both compartments contain an aqueous phase, and the membrane separating them should be permeable to at least one charged species present in the compartments in order that the electric potentials ϕ_1 and ϕ_2 may be defined [3]. The two bars on the extreme left and right hand sides represent reversible electrodes which respond selectively to the ith species (e.g. platinum electrodes which are selective for electrons) while the other two bars represent reference electrodes. The thermodynamic quantities measured by the electrical potential difference between each pair of electrodes are indicated ($\phi' = \phi'_{el} + \phi'_{ref}$). For further explanation see text.

contact the kth phase becomes, in view of Eqns. 27 and 31,

$$\Delta\phi_{er,k} = \phi_{el,k} - \phi_{ref,k} = \mu_{i,k}/(z_i\mathscr{F}) - (\phi'_{el} + \phi'_{ref}) \qquad (33)$$

and thus indicates the *chemical potential* of the ionic species to which the reversible electrode is sensitive. ϕ'_{el} and ϕ'_{ref} are usually not known and therefore the measuring device has to be calibrated by means of a standard. Since the standard potential of a species is independent of composition, we only need a standard with known activity of the species to be assayed and obviously in the same phase where it is contained. Eqn. 33 then lends itself to the direct estimation of the activity of the species in any such phase as is routinely practiced with a H^+-sensitive glass electrode, pH meter and pH-standardized buffer solutions. If the reversible electrode is an inert metal electrode which exchanges electrons with the redox couple R_{ox}/R_{rd}, $\Delta\phi_{er,k}$ is called the *redox potential* of the couple,

$$E_{R,k} = \phi_{el,k} - \phi_{ref,k} + \phi'_{ref} = -\mu_{e(R,k)}/\mathscr{F} \qquad (34)$$

and indicates the chemical potential of the electron. Since ϕ'_{ref} is in general unknown we have to choose a zero point for the redox potential scale. By definition, $E_H = 0$ for the redox couple H_2/H^+ in aqueous solution with $a_H = 1$ M and the partial pressure of H_2 equal to 10^5 Pa (standard hydrogen electrode). This then determines ϕ'_{ref}. For example, a calomel reference electrode with a saturated aqueous KCl solution has $\phi'_{ref} = 0.248$ V at 20°C.

4. Kinetics of chemical reactions and transport

4.1. First-order and second-order chemical reactions

The rate laws for chemical reactions emerge from the view of activation and collision. The molecules of the ith species are stable over a certain period of time because transitions to other conformations which they may adopt (including the event of breaking into fragments) are hampered by energy barriers. A molecule has to gain sufficient energy either due to thermal fluctuations or by collisions with other molecules in order to surmount these energy barriers. These features can be condensed into a temperature dependent rate constant, k_i, which determines the change in mole number, n_i, with time

$$dn_i/dt = -Vk_i a_i(t) \quad \text{(first-order)} \qquad (35)$$

If two molecular species 1 and 2 are ligated in a reaction, they have to collide in the right orientation and with sufficient energy for the bond between them to be formed. Again, a rate constant $k_{1,2}$ can be found which governs the time course of this reaction

$$dn_1/dt = dn_2/dt = -Vk_{1,2}a_1(t)a_2(t) \quad \text{(second-order)} \qquad (36)$$

Equations 35 and 36 give the correct formulation of the rate laws for unidirectional processes. It is customary to assume that the activity coefficients are constant. In this case they can be combined with the rate constants to yield pseudo-constants which depend on conditions, and Eqns. 35 and 36 can be written in terms of concentrations. It should be added that rate constants such as k_i and $k_{1,2}$ may also depend on conditions because of interactions of the transition state complex with its surroundings, in full analogy with the interactions which give rise to the activity coefficients of the reacting species [9].

Equations 35 and 36 pertain only to *unidirectional processes*. These processes can *always* be reversed. Hence, the net change in mole number of a species caused by a given reaction is the sum of the changes arising from two unidirectional processes: one being the process in which the species itself takes part, usually called the "forward reaction", the other being the inverse of this process, usually called "backward reaction". This net change is then the flow of the given reaction (cf. Eqn. 11). Thus, the flow J_i pertinent to the ith reaction where, say, two initial reactants $S_{i,1}$ and $S_{i,2}$ are converted into the final reactant P_i,

$$S_{i,1} + S_{i,2} \rightleftarrows P_i \qquad (37)$$

becomes

$$J_i = V[k_i a_{Si,1} a_{Si,2} - k_{-i} a_{Pi}] \qquad (38)$$

Here k_i and k_{-i} denote, respectively, the (unidirectional) rate constants for the second-order forward and the first-order backward reaction. Note that, in this case, "forward" and "backward" refer to the direction introduced arbitrarily when assigning the attributes initial and final to the reacting species (cf. section 3.3.2).

The final reactant(s) of the ith reaction can in turn become the initial reactant(s) of the jth, kth, $\cdots$ reaction, or they may also be formed as final reactants of other reactions. The change in mole number of any species R involved in one or several such reactions is then given by (cf. Eqn. 18)

$$dn_R/dt = \sum_k v_k J_k \qquad (39)$$

where the sum has to be taken over all reactions in which R participates, and v_k is its stoichiometric coefficient in the kth reaction (the sign of v_k according to the convention introduced in section 3.3.2).

Suppose the final reactant P_i in the reaction of Eqn. 37 is the initial reactant $S_{j,1}$ of the subsequent reaction

$$S_{j,1} + S_{j,2} \rightleftarrows P_{j,1} + P_{j,2} \qquad (40)$$

then, according to Eqn. 39,

$$dn_{Pi}/dt = V[k_i a_{Si,1} a_{Si,2} - (k_{-1} + k_j a_{Sj,2})a_{Pi} + k_{-j} a_{Pj,1} a_{Pj,2}] \quad (41)$$

When a *steady state* for P_i is attained $dn_{Pi}/dt = 0$ and from Eqn. 41

$$a_{Pi} = (k_i a_{Si,1} a_{Si,2} + k_{-j} a_{Pj,1} a_{Pj,2})/(k_{-i} + k_j a_{Sj,2}) \quad (42)$$

Introducing Eqn. 42 into Eqn. 39 written for $R = P_{j,1}$ yields

$$dn_{Pj,1}/dt = V[k_i k_j a_{Si,1} a_{Si,2} a_{Sj,2} - k_{-i} k_{-j} a_{Pj,1} a_{Pj,2}]/[k_{-1} + k_j a_{Sj,2}] \quad (43)$$

The same procedure repeated for $R = P_{j,2}$, $S_{i,1}$, $S_{i,2}$ and $S_{j,2}$ always yields the right hand side of Eqn. 43 for $dn_{Pj,2}/dt$, $-dn_{Si,1}/dt$, $-dn_{Si,2}/dt$ and $-dn_{Sj,2}/dt$. If $k_{-i} \gg k_j a_{Sj,2}$ for all reasonable values of $a_{Sj,2}$, the denominator in Eqn. 43 can be approximated by k_{-i}, and the two consecutive reactions in Eqns. 37 and 40 can be combined into one overall reaction

$$S_{i,1} + S_{i,2} + S_{j,2} \rightleftarrows P_{j,1} + P_{j,2} \quad (44)$$

with

$$J_{i,j} = V[(k_i k_j/k_{-i})a_{Si,1} a_{Si,2} a_{Sj,2} - k_{-j} a_{Pj,1} a_{Pj,2}] \quad (45)$$

which mimics a third-order reaction in the forward direction. Note that the steady state for P_i ($=S_{j,1}$) is reached faster the larger k_{-i} is with respect to $k_i a_{Si,1} a_{Si,2}$ and $k_{-j} a_{Pj,1} a_{Pj,2}$. By the same token, the activity $a_{Pi,1}$ attained in this state (Eqn. 42) decreases and may become negligibly small.

It is generally accepted that unidirectional reactions with order higher than 2 are rather unlikely since they would require a simultaneous collision of three or more species. However, depending on the rate constants of sequential reaction steps and the actual activities of certain reactants, the formulation of higher order reactions may be legitimate as outlined above.

4.2. Enzyme-catalyzed reactions

A catalyst reduces the energy barrier of a reaction and thus increases the rate constants in both directions. In second-order processes, it can also increase the collision probability and promote the proper orientation of the reactants. In order to fulfill this task, the catalyst has to interact with the reactants and may even undergo chemical reactions itself. However, after completion of the catalyzed reaction, the original state of the catalyst has to be regenerated so that it can start a new cycle. An additional aspect of catalysis concerns specificity. The interaction of the catalyst with different members of a group of reactants all able to perform the same type of reaction can greatly vary, which causes a variable increase in reaction rate of the different members. An enzyme is a specialized protein optimally tailored to act as a catalyst for a particular reaction (or reactions) of a selected number of reactants and thus displays all aspects of a catalyst outlined above. The interaction gives rise to binding of the reactants to the enzyme and lowers the energy barriers for the reactions of bound reactants.

4.2.1. Chemical kinetics and the cycle diagram method for the description of enzyme catalysis: In terms of kinetics, the enzyme can be considered as if it were an additional reactant which, however, is present at a much lower concentration than the real reactants and which is restored at the end of a sequence of reactions. Thus, the simplest scheme for the enzyme-catalyzed reaction of S_1 plus S_2 to P is obtained when the following substitutions are made in Eqns. 37, 40, and 44: S_1 for $S_{i,1}$, the free enzyme E for $S_{i,2}$ and $P_{j,2}$, the enzyme-reactant complex ES_1 for P_i, S_2 for $S_{j,2}$, and P for $P_{j,1}$.

Due to the low concentration of the enzyme, the steady state condition $dn_{Pi}/dt = 0$ (i.e. $dn_{ES1}/dt = 0$) is soon reached and is independent of the condition $k_{-i} \gg k_j a_{Sj,2}$. The rate of the reaction is then given by Eqn. 43. It is found to be proportional to Va_E, where a_E is the activity of the free enzyme. This can be replaced by the total mole number of enzyme by means of the mass balance for the enzyme, $n_{E,tot} = n_E + n_{ES1}$. In doing this account is taken of the general relation $n_i = Va_i/y_i$ (cf. Eqn. 22), and of $a_{ES1} = a_{Pi}$ (the latter being given by Eqn. 42). Following this procedure, King and Altman [10] have worked out an algorithm for deriving the rate of enzyme-catalyzed reactions, which was used by Cleland [11], Segel [12] and others to treat a large number of different kinetic schemes.

A substantial extension of the algorithm according to King and Altman was presented by Hill [13]. This author focuses attention on the enzyme whose role is considerably upgraded from just being an additional reactant. Moreover, the cyclic behavior of the enzyme appears explicitly. The enzyme is considered to adopt different states during catalysis. Hence, binding of a reactant to the enzyme causes a transition to a new state in which the bound reactant becomes an integral part of the enzyme. Further transitions can occur due to changes in enzyme conformation. A chemical reaction between bound reactants also gives rise to a change in conformation since bound reactants are taken as integral parts of the enzyme. An appropriate sequence of transitions has to end at the state where it started from, thus constituting a given catalytic cycle of the enzyme. Moreover, dead end branches may start from a given state which evidently do not contribute to catalysis but modify the overall activity of the enzyme, as is the case with inhibitors.

A remark on the definition (or recognition) of different states in an enzyme diagram is in place. An enzyme being a protein can adopt a multitude of conformations even in its native state, i.e. without bound reactants. These conformations are usually not identical to those best adapted to a bound reactant (cf. the notion of induced fit). Moreover, the conformations best adapted to a given set of bound species are usually no longer adapted to a new set of bound species formed in a chemical reaction. As a consequence, a whole sequence of conformational changes may set in, which eventually relaxes the constraints exerted on the enzyme due to binding or alteration of species. In general, these transitions are much faster than the processes of binding or chemical transformation of species. A discernible state in an enzyme cycle thus appears as a collection of conformational states (or better substates) which however, are in equilibrium among each other and therefore need not be discriminated [13]. The fast relaxation of unfavorable conformations evoked by reactants should not be taken as an argument that the true conformational changes of an enzyme need not be considered at all. In fact, a transition between two sets of substates may occur only at a rate comparable to the rates of binding and chemical reaction if it is associated with a substantial rearrangement of protein moieties.

4.2.2. Probabilities of states and transitions in enzyme cycles: The probability P_i of the ith state of an enzyme is defined as

$$P_i = n_{E,i}/n_{E,tot} \tag{46}$$

where $n_{E,i}$ and $n_{E,tot}$ are the mole numbers for the enzyme in state i and for the total enzyme, respectively. Obviously,

$$\sum_i P_i = 1 \tag{47}$$

The transition between states i and j is governed by transition probabilities $\alpha_{i,j}$ and $\alpha_{j,i}$ for the "forward" ($i \rightarrow j$) and "backward" ($j \rightarrow i$) directions, respectively, in close analogy to unidirectional chemical kinetics (cf. Eqn. 35). Forward and backward transitions are true first-order processes for a conformational change. For a transition arising from the binding of reactant R_r to the enzyme, the unidirectional process describing dissociation (say $j \rightarrow i$) is still truly first-order, but the association process is only pseudo-first-order since its transition probability comprises the activity of the reactant, a_{R_r}, according to

$$\alpha_{i,j} = \alpha_{i,j}^0 a_{R_r} \quad \text{for association of } R_r \text{ with enzyme} \tag{48}$$

where $\alpha_{i,j}^0$ is a true second-order transition probability. Note that Eqn. 48 is written in terms of activities in line with Eqn. 36 for unidirectional chemical kinetics. In Ref. 13 activity coefficients of unity are assumed throughout, and hence activities are always replaced by concentrations.

When analyzing a kinetic scheme, a positive direction is arbitrarily chosen which applies to all cycles in a scheme.

4.2.3. Effect of electric fields on transition probabilities: The change in conformation of the enzyme in the course of a cycle is in many cases accompanied by displacements of charges. An obvious example is the binding of a charged reactant to the enzyme. The charge carried by the reactant is displaced from the aqueous phase "outside" the enzyme to a position "inside" the enzyme when the reactant is bound. Displacements within the enzyme occur upon reorientation of charged reactants either due to chemical reactions or by shifting the accessibility of a binding site to a different domain in the enzyme. Less obvious are the cases in which the enzyme itself carries charged groups. A conformational change can then cause a displacement of charges.

The displacement of charges in a transition contributes to the energy barrier separating the two conformations and thus appears implicitly in the transition probabilities. Since electric fields strongly affect charge displacements, we have to expect an alteration of these probabilities when the enzyme is exposed to an electric field. No external electric field is present when the enzyme is suspended in an aqueous phase. However, the rearrangement of charges in the enzyme (or the change of the intrinsic electric field within the enzyme) associated with the transitions may find its expression in a dependence of the transition probabilities on the ionic strength in the aqueous phase [14].

Membrane-bound enzymes are exposed to an electric field whenever a difference in electrical potential across the membrane exists. In the following, we assume that this electric field is constant, i.e. that the potential changes linearly within the membrane. The effect of the field on the transition probabilities follows from the theory of absolute reaction rates [15] and yields a correction factor determined by $\Delta\phi_m$, the membrane potential,

$$\alpha_{i,j}^* = \alpha_{i,j}\exp\{\bar{z}_{i,j}\mathscr{F}\,\Delta\phi_m/(RT)\} \text{ and } \alpha_{j,i}^* = \alpha_{j,i}\exp\{\bar{z}_{j,i}\mathscr{F}\,\Delta\phi_m/(RT)\} \quad (49)$$

where $\alpha_{i,j}^*$ and $\alpha_{j,i}^*$ are the transition probabilities for $i\to j$ and $j\to i$, respectively, in the presence of the electric field.

The quantities $\bar{z}_{i,j}$ and $\bar{z}_{j,i}$ in Eqn. 49 are operationally defined charge numbers effective for the transitions $i\to j$ and $j\to i$, respectively. Suppose that the conformation of the enzyme in state i has the kth charge with charge number z_k at a position with coordinate $x_{k,i}$ on the x-axis shown in Fig. 1D. The conformational change associated with the transition $i\to j$ moves this charge to position $x_{k,j}$ and thus displaces it by $x_{k,j} - x_{k,i}$. The effective charge numbers $\bar{z}_{i,j}$ and $\bar{z}_{i,j}$ are then defined as (cf. Ref. 16)

$$\bar{z}_{i,j} = \gamma_{i,j}\sum_k z_k(x_{k,j} - x_{k,i})/d \text{ and } \bar{z}_{j,i} = \gamma_{j,i}\sum_k z_k(x_{k,j} - x_{k,i})/d \quad (50)$$

where $\gamma_{j,i} = \gamma_{i,j} - 1$, and d is the thickness of the membrane. The sum in Eqn. 50 has to be taken over all charges k including those associated with charged reactants which are bound or released or displaced when going from state i to state j or vice versa. The quantity $\gamma_{i,j}$ expresses the average contribution of charges to the energy barrier separating the two states i and j, and in general is taken to be 0.5 [13, 15, 16] (for further explanation see Appendix B).

4.2.4. The isomerization reaction: The simplest chemical reaction is the isomerization of an initial reactant S to a final reactant P

$$S \rightleftarrows P \tag{51}$$

The minimal kinetic scheme of an enzyme catalyzing this reaction consists of three states as depicted in Fig. 3A. Analysis of the pertinent cycle diagram (Fig. 3B) yields, for the flow of the reaction,

$$J = n_{E,tot}\alpha_{3,2}[K_{2,3}a_S/(a^0K_S) - a_P/(a^0K_P)]/\Sigma_r \tag{52a}$$

where

$$\Sigma_r = (1 + \beta_1 K_{2,3} + \beta_2) + [1 + \beta_2(1 + K_{2,3})]a_S/(a^0K_S)$$

$$+ [1 + \beta_1(1 + K_{2,3})]a_P/(a^0K_P) \tag{52b}$$

The quantity $n_{E,tot}$ denotes the mole number of total enzyme. $K_{2,3}$ is the equilibrium constant (cf. Eqn. 72) for the transition at which the actual isomerization takes place, while $\alpha_{3,2}$ is its transition probability from state 3 to state 2. K_S and K_P are the dissociation constants (cf. Eqn. 73) for the binding of S and P, respectively. The quantities β_i are relative transition probabilities defined as

$$\beta_1 = \alpha_{3,2}/\alpha_{2,1}, \quad \beta_2 = \alpha_{3,2}/\alpha_{3,1} \tag{53}$$

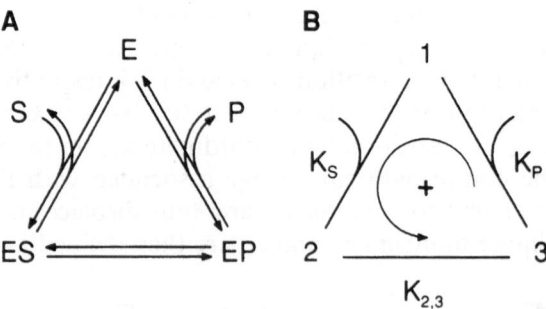

Fig. 3. Kinetic scheme (A) and diagram (B) for the enzyme catalyzed isomerization reaction. E, S, and P in (A) denote the enzyme, the substrate, and the product, respectively. In (B) the different states of the enzyme in the kinetic scheme are numbered. K_S and K_P are the dissociation constants for binding of substrate and product, respectively, while $K_{2,3}$ is the equilibrium constant for the transition $2 \rightleftarrows 3$. The positive direction for the cycle is indicated.

The terms pertinent to S and P in Eqns. 52 can be condensed into quantities known as *Michaelis-Menten* parameters [12]. A pseudo-dissociation constant $K_{m,S}$ (*Michaelis constant*) and an overall first-order rate constant[5] $k_{cat,S}$ (also known as *turnover number*) can be defined for the initial reactant S,

$$K_{m,S} = K_S(1 + \beta_1 K_{2,3} + \beta_2)/[1 + \beta_2(1 + K_{2,3})] \tag{54a}$$

and

$$k_{cat,S} = \alpha_{3,2} K_{2,3}/[1 + \beta_2(1 + K_{2,3})] \tag{54b}$$

Similarly for the final reactant P,

$$K_{m,P} = K_P(1 + \beta_1 K_{2,3} + \beta_2)/[1 + \beta_1(1 + K_{2,3})] \tag{55a}$$

and

$$k_{cat,P} = \alpha_{3,2}/[1 + \beta_1(1 + K_{2,3})] \tag{55b}$$

In terms of the Michaelis-Menten parameters Eqns. 52 read

$$
\begin{aligned}
J &= n_{E,tot} \frac{k_{cat,S} a_S/K_{m,S} - k_{cat,P} a_P/K_{m,P}}{a^0 + a_S/K_{m,S} + a_P/K_{m,P}} \\
&= n_{E,tot}\left[\frac{k_{cat,S} a_S}{a_S + K_{m,S}(a^0 + a_P/K_{m,P})} - \frac{k_{cat,P} a_P}{a_P + K_{m,P}(a^0 + a_S/K_{m,S})}\right]
\end{aligned} \tag{56}
$$

The Michaelis-Menten parameters are conveniently estimated from the initial flow of the reaction, i.e. when only initial reactant ($a_P \approx 0$) or only final reactant ($a_S \approx 0$) is present. Then, Eqn. 56 reduces to the form

$$|J| = n_{E,tot} k_{cat,X} a_X/(a^0 K_{m,X} + a_X), \quad X = S \text{ or } P \tag{56a}$$

Performing such experiments for different substrate or product activities yields $K_{m,S}$ and $k_{cat,S}$ or $K_{m,P}$ and $k_{cat,P}$ by well-known techniques based on linear plots (see, for example, Ref. 12).

The terms $K_{m,S}(a^0 + a_P/K_{m,P})$ and $K_{m,P}(a^0 + a_S/K_{m,S})$ in the denominators of the fractions in square brackets of Eqn. 56 are reminiscent of the dependence of K_m on the concentration of an inhibitor for competitive inhibition. This has led to the rather unfortunate notion that "the reaction $S \rightarrow P$ is inhibited by the product" (which, by the same token, could also read "the reaction $P \rightarrow S$ is inhibited by the substrate") [11, 12]. In fact, if and only if $k_{cat,P} \approx 0$, the flow of the reaction is uniquely determined by the first term comprising $K_{m,S}(a_0 + a_P/K_{m,P})$, and P behaves as if it were a competitive inhibitor for S. In all other cases, however, the slowing down of the net rate of the reaction with increasing a_P is due to an increasing back reaction $P \rightarrow S$ which counteracts the

---

[5]If the mole number of enzyme $n_{E,tot}$ is not known, the maximal velocity $v_{max,X} = n_{E,tot} k_{cat,x}/V$ ($X = S$ or P) is used.

forward reaction $S \to P$ and eventually balances it at equilibrium. Hence, the above statement should be avoided, and the notion of product or substrate inhibition should be restricted to those possible cases where P or S, in addition to being the final or the initial reactant of the reaction, indeed take over the role of an inhibitor.

4.2.5. Redox reactions: Two cases can be distinguished for redox reactions (cf. Eqn. 14) catalyzed by an enzyme. If the enzyme itself cannot undergo a redox reaction its role is restricted to bringing the reactants together and lowering the activation barrier for the electron exchange. In this case, redox reactions do not differ from any other reaction in which two final reactants are formed from two initial reactants. If the enzyme itself is a redox couple, D_{rd} can bind to the enzyme, transfer its electron(s) to the enzyme and leave it as D_{ox} (see Fig. 4A which pertains

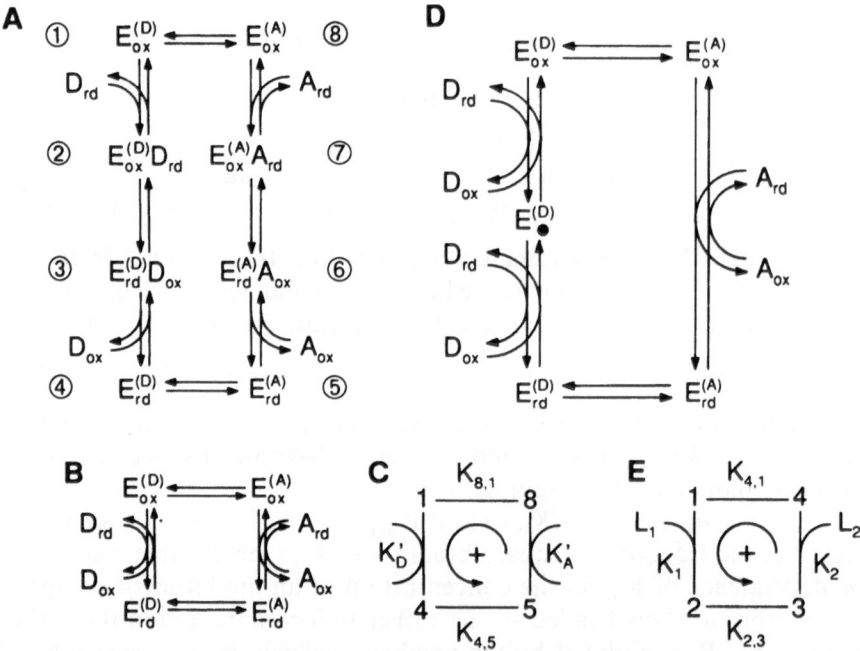

Fig. 4. Kinetic schemes (A, B, D) and diagram (C) for enzyme-catalyzed redox reactions as well as diagram (E) for a carrier-mediated transport. The symbols E, D and A represent the redox enzyme, the electron donating and the electron accepting redox couple, respectively, while subscripts rd and ox indicate the reduced and oxidized species of each couple, respectively. The redox enzyme can adopt different conformations which react only with either D or A as indicated by the corresponding superscripts. The dot on E in (D) means "half reduced" enzyme (semiquinoid form). The scheme in (B) and its diagram in (C) are obtained from the scheme in (A) by a reduction of states. The diagram in (E) pertains to a carrier-mediated transport of species L between the compartments indicated by the subscripts 1 and 2 (cf. Fig. 1D). States 1 and 2 represent the carrier whose binding site is accessible only from compartment 1 without and with bound species, respectively, while states 4 and 3 are the corresponding states for the carrier facing compartment 2.

to a redox reaction with $v_{e(D)} = v_{e(A)}$). In a subsequent sequence of transitions, A_{ox} can then bind to the enzyme and get the electron(s) whereby A_{rd} is formed which dissociates from the enzyme. The scheme in Fig. 4A includes an additional feature, viz. a transition for the oxidized and the reduced enzyme which represents an adaptation of the enzyme's conformation to the binding of D and A.

Electron transfer reactions in a complex of redox couples are usually much faster than ordinary chemical reactions or binding [17]. As a consequence, the probabilities of the states 2, 3, 6, and 7 become rather small, which allows us to condense the sequence of transitions $1 \rightleftarrows 2 \rightleftarrows 3 \rightleftarrows 4$ and $5 \rightleftarrows 6 \rightleftarrows 7 \rightleftarrows 8$ into *one* transition $1 \rightleftarrows 4$ and $5 \rightleftarrows 8$, respectively (see Hill [13]). The analysis of the resulting scheme (Fig. 4B) yields

$$J_e = n_{E,tot} v_{e(E)} \alpha_{5,4} [K_{4,5} K_{8,1} \rho_D / K'_D - \rho_A / K'_A] / \Sigma_e \qquad (57a)$$

where

$$\Sigma_e = (1 + K_{8,1})[1 + \beta_2 K_{Aox}/a_{Aox} + \beta_1 K_{4,5} K_{Dox}/a_{Dox}]$$

$$+ K_{8,1}[1 + \beta_2(1 + K_{4,5})K_{Aox}/a_{Aox} + \beta_3 K_{4,5}]\rho_D / K'_D$$

$$+ [1 + \beta_1(1 + K_{4,5})K_{Dox}/a_{Dox} + \beta_3 K_{8,1}]\rho_A / K'_A$$

$$+ \beta_3 K_{8,1}(1 + K_{4,5})\rho_D \rho_A / (K'_D K'_A) \qquad (57b)$$

Here, $n_{E,tot}$ and $v_{e(E)}$ denote, respectively, the mole number of the enzyme and the number of electrons which it exchanges with D or A; ρ_D and ρ_A indicate the reduction state (see Eqn. 24b) of the donating and accepting redox couple, respectively. $K_{4,5}$ and $K_{8,1}$ are the equilibrium constants for the conformational change of the reduced and the oxidized enzyme. K'_D and K'_A are combined equilibrium constants defined as

$$K'_D = K_{3,2} K_{Drd}/K_{Dox}, \quad K'_A = K_{6,7} K_{Ard}/K_{Aox} \qquad (58a)$$

where K_{Xrd} and K_{Xox} (X = D or A) are the dissociation constants of the reduced and oxidized species, respectively, for the donating (X = D) or the accepting (X = A) redox couple. The relative transition probabilities β_i are

$$\beta_1 = \alpha_{5,4}/(\alpha'_{2,1} K_{3,2}), \quad \beta_2 = \alpha_{5,4}/(\alpha'_{7,8} K_{6,7}), \quad \beta_3 = \alpha_{5,4}/\alpha_{8,1} \qquad (58b)$$

where $\alpha'_{2,1}$ and $\alpha'_{7,8}$ are condensed transition probabilities arising from the reduction of states [1, 13]:

$$\alpha'_{2,1} = \alpha_{2,1}/[1 + \alpha_{2,1}/\alpha_{2,3} + K_{3,2}\alpha_{2,1}/\alpha_{3,4}],$$
$$\alpha'_{7,8} = \alpha_{7,8}/[1 + \alpha_{7,8}/\alpha_{7,6} + K_{6,7}\alpha_{7,8}/\alpha_{6,5}] \qquad (58c)$$

Inspection of the quantity Σ_e reveals that an enzyme-catalyzed redox reaction behaves like a carrier-mediated transport (compare Eqn. 64b) if $\beta_1 = \beta_2 \approx 0$ or if $a_{Dox}/K_{Dox} \gg 1$ and $a_{Aox}/K_{Aox} \gg 1$. Electrons are transported between two "activities" which are expressed by the reduction states ρ_D and ρ_A. It should be noted that this analogy is complete for a membrane-bound redox enzyme with the donating redox couple in a different compartment than the accepting couple (compare diagrams in Fig. 4C and 4E). The complexity encountered with the transport of a charged species across the membrane to be discussed in section 4.3.2 then applies to the transport of electrons as well.

A redox enzyme catalyzing the reaction where $v_{e(D)} \neq v_{e(A)}$ has to be able to exchange $v_{e(D)}$ and $v_{e(A)}$ electrons in the conformations adapted to D and A, respectively. A scheme describing such a situation for $v_{e(D)} = 1$ and $v_{e(A)} = 2$ is depicted in Fig. 4D. Obviously, two sequential transitions with $D_{rd} \rightleftarrows D_{ox}$ are required in order to collect the two electrons subsequently transferred to A_{ox}. Alternatively, the enzyme may be able to transfer only one electron in all transitions. Two redox centers in the enzyme are then required, and A must be able to accept electrons in a sequence of two transitions each transferring one electron. The corresponding scheme can be converted to the scheme in Fig. 4D by a possible reduction of states (see Ref. 13). The situation becomes more complicated if the half reduced enzyme created in the first transition involving D_{rd} can change to a conformation which binds D_{rd} as well as A_{ox}. Additional branches are then introduced which yields a diagram with four cycles. Its analysis, though straightforward, will be omitted since it results in relations too complex to be reproduced here.

4.3. Transport

Transport of species within the aqueous phases need not be considered because of stirring or rapid diffusion in the internal unstirred spaces (see section 2). It is worth mentioning, however, that certain membrane topologies such as tight folding may lead to unequal distributions of solutes despite stirring [18]. Transport through membranes is usually assumed to occur by diffusion. If present, pores formed by specialized proteins or carriers (which bind and transfer selected species) can substantially increase the transport rates. The transport through pores has been comprehensively treated by Läuger [19] and is therefore omitted here.

4.3.1. Permeation through membranes: The flow of the ith species, J_i, in the membrane arises from a unidirectional gradient in its electrochemical potential $\tilde{\mu}_i$ (Nernst-Planck equation)

$$J_i(x)/A = u_i a_i(x) \, d\tilde{\mu}_i/dx = -u_i[RT \, da_i/dx + z_i a_i(x) \, d\phi/dx] \quad (59)$$

with u_i and A denoting, respectively, the mobility of the species in the membrane phase and the total membrane area. For a steady state of transport J_i = constant, and Eqn. 59 can be integrated between x_1 and x_2 (see Fig. 1D) which yields

$$J_i = AP_i \frac{z_i \mathscr{F} \Delta\phi_m}{RT} \frac{a_{i,1} \exp\{z_i \mathscr{F} \Delta\phi_m/(RT)\} - a_{i,2}}{\exp\{z_i \mathscr{F} \Delta\phi_m/(RT)\} - 1} \tag{60}$$

Here, $\Delta\phi_m = \phi_1 - \phi_2$ is the membrane potental (see section 3.5.2). P_i is the permeability of the ith species,

$$P_i = RTu_i K_{pi}/d \tag{61}$$

where K_{pi} and d denote the partition coefficient of the species between membrane and aqueous phase (see Eqn. 25) and the thickness of the membrane, respectively. Note that for an uncharged species or in the absence of a membrane potential Eqn. 60 is reduced to

$$J_i = AP_i(a_{i,1} - a_{i,2}) \quad \text{for } z_i = 0 \quad \text{or } \Delta\phi_m = 0 \tag{60a}$$

When we insert Eqn. 60 written for different species into Eqn. 20 we realize that transcendental equations emerge which comprise sums of exponentials of the membrane potential. Hence, an explicit solution does not exist in general but is possible for special cases. If only one charged species (which is given the index p for permeable) can move across the membrane Eqn. 20 is simplified to $J_p = 0$, and

$$\Delta\phi_m = RT/(z_i \mathscr{F}) \ln\{a_{p,2}/a_{p,1}\} \tag{62}$$

This relation is known as the Nernst equation and could as well be derived from the equilibrium condition $X_p = 0$ since $J_p = 0$ (cf. Eqns. 17 and 21). If only univalent ions permeate through the membrane we obtain, by virtue of Eqn. 60, what is known as the Goldman equation,

$$\Delta\phi_m = (RT/\mathscr{F}) \ln\left\{\left[\sum_i P_i a_{i,2} + \sum_j P_j a_{j,1}\right] \middle/ \left[\sum_i P_i a_{i,1} + \sum_j P_j a_{j,2}\right]\right\}$$
$$\text{for } z_i = 1, z_j = -1 \quad (63)$$

Note that Eqn. 63 becomes identical to Eqn. 62 if the permeabilities of all species except one are set to zero. Moreover, Eqn. 63 is usually given in the literature in terms of concentrations rather than activities which means that activity coefficients are set to unity.

4.3.2. Carrier-mediated transport:

4.3.2. Carrier-mediated transport: A carrier is usually envisaged as a molecule which shuttles back and forth in the membrane and which binds a given species from the aqueous phases on either side of the membrane. It thus enables this species to cross the membrane without being exposed to the hydrophobic environment in the membrane. Valinomycin transporting K^+ or Rb^+ ions is the classical example of this mechanism [20]. However, a relatively small carrier molecule which actually moves within the membrane is a less frequent case and is only suitable for the transport of small species. Most carriers are integral membrane proteins able to adopt at least two conformations. The conformations differ in the accessibility (and most likely the position) of a binding site for the transported species. This site must face and be open to the aqueous phases on either side of the membrane. Note that the transition between the two conformations with bound species requires the displacement of the latter within the protein. This may cause substantial strains in the protein, particularly for bulky species, which translates into a high energy barrier and low transition probabilities (see Appendix B).

Both schemes, i.e. mobile carrier and conformational model, result in the same diagram (Fig. 4E). Its analysis yields, for the flow of the ith species catalyzed by a carrier with $\bar{z}_{i,j} = \bar{z}_{j,i} = 0$ (see section 4.2.3), or in the absence of a membrane potential,

$$J_i = n_c \alpha_{3,2} [K_{2,3} K_{4,1} a_{i,1}/(a^0 K_1) - a_{i,2}/(a^0 K_2)]/\Sigma_c,$$

$$\bar{z}_{i,j} = 0 \text{ or } \Delta\phi_m = 0 \tag{64a}$$

where

$$\Sigma_c = (1 + K_{4,1})(1 + \beta_1 K_{2,3} + \beta_2) + K_{4,1}[1 + \beta_2(1 + K_{2,3})$$

$$+ \beta_3 K_{2,3}] a_{i,1}/(a^0 K_1) + [1 + \beta_1(1 + K_{2,3}) + \beta_3 K_{4,1}] a_{i,2}/(a^0 K_2)$$

$$+ \beta_3 K_{4,1}(1 + K_{2,3}) a_{i,1} a_{i,2}/(a^{02} K_1 K_2) \tag{64b}$$

In Eqns. 64, n_c denotes the mole number of the carrier. K_1 and K_2 are the dissociation constants for binding of the transported species to the carrier from compartments 1 and 2 with activities $a_{i,1}$ and $a_{i,2}$, respectively. $K_{2,3}$, $K_{4,1}$, and $\alpha_{3,2}$ denote the equilibrium constants and the transition probability for the transitions indicated by the indices, respectively. The relative transition probabilities are defined as

$$\beta_1 = \alpha_{3,2}/\alpha_{2,1}, \quad \beta_2 = \alpha_{3,2}/\alpha_{3,4}, \text{ and } \beta_3 = \alpha_{3,2}/\alpha_{4,1} \tag{65}$$

Inspection of the quantity Σ_c given in Eqn. 64b shows that it is formally identical with Σ_r in Eqn. 52b for the isomerization reaction except for the additional term comprising $a_{i,1} a_{i,2}$. This term vanishes if $\beta_3 \approx 0$, i.e. if the transition of the free carrier is much faster than that of the loaded carrier (cf. Eqn. 65), and the four-state diagram can then be reduced to a three-state model like that in Fig. 3B. The term with the product of both activities also vanishes when one of the activities is zero, which enables us to define Michaelis-Menten parameters in the usual way.

They are, for the conditions $a_{i,2} \approx 0$ and $a_{i,1} \approx 0$, respectively,

$$K_{m,1} = K_1(1 + K_{4,1})(1 + \beta_1 K_{2,3} + \beta_2)/(K_{4,1}D_1),$$

$$k_{cat,1} = \alpha_{2,3}/D_1 \quad \text{with} \quad D_1 = 1 + \beta_2(1 + K_{2,3}) + \beta_3 K_{2,3}$$

(66a)

and

$$K_{m,2} = K_2(1 + K_{4,1})(1 + \beta_1 K_{2,3} + \beta_2)/D_2,$$

$$k_{cat,2} = \alpha_{3,2}/D_2 \quad \text{with} \quad D_2 = 1 + \beta_1(1 + K_{2,3}) + \beta_3 K_{4,1}$$

(66b)

The situation becomes considerably more complex if the *transported species and/or the carrier are charged*. Eqn. 64 is then converted to

$$J_i = n_c \alpha_{3,2}[K_{2,3}K_{4,1}a_{i,1} \exp\{z_i \mathscr{F} \Delta\phi_m/(RT)\}/(a^0 K_1) - a_{i,2}/(a^0 K_2)]/\Sigma_c^*$$

(67a)

with

$$\begin{aligned}
\Sigma_c^* = \exp\{(\bar{z}_{2,3} + \bar{z}_{3,4} - \bar{z}_{4,3})\mathscr{F} \Delta\phi_m/(RT)\}[&(1 + K_{4,1}^*)(1 + \beta_1^* K_{2,3}^* + \beta_2^*) \\
&+ K_{4,1}^*[1 + \beta_2^*(1 + K_{2,3}^*) + \beta_3^* K_{2,3}^*]a_{i,1}/(a^0 K_1^*) \\
&+ [1 + \beta_1^*(1 + K_{2,3}^*) + \beta_3^* K_{4,1}^*]a_{i,2}/(a^0 K_2^*) \\
&+ \beta_3^* K_{4,1}^*(1 + K_{2,3}^*)a_{i,1}a_{i,2}/(a^{02} K_1^* K_2^*)]
\end{aligned}$$

(67b)

where

$$\beta_1^* = \beta_1 \exp\{(\bar{z}_{3,2} - \bar{z}_{2,1})\mathscr{F} \Delta\phi_m/(RT)\},$$

$$\beta_2^* = \beta_2 \exp\{(\bar{z}_{3,2} - \bar{z}_{3,4})\mathscr{F} \Delta\phi_m/(RT)\},$$

$$\beta_3^* = \beta_3 \exp\{(\bar{z}_{3,2} - \bar{z}_{4,1})\mathscr{F} \Delta\phi_m/(RT)\}$$

(68a)

and

$$K_1^* = K_1 \exp\{(\bar{z}_{2,1} - \bar{z}_{1,2})\mathscr{F} \Delta\phi_m/(RT)\},$$

$$K_2^* = K_2 \exp\{(\bar{z}_{3,4} - \bar{z}_{4,3})\mathscr{F} \Delta\phi_m/(RT)\},$$

$$K_{2,3}^* = K_{2,3} \exp\{(\bar{z}_{2,3} - \bar{z}_{3,2})\mathscr{F} \Delta\phi_m/(RT)\},$$

$$K_{4,1}^* = K_{4,1} \exp\{(\bar{z}_{4,1} - \bar{z}_{1,4})\mathscr{F} \Delta\phi_m/(RT)\},$$

(68b)

In the case of a constant membrane potential $\Delta\phi_m$, the discussion of Eqns. 67 is straightforward and similar to that of Eqns. 64. However, the Michaelis-Menten parameters are dependent on the membrane potential. If $\Delta\phi_m$ varies, the behavior of J_i becomes too complex to be discussed here. As a simplification, the analysis could be restricted to small membrane potentials where exponentials can be linearized.

5. Connection between thermodynamics and kinetics

Thermodynamics tells us in which direction a process proceeds spontaneously, but it does not indicate how fast the process runs. On the other hand, a kinetic scheme of a process comprises all information about rates and seemingly does not require any thermodynamic elements. In fact, a *properly formulated kinetic scheme* takes care of thermodynamics by itself. The attribute "properly formulated" implies that the kinetic scheme complies with the frame set by thermodyanamics or, in other words, does not violate thermodynamic laws. As we shall see in the following section, this condition gives rise to restrictions imposed on the rate constants or transition probabilities in enzyme kinetics.

5.1. Microscopic reversibility, detailed and thermokinetic balancing

The twin concepts of microscopic reversibility and detailed balancing pertain to the same phenomenon and are often used interchangeably. They relate to the fact that all flows have to vanish in the equilibrium state where all parameters of a system no longer change in time (see section 3.5). The conditions derived from this fact depend on the level of description of the system. In order to avoid confusion, it was therefore suggested [21] that the concept of *microscopic reversibility* should be used exclusively in a microscopic description of the system in terms of statistical mechanics. *Detailed balancing* then applies to the macroscopic (or phenomenological) level of description in terms of rate constants which arise from but are not identical to the transition probabilities pertinent to transitions between states of a system in the microscopic description [21]. These transition probabilities should not be confused with the transition probabilities $\alpha_{i,j}$ pertinent to the transitions between states of an enzyme (see section 4.2.2).

5.1.1. Detailed balancing and thermokinetic balancing:

5.1.1. Detailed balancing and thermokinetic balancing: The vanishing of all flows at equilibrium required by detailed balancing leads to a relation between forward and backward rate constants and the equilibrium constant, K_r, of a chemical reaction. Thus, for the ith reaction in Eqn. 37 (cf. Eqns. 38 and 26)

$$k_i/k_{-i} = K_{ri}/a^0 = [a_{Pi}/(a_{Si,1} a_{Si,2})]_{eq} \qquad (69a)$$

where the subscript eq indicates the activities of the final and initial reactants at equilibrium. The left hand side of Eqn. 69a is valid at, close to, or far from the equilibrium state. It was termed *thermokinetic balancing* [21] to distinguish it from detailed balancing. Similarly for the reaction in Eqn. 40

$$k_j/k_{-j} = K_{rj} = [a_{Pj,i}a_{Pj,2}/(a_{Sj,1}a_{Sj,2})]_{eq} \qquad (69b)$$

For a sequence of two reactions where the final reactant of the first reaction is an initial reactant of the second reaction

$$k_i k_j/(k_{-i}k_{-j}) = K_{ri}K_{rj}/a^0 = K_{ri,j}/a^0 = [a_{Pj,1}a_{Pj,2}/(a_{Si,1}a_{Sj,1}a_{Sj,2})]_{eq} \qquad (70)$$

where $K_{ri,j}$ denotes the equilibrium constant for the overall reaction in Eqn. 44. This can easily be extended to sequences of several reactions. Detailed balancing then imposes the condition that the product of forward rate constants divided by the product of backward rate constants is equal to the product of the equilibrium constants for each reaction in the sequence, which in turn is equal to the equilibrium constant for the overall reaction emerging from the sequence. It is important to note that the relations between rate constants and equilibrium constants always hold irrespective of the reactions being at equilibrium or a sequence of reactions being in a steady state.

5.1.2. Thermokinetic balancing for enzymes: Detailed balancing applied to the kinetic scheme of an enzyme requires that the flows through all transitions vanish. This condition yields a relation between the transition probabilities $\alpha_{i,j}$ (see section 4.2.2) or the equilibrium constants for the transitions in a given cycle and the equilibrium constant for the reaction catalyzed by this cycle:

$$\prod_{c+} \alpha_{i,j}\alpha_{k,1}^0\alpha_{n,m} \Big/ \prod_{c-} \alpha_{j,1}\alpha_{1,k}\alpha_{m,n}^0 = \prod K_{i,j} \prod K_{Rr(mn)} \Big/ \prod K_{Rr(lk)} = K_c \quad (71)$$

The term $\prod_{c+} \alpha_{i,j}\alpha_{k,1}^0\alpha_{n,m}$ means the product of all transition probabilities in cycle c taken in the positive direction, where $\alpha_{i,j}$, $\alpha_{k,1}^0$, and $\alpha_{n,m}$ stand for transitions with a conformational change, transitions with association of a reactant R_r, and transitions with association of a reactant R_r in the opposite direction, respectively. The term $\prod_{c-} \alpha_{j,i}\alpha_{1,k}\alpha_{n,m}^0$ is the corresponding product taken in the negative direction.

The quantities $K_{i,j}$ and K_{Rr} in Eqn. 71 are equilibrium constants, defined as

$$K_{i,j} = \alpha_{i,j}/\alpha_{j,i} = [P_j/P_i]_{eq} \qquad (72)$$

for conformational transitions, and

$$K_{Rr}a^0 = \alpha_{j,i}/\alpha_{i,j}^0 = [P_i a_{Rr}/P_j]_{eq} \qquad (73)$$

for transitions pertinent to binding. They also relate the probabilities P_i and P_j of the states (see section 4.2.2) at equilibrium as indicated on the right hand side of Eqns. 72 and 73. The second part of Eqn. 71 then means the product of all $K_{i,j}$ pertinent to conformational changes times the product of all dissociation constants for binding transitions whose dissociation coincides with the positive direction, divided by the product of dissociation constants for dissociation of R_r in the negative direction.

K_c in Eqn. 71 is the equilibrium constant for the overall process catalyzed by the given cycle c. Note that $K_c = 1$ if no chemical reaction is associated with a full turn around the cycle. $K_c = 1$ also holds if one turn of the cycle results only in a transport of species from one compartment to another, in line with the notion that the partition coefficient between two equal phases is unity (see section 3.5.1). It should be added that the Haldane relations [22] also arise from thermokinetic balancing but are expressed in terms of the Michaelis-Menten parameters.

In the presence of a membrane potential and charged species, the effect of electric field on the transition probabilities discussed in section 4.2.3 gives rise to an auxiliary relation which is obligatory for thermokinetic balancing. The quantities $\bar{z}_{i,j}$ and $\bar{z}_{j,i}$ (cf. Eqn. 50) for a given cycle c are subject to the condition

$$\sum_c (\bar{z}_{i,j} - \bar{z}_{j,i}) = \sum_{c+} z_t \qquad (74)$$

where z_t is the charge number of a species transported across the membrane when going around the cycle in the positive direction. $\sum_{c+} z_t$ then means that the sum has to be taken for all species transported by the given cycle and z_t has to be multiplied by 1 or -1 if the transport occurs in the positive or negative direction of the flows (see section 3.3.2), respectively. The sum on the left hand side of Eqn. 74 has to include all transitions of the given cycle c, where the index i, j indicates the positive direction of the cycle. Applying detailed balancing to the kinetic cycles of an enzyme subject to the influence of an electric field, and introducing Eqns. 48, 49 and 74, leads once again to Eqn. 71, where all the transition probabilities (of whatever type) are the *uncorrected* (or *intrinsic*) quantities. It should be noted that other force fields which can influence the transition probabilities of an enzyme, such as mechanical stress, will also give rise to correction factors (cf. Eqn. 49) governed by fractional relations analogous to Eqn. 50 [23]. These relations would be subject to a condition similar to Eqn. 74. In all cases it is the uncorrected or intrinsic transition probabilities that must satisfy thermokinetic balancing.

Thermokinetic balancing for the cycle in the diagram of the enzyme-catalyzed isomerization reaction (Fig. 3B) yields

$$K_{2,3} K_P / K_S = K_c \quad \text{for } S \rightleftarrows P \qquad (75)$$

where K_c denotes the equilibrium constant of the reaction in Eqn. 51. The full cycle in Fig. 4A for a redox reaction complies with

$$K_{4,5} K_{8,1} K_{6,7} K_{Dox} K_{Ard} / (K_{3,2} K_{Drd} K_{Aox}) = K_c \qquad (76a)$$

The reduction of states and the definition of the condensed constants K'_D and K'_A in Eqn. 58a reduces Eqn. 76a to

$$K_{4,5} K_{8,1} K'_A / K'_D = K_c \quad \text{for } D_{rd} + A_{ox} \rightleftarrows D_{ox} + A_{rd} \qquad (76b)$$

For the cycle in the diagram for a carrier (Fig. 4E)

$$K_{2,3} K_{4,1} K_2 / K_1 = 1 \qquad (77)$$

The ratio of equilibrium constants in Eqn. 77 is unity because the catalyzed reaction is the partitioning of a species between two aqueous phases. Moreover, the auxiliary relation in Eqn. 74 applies to all cases if charged, membrane-bound enzymes are involved and/or if the transported species are charged.

5.2. Flow-force relations

The analysis of the kinetic scheme of a process results in a relation indicating how the flow of the process depends on the activities of the reactants or the transported species. On the other hand, the thermodynamic force of a process is defined in terms of electrochemical potentials which implicitly comprise these activities. Making the activities explicit by virtue of Eqn. 21 yields from Eqn. 12, in view of Eqns. 26a and 26b,

$$\mathscr{A}_{j,k} = RT \ln \left\{ K_{rj,k} \prod_r [a_{Rr(j,k)} / a^0]^{-\nu_{Rr(j,k)}} \right\} \qquad (78a)$$

for a chemical reaction. Note that the terms $z_{Rr(j,k)} \mathscr{F} \phi$ in Eqn. 21 are cancelled because $\Sigma_r \nu_{Rr(j,k)} z_{Rr(j,k)} = 0$ for a chemical reaction. Moreover, the stoichiometric coefficients $\nu_{Rr(j,k)}$ have to be taken with signs according to the convention introduced in section 3.3.2. Similarly, from Eqn. 15 and by virtue of Eqn. 26c,

$$\mathscr{A}_{ej,k} = RT \ln \{ K_{ej,k} \rho_{Dj,k}^{1/\nu_{e(Dj,k)}} / \rho_{Aj,k}^{1/\nu_{e(Aj,k)}} \} \qquad (78b)$$

For a transport process, we obtain from Eqn. 17,

$$\Delta \tilde{\mu}_i = RT \ln \{ a_{i,1} / a_{i,2} \} + z_i \mathscr{F} \Delta \phi_m \qquad (78c)$$

where $\Delta \phi_m$ is the membrane potential.

It is seen that thermodynamic forces are determined by the equilibrium constant of a process (which is frequently unity for transport between compartments), the activities of the reactants, and the membrane potential where appropriate. In view of the thermodynamic equivalence of forces (see section 3.3.2) all these contributions must be equivalent. Therefore the assignment of different names such as "enthalpic" for the contribution of the equilibrium constant or the membrane potential and "entropic" for contribution of the activities, as has become common practice (see, for example, ref. 24) is meaningless.

Inspection of the expressions for the flows given in section 4 (Eqns. 38, 45, 52, 57, 60, 64, 67) shows that, in view of the pertinent relations for detailed balancing (Eqns. 69 and 70) and thermokinetic balancing (Eqns. 75–77), the flow is in every case related to the force by

$$J_i = [\exp\{X_i/(RT)\} - 1]/f(a_i) \tag{79}$$

Here, $f(a_i)$ is the abbreviation for a generally quite complex function which comprises the activities of the reactants ($a_i = a_{Rr}$) or the transported species ($a_i = a_{i,k}$). It is equivalent to a reciprocal flow and can be read from the pertinent equations listed above. Eqn. 79 evidently fulfils the condition $J_i = 0$ for $X_i = 0$ as required by thermodynamics for the equilibrium state (see section 3.5). This is not surprising because we have taken care of this condition by means of detailed or thermokinetic balancing.

The flow of a process is not unequivocally defined by its thermodynamic force. As is evident from Eqns. 78 there are a multitude of combinations of activities a_i which constitute the same force but yield different $f(a_i)$. However, constraints almost invariably introduced by the design of an experimental system as discussed in section 2 create an interrelation between the concentrations of some of the species involved in the processes. In fact, due to mass balance (cf. Eqns. 10 and 16), the mole numbers of initial and final reactants in a chemical reaction or those of a transported species in the two compartments cannot vary independently in an isolated system. Moreover, when choosing such conditions that a steady state of the system can be attained (see section 3.4) we have to clamp the concentration of some of the reactants. Alternatively, we set the initial concentration and the chemical capacity for some reactants high enough in order to prevent an appreciable change in their concentration.

5.2.1. Chemical reactions under constraints: The above mentioned constraints applied to the isomerization reaction specified in Eqn. 51 read

$$c_S + c_P = c_{tot} = \text{const} \tag{80a}$$

which, by virtue of Eqns. 22 and 78a, is transformed into

$$a_P = y_P c_{tot}/[1 + \exp\{X/(RT)\}y_P/(K_r y_S)] \tag{80b}$$

The function f in Eqn. 79 then becomes

$$f(a_i) = \exp\{X/(RT)\}/J_+ + 1/J_- \tag{81}$$

where the flows J_+ and J_- are the extreme flows for $a_S = y_S c_{tot}$, $a_P = 0$ and $a_S = 0$, $a_P = y_P c_{tot}$, respectively. For a reaction not catalyzed by an enzyme they amount to

$$J_+ = V k_1 y_S c_{tot} \quad \text{and} \quad J_- = V k_{-1} y_P c_{tot} \tag{82a}$$

with k_1 and k_{-1} denoting the forward and backward rate constants, respectively. For an enzyme-catalyzed reaction the extreme flows can be expressed in terms of the Michaelis-Menten parameters (see Eqns. 54 and 55).

$$J_+ = n_{E,tot} k_{cat,S}/[1 + a^0 K_{m,S}/(y_S c_{tot})]$$

and

$$J_- = n_{E,tot} k_{cat,P}/[1 + a^0 K_{m,P}/(y_P c_{tot})] \qquad (82b)$$

By virtue of Eqn. 81, Eqn. 79 can be reformulated as

$$J = \frac{J_+ + J_-}{2} \tanh\left\{\frac{X}{2RT} - \frac{1}{2}\ln\frac{J_+}{J_-}\right\} + \frac{J_+ - J_-}{2} \qquad (83)$$

Figure 5 shows a plot of the now unambiguous flow-force relation in Eqn. 83. It thus appears that even a non-catalyzed chemical reaction displays saturation properties under the constraint specified in Eqn. 80a. The hyperbolic tangent has an *inflection point* when its argument is zero. Hence, from Eqn. 83

$$X_0/(RT) = \ln\{J_+/J_-\} \qquad \text{(inflection point)} \qquad (84)$$

which is equal to $\ln\{K_r y_S/y_P\}$ for a non-catalyzed reaction. In the case of enzyme catalysis, however,

$$\frac{X_0}{RT} = \ln\left\{K_r \frac{y_S}{y_P} \frac{1 + y_P c_{tot}/(a^0 K_{m,P})}{1 + y_S c_{tot}/(a^0 K_{m,S})}\right\} \qquad (85)$$

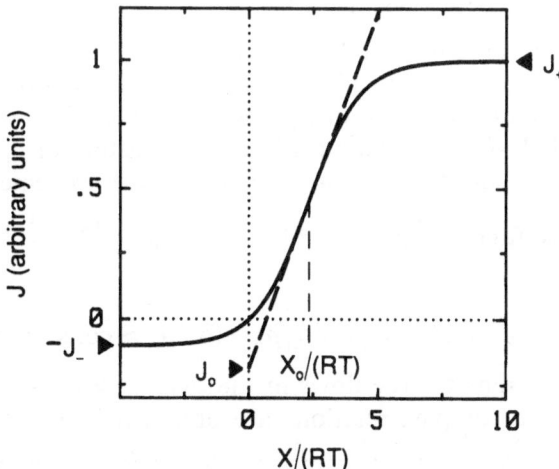

Fig. 5. Dependence of flow, J, on thermodynamic force, X, for a reaction or transport under constraints. The curve is the hyperbolic tangent and the broken line the tangent as its inflection point located at X_0. For large positive and negative forces the curve converges to the limits determined by the extreme flows J_+ and J_-, respectively. J_0 indicates the intercept of the tangent on the ordinate. Flows are given in arbitrary units.

The argument in curly brackets varies between $K_r y_S/y_P$ if $y_X c_{tot}/(a^0 K_{m,X}) \ll 1$ (X = S and P) and $K_r K_{m,S}/K_{m,P} = k_{cat,S}/k_{cat,P} = K_{2,3}[1 + \beta_1(1 + K_{2,3})]/[1 + \beta_2(1 + K_{2,3})]$ (cf. Eqns. 54 and 55) if $y_X c_{tot}/(a^0 K_{m,X}) \gg 1$. The location of the inflection point thus depends on c_{tot}, and a substantial offset of X_0 due to a large or small K_r can be partially or fully balanced by a large enough concentration c_{tot} provided that $K_{2,3}$, β_1 and β_2 have appropriate values.

The hyperbolic tangent deviates less than 10% from its tangent at the inflection point if the argument varies within the range from -0.71 to 0.71 around this point. Hence, we can reasonably approximate Eqn. 83 by the linear relation

$$J \approx \left[\frac{J_+ - J_-}{2} - \frac{J_+ + J_-}{4} \{X_0/(RT)\} \right] + \frac{J_+ + J_-}{4} \{X/(RT)\} \quad (83a)$$

(see Fig. 5) which, in terms of X, covers the range of ± 1.75 kJ/mol around X_0. Note that the intercept on the J-axis, which is enclosed in square brackets in Eqn. 83a, also goes to zero when X_0 approaches zero.

The same flow-force relation (Eqn. 83) emerges if a_S or a_P is constant with $J_- = n_{E,tot} k_{cat,P}$ or $J_+ = n_{E,tot} k_{cat,S}$ while the other extreme flow is given by Eqns. 82b with c_{tot} representing the constant concentration, respectively. However, the non-catalyzed reaction no longer complies with Eqn. 83 under these conditions.

The situation is somewhat different in the case of redox reactions. Mass balance requires only that

$$c_{Xrd} + c_{Xox} = c_{X,tot} = const \quad \text{for X = D and A} \quad (86)$$

which does not provide a relation between ρ_D and ρ_A. Additional constraints are necessary which can arise from clamping or high enough concentrations and capacities for the species of one of the redox couples (e.g. the couple H_2O/O_2) yielding either $\rho_A \approx$ constant or $\rho_D \approx$ constant. In this case $f(a_i)$ adopts the form of Eqn. 81 with extreme flows

$$J_+ = n_{E,tot} v_{e(E)} \alpha_{5,4} K_{4,5}, \quad J_- = n_{E,tot} v_{e(E)} \alpha_{5,4} \rho_A/[K'_A(1 + K_{8,1})]$$

and

$$J_+/J_- = K_{4,5}(1 + K_{8,1}) K'_A/\rho_A \quad \text{for } \rho_A \approx const, \beta_i \approx 0 \quad (87)$$

Analogous equations for the flows in the case $\rho_D \approx$ constant exist, and similar but more complex relations are obtained if $\beta_i \neq 0$. Thus, provided that the affinity of the electron is only varied by changing ρ of one of the couples under the constraints in Eqn. 86 while ρ of the other couple is approximately constant, the flow of electrons depends on the affinity according to the hyperbolic tangent (Eqn. 83) and the inflection point may be near $\mathscr{A}_e = 0$ for suitable values of the fixed ρ (and $c_{X,tot}$ in case of $\beta_i \neq 0$).

5.2.2. Transport under constraints, and the kinetic inequivalence of chemical and electrical potential: Mass balance in the case of transport between two compartments requires that

$$V_1 c_{i,1} + V_2 c_{i,2} = n_{i,tot} = c_{i,tot} V_{tot} = \text{const} \tag{88}$$

where $n_{i,tot}$ is the total mole number of the ith species, and $c_{i,tot}$ is the total concentration defined with respect to the total volume $V_{tot} = V_1 + V_2$. For the *transport of uncharged species* $(z_i = 0)$ *or if* $\Delta\phi_m = 0$, the function $f(a_i)$ adopts the form given in Eqn. 81 under the constraints specified in Eqn. 88 in the case of permeation (cf. Eqn. 60a) and carrier-mediated transport (Eqn. 78). The extreme flows are

$$J_+ = AP_i y_{i,1} n_{i,tot}/V_1 \quad \text{and} \quad J_- = AP_i y_{i,2} n_{i,tot}/V_2 \tag{89a}$$

for permeation, while for a carrier-mediated transport

$$J_+ = n_c k_{cat,1}/[1 + V_1 a^0 K_{m,1}/(y_{i,1} n_{i,tot})]$$
$$J_- = n_c k_{cat,2}/[1 + V_2 a^0 K_{m,2}/(y_{i,2} n_{i,tot}) + D_c] \tag{89b}$$

The additional term D_c in Eqn. 89b reads

$$D_c = \beta_3 \frac{a_{i,1}}{a^0 K_1} \frac{K_{4,1}(1 + K_{2,3})}{1 + \beta_1(1 + K_{2,3}) + \beta_3 K_{4,1}} \tag{89c}$$

and vanishes if $\beta_3 \approx 0$, i.e. if the transition of the empty carrier is much faster than that of the loaded carrier. Otherwise, the activity $a_{i,1}$ should be approximately constant. Since the volume of the suspending medium is much larger than that of the combined interior spaces of the vesicles, i.e. $V_1 \gg V_2$, the chemical capacity in compartment 1, particularly if supported by suitable buffers, is also much larger than that in compartment 2. Hence, $c_{i,1} \approx \text{const}$ and $a_{i,1} = y_{i,1} c_{i,1} \approx \text{const}$ is easily achieved.

Transport under the above specified constraints and conditions thus complies with the unambiguous flow-force relation of Eqn. 83. Its inflection point is rather offset from the origin in the case of permeation since (cf. Eqns. 84 and 89a) $X_0/(RT) = \ln\{V_2/V_1\}$ which is rather negative because of $V_2/V_1 \ll 1$. This is not so for transport mediated by a carrier. In this case, the location of the inflection point is mainly determined by $y_{i,k} n_{i,tot}/V_k$ $(k = 1, 2)$ with respect to the binding constants and the equilibrium constants for the translocation steps. Assuming for the moment that all $\beta_i \approx 0$ and $V_2/V_{tot} \approx 0$ while $V_1/V_{tot} \approx 1$, $X_0/(RT) \approx \ln\{K_{2,3}/[1 + a^0 K_1(1 + 1/K_{4,1})/(y_{i,1} c_{i,tot})]\}$.

Let us now turn to the other extreme where only an electrical potential difference across the membrane exists while the activities in both compartments are equal, i.e. $a_{i,1} = a_{i,2} = a_i$. Obviously, $J_i = 0$ for an uncharged species. For a charged species with charge number z_i, $X_i/(RT) = z_i \mathscr{F} \Delta\phi_m/(RT)$ (cf. Eqn. 78c) and for permeation (cf. Eqn. 60a)

$$J_i = X_i AP_i a_i/(RT) \tag{90}$$

The flow is always proportional to its conjugate force, in strong contrast to the other extreme case (viz. no electrical potential but a difference in activity) where it was at most linear over a limited range of force. We thus encounter the unpleasant situation that the chemical and electrical potentials which are *thermodynamically* equivalent in determining the force are *kinetically* not equivalent. This is in general also true for transport mediated by a carrier because the effect of an electrical potential on the transition probabilities usually concerns many or even all transitions in a cycle while different activities in the compartments affect only the association probabilities in the binding transitions (cf. Eqn. 67b). However, if a carrier is built in such a way that electrical potential and activity operate essentially on the same transitions, the two terms in the force have the same effect on transport, and such devices are said to display kinetic equivalence of electrical and chemical potential.

5.3. Energetics of enzymes

The cycle diagram method of Hill also provides insight into the energetics of the enzyme states. This will allow us to address among other topics the question of "rate-limiting steps".

5.3.1. Standard, basic, and gross free energy levels in enzymes: The equilibrium constants defined in Eqns. 72 and 73 for each transition in a diagram can be interpreted in terms of what Hill calls *standard free energy levels* [13], in full analogy with the equilibrium constants for reactions in solution (cf. Eqns. 26). Each transition from state i to state j is then associated with a change in standard free energy, $G_i^0 - G_j^0$, of the states. Thus, from Eqn. 72,

$$G_j^0 = G_i^0 - RT \ln\{K_{i,j}\} \tag{91a}$$

for a transition due to a conformational change (see section 4.2.1), and from Eqn. 73

$$G_{j(Rr)}^0 = G_i^0 + RT \ln\{K_{Rr}\} \tag{91b}$$

for a transition pertinent to binding of reactant R_r where j is the state with bound reactant as indicated. By virtue of Eqns. 91, we can construct a diagram for the levels of the standard free energies of the states in each cycle. These levels are with respect to an arbitrarily chosen reference state which usually is the state of the enzyme or carrier without bound reactants.

Figure 6B shows an example of standard free energy levels for the Ca^{2+}-transporting ATPase in the sarcoplasmic reticulum, calculated as described in Ref. 21 using data taken from the literature. The corresponding enzyme cycle is depicted in Fig. 6A. The zero of the standard

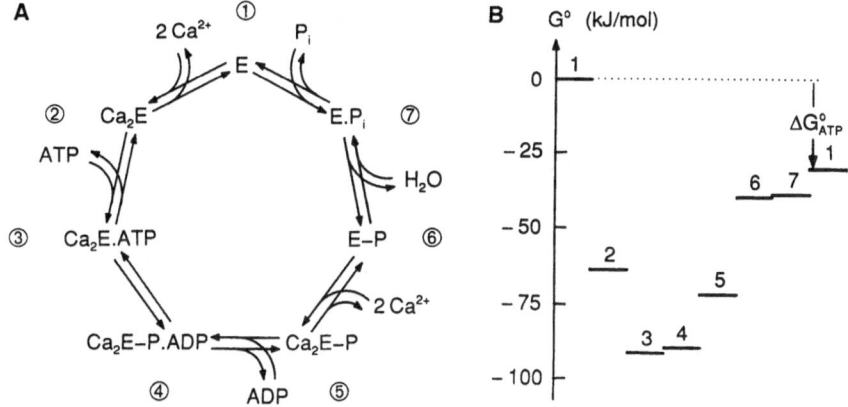

Fig. 6. Kinetic scheme (A) and standard free energy levels (B) for the Ca^{2+}-transporting ATPase of sarcoplasmic reticulum. In (A), $E.P_i$ represents the phosphate-liganded enzyme, E-P the phosphorylated enzyme, and so on. All reactant-binding transitions refer to the cytosol, except the transition $5 \rightleftarrows 6$, which refers to the the lumen. G^0 in (B) was calculated by means of Eqns. 91 using data from the literature (see Ref. 21).

free energy scale is chosen to coincide with the level corresponding to state 1, i.e. the free enzyme. The last level in Fig. 6B again represents the free enzyme and is displaced from the reference level by $-\Delta G^0_{ATP} = RT \ln\{K_c\}$. This is nothing other than thermokinetic balancing (Eqn. 71) which relates the equilibrium constants of the transitions in a cycle (Eqns. 72 and 73) to the equilibrium constant of the reaction associated with a turn around a cycle. In the present case, a rotation of the cycle in the direction which moves Ca^{2+} ions from the cytoplasm into the lumen of the reticulum (counter-clockwise in Fig. 6A) also hydrolyses one ATP molecule to ADP and Pi.

The standard free energy levels set the absolute frame or the 'scaffold' in which the enzyme can move. As is evident from the example in Fig. 6B, they can go up or down and thus do not determine the direction in which an enzyme cycle will turn. In fact, they are modified by the actual activities of the reactants in a steady state of the system. Including these activities leads to what Hill [13] has termed the *basic free energy* G′ of the enzyme states. Conformational transitions are obviously not affected by activities of reactants, hence

$$G'_j = G'_i - RT \ln\{K_{i,j}\} \tag{92a}$$

but for a transition with binding of a reactant,

$$G'_{j(Rr)} = G'_i + RT \ln\{a^0 K_{Rr}/a_{Rr}\} \tag{92b}$$

Figure 7 depicts the basic free energy levels obtained from the standard free energy levels in Fig. 6B for three different steady states. The

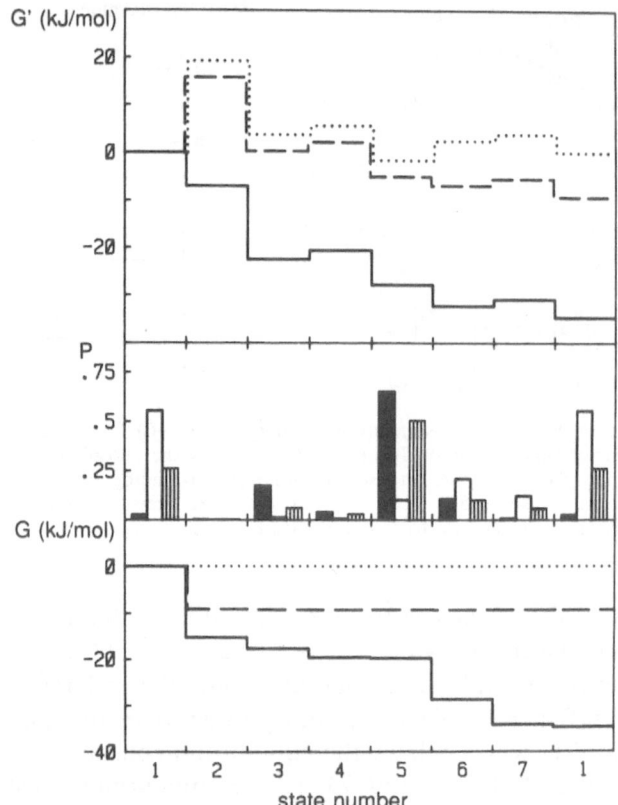

Fig. 7. Basic free energy levels G′, state probabilities P, and gross free energy levels G for the kinetic scheme in Fig. 6A. The Ca^{2+} ion concentrations and the symbols are as follows:

	cytoplasmic Ca^{2+} (μM)	lumenal Ca^{2+} (mM)	lines	bars
(i)	10	0.6	solid	solid
(ii)	0.1	1.0	broken	open
(iii)	0.05	3.308	dotted	hatched

The concentrations of ATP, ADP, and Pi are, respectively, 8 mM, 40 μM and 8 mM. Basic and gross free energy levels were calculated by means of Eqns. 92 and 93 as described in Ref. 21. Values for P were obtained using the simulation program SPICE (cf. chapter 2). Note that the three probabilities corresponding to state 2 are: (i) 8.9×10^{-4}; (ii) 2.4×10^{-5}; (iii) 1.1×10^{-4}.

difference between reference and final level is now equal to the thermo-dynamic force of the reaction $2Ca_{cyt}^{2+} + ATP + H_2O \rightleftarrows 2Ca_{lum}^{2+} + ADP + Pi$, where Ca_{cyt}^{2+} and Ca_{lum}^{2+} denote Ca^{2+} in the cytoplasm and the lumen of the sarcoplasmic reticulum, respectively. Accordingly, this difference is zero at equilibrium (cf. dotted line in top panel of Fig. 7). The basic free energies define the levels attainable by the enzyme states in a given steady state of the system, and still can go up and down. The enzyme so to speak executes a random walk on the basic free energy levels. This causes the probabilities P_i of the states to eventually adopt such values that the flows for each transition in a cycle are the same at the *steady state of the enzyme*.

Just as we have related the thermodynamic force of a reaction to the activities of the reactants by means of the equilibrium constant (see Eqn. 78a), we can define a thermodynamic force for each transition by means of the state probabilities and the pertinent equilibrium constants. The force is equal to the difference in what Hill [13] calls the *gross free energy* G_i of the enzyme states. Hence, in the absence of a membrane potential,

$$G_j = G_i - RT \ln\{K_{i,j} P_i / P_j)\} \tag{93a}$$

and

$$G_{j(Rr)} = G_i + RT \ln\{a^0 K_{Rr} P_j / (a_{Rr} P_i)\} \tag{93b}$$

Fig. 7 shows the gross free energy levels for the three steady states. Note that the difference between final and reference level is still equal to the force of the reaction, in line with the notion that catalysis does not change the energetics of the catalyzed reaction. The force for each transition now has the same sign, and indicates the net turnover direction of the cycle resulting from stochastic backward and forward movements of the enzyme between the states.

The above considerations apply to enzymes in solution as well. In the case of membrane-bound enzymes in the presence of a membrane potential the effect of the electric field on the transition probabilities has to be included. This adds the term $(\bar{z}_{i,j} - \bar{z}_{j,i})\mathcal{F}\Delta\phi_m / (RT)$ to Eqns. 92 and 93 (for $\bar{z}_{i,j}$ and $\bar{z}_{j,i}$ see Eqn. 50). The membrane potential shifts the basic and gross free energy levels for those transitions which involve a charge displacement (see section 4.2.3) and contributes to the thermodynamic force of the transmembrane processes (cf. Eqn. 78c). It does not, however, affect the standard free energy levels.

5.3.2. Relevance of free energies to enzyme performance: The standard free energy levels reflect the interaction of the enzyme with the reactants. Thus, in the example shown in Fig. 6, there is a strong interaction between Ca^{2+} ions and the cytoplasmic face of the enzyme while the opposite is true for the luminal face of the enzyme. These interaction energies should be considered in the context of the conditions the enzyme is designed to encounter, i.e. the activities of Ca^{2+} ions in the cytosol and lumen of the sarcoplasmic reticulum. The respective dissociation constants are roughly 2×10^{-6} and 2×10^{-3} for the release of these ions, well adjusted to the Ca^{2+} concentrations in molar units typical for the two spaces (see legend to Fig. 7). The basic free energy levels for the three steady states show that in the case of rapid turnover of the enzyme (solid line) both the association of Ca^{2+} from the cytoplasm and its dissociation into the lumen are favorable.

A frequently used approach is to declare one transition in a kinetic scheme to be the "rate-limiting" step. This is supposed to justify the assumption that all other transitions are either at or very close to equilibrium. Such close-to-equilibrium transitions are recognizable by a vanishing difference between the gross free energy levels of the pertinent states, which immediately brings us to the following conclusion. *The rate-limiting step in a scheme practically does not exist because G is determined by G′ which in turn depends on the activities of the reactants* (cf. Eqns. 92 and 93). Hence, only in exceptional cases will the variation in G′ be balanced by appropriate adjustments of the state probabilities. These adjustments must be such that G remains constant prior to the rate-limiting step and again subsequent to that step, while the varying thermodynamic force of the reaction appears entirely as a difference in G between the states adjacent to the rate-limiting step. A situation of this kind only occurs if all the β_i in a given scheme are essentially zero. Otherwise, transitions with a very small difference in G can occur (see Fig. 7, bottom panel) but this condition does not have to persist when the activities of the reactants vary. Under no circumstances can a rate-limiting step be inferred from the standard free energy levels.

Acknowledgements

D.W. acknowledges financial support by the Swiss National Science Foundation and by the Julius Bär Foundation (Zürich). S.R.C. was supported by the Basic Research Foundation administered by the Israel Academy of Sciences and Humanities.

References

1. D Walz, Biochim. Biophys. Acta 1019 (1990) 171–224.
2. D Walz and SR Caplan, Bioelectrochem. Bioenergetics 28 (1992) 5–30.
3. PW Atkins, *Physical Chemistry*, Oxford University Press, Oxford, 1994.
4. SR Caplan in *Current Topics in Bioenergetics*, DR Sanadi (ed), Academic Press, New York, 1971, Vol. 4, pp. 2–77.
5. I Prigogine, *Introduction to Thermodynamics of Irreversible Processes*, Interscience Publ., New York, 1967.
6. D Walz Biochim. Biophys. Acta 505 (1979) 279–353.
7. O Kedem and A Katchalsky, Biochim. Biophys. Acta 27 (1958) 229–246; Trans. Farad. Soc. 59 (1963) 1918, 1931, 1941.
8. D Walz and SR Caplan, Biochim. Biophys. Acta 859 (1986) 151–164.
9. K Denbigh, *The Principles of Chemical Equilibrium*, Cambridge University Press, Cambridge, 1964.
10. EL King and C Altman, J. Phys. Chem. 60 (1956) 1375–1378.
11. WW Cleland, Biochim. Biophys. Acta 67 (1963) 104–137.
12. IH Segel, *Enzyme Kinetics*, John Wiley & Sons, New York, 1975.
13. TL Hill, *Free Energy Transduction in Biology*, Academic Press, New York, 1977.
14. G Tollin, TE Meyer and MA Cusanovich, Biochim. Biophys. Acta 853 (1986) 29–41.

15. S Glasstone, KJ Laidler and H Eyring, *The Theory of Rate Processes*, McGraw-Hill, New York, 1941.
16. P Läuger, Biochim. Biophys. Acta 779 (1984) 307–341.
17. RA Marcus and N Sukin, Biochim. Biophys. Acta 811 (1985) 265–322.
18. IW Richardson, V Litcko and E Bartoli J. Membrane Biol. 11 (1973) 293–308.
19. P Läuger, Biochim. Biophys. Acta 311 (1973) 423–441.
20. G Stark, B Ketterer, R Benz and P Läuger, Biophys. J. 11 (1971) 981–994.
21. D Walz and SR Caplan, Cell Biophys. 12 (1988) 13–28.
22. JBS Haldane, *Enzymes*, Longmans, Green & Co., London, 1930 (reprinted by M.I.T. Press, Cambridge, MA, 1965).
23. SR Caplan and M Kara-Ivanov, Int. Rev. Cytology 147 (1993) 97–164.
24. RM Macnab in *Biological Structures and Coupled Flows*, A Oplatka and M Balaban (eds), Academic Press, New York, 1983, pp. 147–160.

Appendix A. Glossary

$a_{i,k}$	activity of ith species in kth compartment or phase
a^0	standard activity (1 M)
A	total membrane area
A_{ox}, A_{rd}	oxidized, reduced species of acceptor redox couple
$\mathscr{A}_{j,k}$	affinity of jth chemical reaction in kth compartment
$\mathscr{A}_{ej,k}$	affinity of jth redox reaction in kth compartment
$c_{i,k}$	concentration of ith species in kth compartment or phase
d	thickness of membrane
D_{ox}, D_{rd}	oxidized, reduced species of donor redox couple
$E_{R,k}$	redox potential of couple R_{ox}/R_{rd} in kth compartment
$E^0_{R,k}$	standard redox potential of couple R_{ox}/R_{rd} in kth compartment
$\mathscr{F}$	Faraday constant
G	free enthalpy (Gibbs free energy)
G^0_i	standard free energy of ith enzyme state
G'_i	basic free energy of ith enzyme state
G_i	gross free energy of ith enzyme state
$G^{\#}_{ij}$	standard free energy of the transient state between states i and j of an enzyme
$\Delta G^0_{j,k}$	standard free enthalpy (standard Gibbs free energy) of jth chemical reaction in kth compartment
$\Delta H^0_{j,k}$	standard enthalpy of jth chemical reaction in kth compartment
$J_{j,k}$	flow for jth chemical reaction in kth compartment
$J_{ej,k}$	flow for jth redox reaction in kth compartment
J_i	flow for transport of ith species
J_+, J_-	extreme flows for processes under constraint
$k_{cat,X}$	turnover number for reactant X of an enzyme-catalyzed process
k_i	unidirectional first-order or second-order rate constant for forward reaction of ith chemical reaction

k_{-i} unidirectional first-order or second-order rate constant for backward reaction of ith chemical reaction

K_c equilibrium constant for overall process catalyzed by given cycle c in a kinetic scheme of an enzyme

$K_{ej,k}$ equilibrium constant for jth redox reaction in kth compartment

$K_{i,j}$ equilibrium constant for the transition between states i and j of an enzyme

$K_{i,j}^*$ apparent equilibrium constant in the presence of an electric field

$K_{m,X}$ Michaelis constant for reactant X of an enzyme-catalyzed process

$K_{pi,kk'}$ equilibrium constant for partitioning of ith species between phase or compartment k and k' (partition coefficient)

$K_{ij,k}$ equilibrium constant for jth chemical reaction in kth compartment

K_X dissociation (equilibrium) constant for binding of reactant X to an enzyme

K_X^* apparent dissociation constant in the presence of an electric field

K_X' combined dissociation constant resulting from a reduction of states

l elongation

$n_{E,i}$ mole number of enzyme in ith state

$n_{E,tot}$ total mole number of enzyme

$n_{i,k}$ mole number of ith species in kth compartment or phase

p pressure

P_i probability of ith state of an enzyme; permeability of ith species

$P_{p(j,k)}$ pth final reactant (or product) of jth chemical reaction in kth compartment

R gas constant

R_{ox}, R_{rd} oxidized, reduced species of redox couple R

$R_{r(j,k)}$ rth reactant (initial or final) of jth chemical reaction in kth compartment

S entropy

$\Delta S_{j,k}^0$ standard entropy of jth chemical reaction in kth compartment

$S_{s(j,k)}$ sth initial reactant (or substrate) of jth chemical reaction in kth compartment

t time

T absolute temperature

u_i mobility of ith species in membrane phase

U internal energy

V volume

x   cartesian coordinate

X   force (thermodynamic or mechanical)

X_0 thermodynamic force at inflection point of a flow-force relation

$y_{i,k}$ activity coefficient of ith species in kth compartment or phase

z_i charge number of ith species

$\bar{z}_{i,j}$ effective charge number for transition from state i to state j of an enzyme

$\alpha_{i,j}$ first-order or pseudo-first-order transition probability for transition from state i to state j of an enzyme

$\alpha_{i,j}^*$ transition probability in the presence of an electric field

$\alpha'_{i,j}$ combined transition probability resulting from a reduction of states

$\alpha_{i,j}^0$ second-order transition probability for transition from state i to state j of an enzyme

β relative transition probability

β^* relative transition probability in the presence of an electric field

$\gamma_{i,j}$ factor expressing average contribution of charges to energy barrier separating states i and j of an enzyme

$\tilde{\mu}_{i,k}$ electrochemical potential of ith species in kth compartment or phase

$\mu_{i,k}$ chemical potential of ith species in kth compartment or phase

$\mu_{i,k}^0$ standard chemical potential of ith species in kth compartment or phase

$\tilde{\mu}_{e(R,k)}$ electrochemical potential of electron associated with R_{ox}/R_{rd} in kth compartment

$\mu_{e(R,k)}$ chemical potential of electron associated with R_{ox}/R_{rd} in kth compartment

$\mu_{e(R,k)}^0$ standard chemical potential of electron associated with R_{ox}/R_{rd} in kth compartment

$\nu_{e(R)}$ number of electrons exchanged by redox couple R_{ox}/R_{rd}

$\nu_{Rr(j,k)}$ stoichiometric coefficient for reactant R_r of jth chemical reaction in kth compartment

$\xi_{j,k}$ degree of advancement of jth chemical reaction in kth compartment

$\rho_{R,k}$ reduction state of redox couple R_{ox}/R_{rd} in kth compartment

Σ_x abbreviation for denominator in relation describing flow of an enzyme-catalyzed chemical reaction ($x = r$), redox reaction ($x = e$), or transport ($x = c$)

Σ_c^* Σ_c in the presence of an electric field

Φ dissipation function

ϕ_k electrical potential (Galvani potential) of kth compartment
            or phase
$\Delta\phi_m$ electrical potential difference across membrane (membrane
            potential)
[ ]$_{eq}$ equilibrium value of quantities given between brackets

Appendix B: Energy barriers in transitions

The transition from a given state i to a given state j usually involves a
geometrical rearrangement of some or all of the enzyme components.
This is accompanied by a transient increase in the standard free energy of the enzyme (see section 5.3.1). According to the absolute rate
theory [15] the transition probabilities $\alpha_{i,j}$ and $\alpha_{j,i}$ are governed by
$\exp\{(G_i^0 - G_{ij}^{\#})/(RT)\}$ and $\exp\{(G_j^0 - G_{ij}^{\#})/(RT)\}$, respectively, where
$G_{ij}^{\#}$ denotes the maximum of the transient energy increase. In general
the value of $G_{ij}^{\#}$ is not known and not even required, since the intrinsic
transition probabilities are given.

If the geometrical rearrangement includes movements of charged
groups in an electric field, an electrical contribution must be added to
$G_{ij}^{\#}$. It is to be expected that each charge adds a different contribution
to the maximal value. Since there is no way of knowing what these
individual contributions are, a reasonable assumption is that the contribution of each charge to the maximal value is, on the average, one half
of its possible total contribution. Accordingly, $\gamma_{i,j}$ is invariably chosen to
be 0.5. Any other choice requires rigorous justification.

Bioelectrochemistry: General Introduction
ed. by S. R. Caplan, I. R. Miller and G. Milazzo†
© 1995 Birkhäuser Verlag Basel/Switzerland

CHAPTER 2
Methods of mathematical modelling

Dieter Walz[1], S. Roy Caplan[2], David R. L. Scriven[3], and
Donald C. Mikulecky[4]

[1]*Biozentrum, University of Basel, Basel, Switzerland*
[2]*Department of Membrane Research and Biophysics, The Weizmann Institute of Science, Rehovot, Israel*
[3]*Department of Physiology, University of the Witwatersrand Medical School, Parktown, Johannesburg, South Africa*
[4]*Department of Physiology and Biophysics, Medical College of Virginia, Virginia Commonwealth University, Richmond, VA, USA*

1. Introduction

Nonequilibrium thermodynamics and kinetics as presented in chapter 1 are adequate and sufficient for the analysis of systems or processes in the steady state. On the other hand, they only provide necessary tools for the analysis of a system on its way to a steady state, but not the actual means of performing the analysis. To carry out such an analysis one must invoke mathematical modelling. Moreover, even in the steady state modelling proves to be a valuable means of exploring the behavior of complex systems. Thus mathematical modelling is an important subject in bioelectrochemistry and hence is dealt with in this chapter. Furthermore, this chapter includes a number of carefully-chosen worked examples. These examples are not only illustrative and designed to cover most features of modelling, but they are also biologically realistic, and in an important sense extend the treatment given in chapter 1. In addition, they present novel aspects of bioelectrochemistry which are difficult to display without the aid of simulation.

1.1. What is mathematical modelling?

The aim of mathematical modelling is to simulate the behavior of a given system. The system may derive from a translation of an experimental set-up into thermodynamic terms (cf. section 2 in chapter 1). In this case modelling enables us to understand and interpret data collected in experiments on the basis of the mechanisms and processes assumed to occur in the system. Alternatively, the system may be hypothetical and emerge from the modeller's fantasy, in which case modelling will tell how the features implanted into the system show up in its parameters when the system is subjected to a Gedankenexperiment.

The first stage of modelling consists in designing a suitable model. Although this statement may sound almost trivial it is far from being so. In fact it means that the system is specified to such a level of detail that an *umambiguous description in mathematical terms* is possible. Neither undetermined elements nor redundancies are permissible. The mathematical description of the model forms the second stage of modelling. It implies that the modeller has a precise idea of which mechanisms or which relations govern the processes assumed to occur in the system. Moreover, the effect of these processes on the parameters of the system has to be assessed by mathematical relations. The result of a consistent and unambiguous mathematical description is a set of usually non-linear algebraic and/or differential equations. In most cases these equations contain the system's parameters in a form which is not suitable for judging the system's behavior. Hence, in the third stage of modelling, the equations have to be solved for the parameters of interest, which often is the most tedious and difficult step to perform.

1.2. An example which serves to illustrate the essentials of modelling

Consider the system shown in Fig. 1A which consists of two compartments separated by a membrane. The two compartments are initially filled with an aqueous solution which contains a uni-univalent salt. The

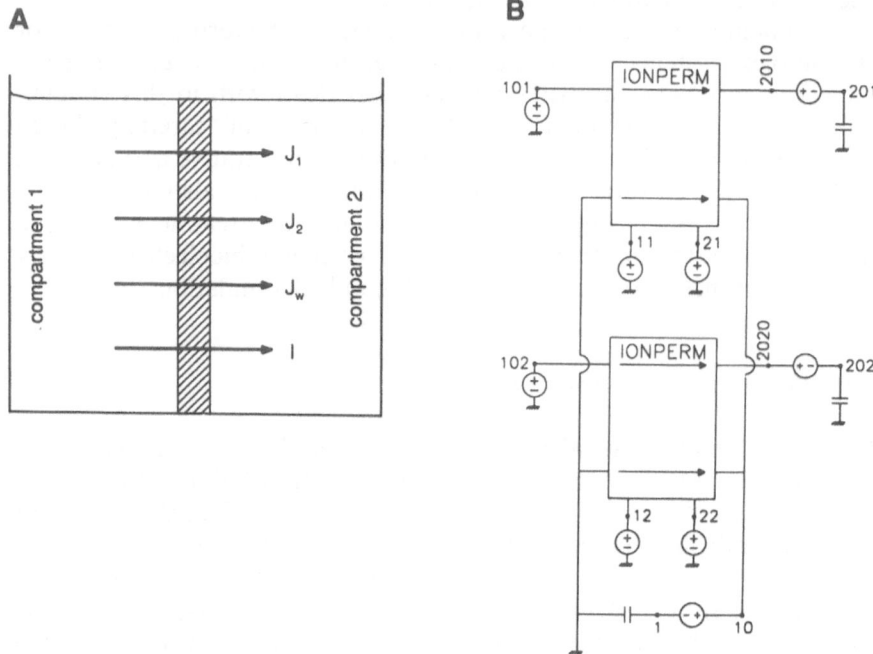

Fig. 1. Permeation of salt through a membrane. A. Topological map of the system with two compartments separated by a membrane. The arrows indicate the flows of ionic species (J_1, J_2), water (J_w), and charges (I), respectively. B. Representation of the model describing the system in terms of an equivalent electrical network. This network allows one to simulate the behavior of the system by means of the program SPICE. The names of the elements are omitted due to lack of space, but they can be found in the listing of the pertinent SPICE code in section C1. The symbol $\overline{7\!7\!7}$ denotes the "electrical ground" which has zero potential. For further explanation see text.

concentration of the salt is known but different in the two compartments. The membrane is permeable to both ions of the salt, and the pertinent permeabilities are known. Let us assume that the transport of the ions through the membrane and that of the charges associated with the ions have reached a steady state so that the condition of electroneutrality applies (cf. Eqn. 20 in chapter 1). Under this condition the electrical potential difference between the compartments (i.e. the membrane potential) $\Delta\phi_m$, which evolved while approaching the steady state of the processes, can be calculated by means of the Goldman equation (Eqn. 63 in chapter 1). The parameters required in this equation are the permeabilities (which are known) and the activities of the ions in the two compartments. For the sake of simplicity, the activities are equated to the concentrations (i.e. activity coefficients are set to unity, cf. Eqn. 22 in chapter 1).

As a first approximation the initial values for the concentrations could be used. This, however, implies the assumption that these concentrations hardly change during the transient leading to steady state. A more realistic description of the system during this transient includes the effect of the flow of cations, J_1, and anions, J_2 (see Fig. 1A) on the concentrations of these species, $c_{1,k}$ and $c_{2,k}$, in both compartments ($k = 1$ and 2). Moreover, it takes into account that ion flows give rise to an electrical current, I, across the membrane which vanishes only when the membrane has reached a steady state where the condition of electroneutrality applies. In general the membrane is also permeable to water which causes a flow of water (marked by J_w in Fig. 1A) whenever the concentrations of solutes in the two compartments differ. In the present example the permeability of the membrane for water is assumed to be so small that J_w can be neglected.

In order to transform this description into a set of mathematical relations a sign convention for transport has to be adopted first (see section 3.3.2 in chapter 1). Here flows from compartment 1 to compartment 2 are counted positive. Next it is recognized that the flow J_i of the ith species ($i = 1$ for cation, $i = 2$ for anion) causes a change in the mole number, $n_{i,k}$, of the ith species in the kth compartment, which is given by Eqn. 16 in chapter 1. The change in mole number is related to the change in concentration by the chemical capacity $C_{ci,k}$ for species i in comparment k:

$$C_{ci,k} = dn_{i,k}/dc_{i,k} \tag{1}$$

In the absence of buffering $C_{ci,k}$ is equal to the volume, V_k, of the compartment [1]. It then follows that

$$dc_{i,1}/dt = -J_i/V_1 \quad \text{and} \quad dc_{i,2}/dt = J_i/V_2 \tag{2}$$

Each ion flow carries an electrical current, I_i, which amounts to

$$I_i = z_i \mathscr{F} J_i \tag{3}$$

Here z_i is the charge number of the ith species, while $\mathscr{F}$ denotes the Faraday constant. The total electric current, I, is the sum of all I_i, hence

$$I = \mathscr{F} \sum_i z_i J_i = dQ/dt \tag{4}$$

It transfers the charge increment dQ per time increment dt from compartment 1 to compartment 2, which in turn causes a change in the difference in electrical potential $\Delta\phi_m = \phi_1 - \phi_2$ between the compartments. Since the electrical capacity of the membrane is defined as $C_m = dQ/d\Delta\phi_m$, the time derivative of $\Delta\phi_m$ becomes, in view of Eqn. 4,

$$d\Delta\phi_m/dt = (-1/C_m)\, dQ/dt = (-\mathscr{F}/C_m) \sum_i z_i J_i \tag{5}$$

If the transport of ions in the membrane is assumed to be in a steady state the flow of the ith species is described by the integrated Nernst-

Planck equation (Eqn. 60 in chapter 1) which reads, with activity coefficients set to unity,

$$J_i = AP_i \frac{z_i \mathscr{F} \Delta\phi_m}{RT} \frac{c_{i,1} \exp\{z_i \mathscr{F} \Delta\phi_m/(RT)\} - c_{i,2}}{\exp\{z_i \mathscr{F} \Delta\phi_m/(RT)\} - 1} \tag{6}$$

where P_i denotes the permeability of the ith species and A is the total membrane area. Equations 2, 5, and 6 constitute a set of nonlinear differential equations which fully describes the system. The actual behavior of the system is obtained upon integration of the differential equations. This yields the time course of the relevant parameters of the system which, in the present example, are the concentrations of the species in the compartments and the membrane potential. The integration is most conveniently done by means of a technique which makes use of an equivalent electrical network representation of the system (see Fig. 1B). The results to be presented in section 5.1 provide detailed information about the system, in particular how the membrane approaches a steady state and how much the concentrations in the compartments change during this initial phase.

1.3. Goal of this chapter

This chapter is intended to provide a reader having little or no previous experience in mathematical modelling with a tool-kit for the construction of both steady-state and time-dependent models. The systems to be modelled may be macroscopic, such as those encountered in electrophysiology, or they may be microscopic (i.e., cellular or molecular), such as those that crop up repeatedly in biochemistry or bioelectrochemistry. The techniques to be described have been routinely used by the authors for several years. Not only are they convenient, but they have the advantage of providing the user with an algorithmic method which has considerable intuitive appeal, since it relates in a fairly direct way to the user's physical or conceptual picture of the system. Manipulating the model can provide a great deal of insight because of the model's one-to-one correspondence with the elements of the actual system under investigation. In this sense the model is highly transparent, and lends itself readily to the performance of Gedankenexperiments.

But this is not to say that we are offering a technique here which is trivially easy to master. Most biological systems are fairly complex, and hence a successful model must necessarily be a somewhat complex affair. However, by carefully following the procedures set out below, which are less formidable than they may seem at first sight, we believe that within a reasonably short period the prospective modeller should have a working model available. This will enable him or her to simulate the behavior of his or her system under all possible conditions which can be (and even some which cannot be) realized in the laboratory.

2. Design of a model and its use for simulations

In this section the procedure for designing a model is outlined in the form of a set of general rules. Each rule is applied to the example given in section 1.2 in order to illustrate its meaning. Since the major goal of modelling is the simulation of the system's behavior, the options available to the modeller for performing this task are discussed. Special emphasis is placed on the network representation of models since this approach paves the way to the most easily handled and yet most versatile simulation procedure.

2.1. How to devise a model

The development of a model comprises the following sequence of steps:

(1) Enumerate all the elements of the system (compartments, barriers, pumps, carriers, contractile elements, etc.).
   The elements of the system shown in Fig. 1A are the two compartments 1 and 2, and the membrane.
(2) Make a topological map (i.e., a schematic sketch) showing how the elements are interconnected. The elements may be morphological or the sequential steps of a sequence of reactions or both.
   Fig. 1A displays the topological arrangement of the two compartments and the membrane.
(3) Enumerate all the elemental processes occurring (material flows of all species, volume flows, electrical flows, heat flows, chemical reactions) which could conceivably be anticipated from the topological map. Some of these processes may be eliminated (either permanently or as a first approximation) on the basis of experimental data or the modeller's intuition.
   The elemental processes in the example given are the flows of ions and water across the membrane (indicated by the arrows marked with a J in Fig. 1A), as well as the electric current (arrow marked with an I) which charges the membrane capacity. Additional elemental processes (not shown) are the partitioning of ions and water into the membrane.
(4) Describe the remaining elemental processes in mathematical terms (i.e. flows as functions of the parameters of the system). This forces one to decide on which level the problem can be approached: the phenomenological level or a more or less detailed molecular kinetic

level. If both options are available, a decision must be made as to which level is appropriate.

The description chosen for the example given is phenomenological. It uses the integrated Nernst-Planck equation (Eqn. 6) for ion transport and a constant capacity to represent the electrical properties of the membrane. Moreover, the flow of water is neglected. A more detailed molecular kinetic level would consider the partitioning of the ions into the membrane and their diffusion within the membrane (as is done in ref. 2). It would further take into account that the storage property of the membrane for charges is also determined by the diffuse double layers of ions adjacent to the membrane surfaces and the fixed charge possibly present on the membrane surfaces (see ref. 1 and chapters 4 and 5).

(5) Decide whether it is sufficient to consider steady states only, or also time-dependent behavior. In the former case all the necessary equations have already been collected, while in the latter the effects of the flows on the parameters of the system must be taken into consideration, giving rise to their dependence on time (e.g. changes in time of concentrations, electrical potentials, tension in elastic elements, etc.).

The time behavior of the example given is of interest. The effects of the flows on the parameters of the system are expressed by Eqns. 2 and 5.

(6) Ascertain which parameters are experimentally controllable (in an actual or a Gedankenexperiment) and which adjust themselves to the controlled parameters. The uncontrolled parameters (i.e., unknowns) should now be expressed in terms of the controlled parameters, which implies that the number of equations must be equal to the number of uncontrolled parameters. If this is true the model is fully described mathematically.

The only controllable parameters in the example given are the volumes of the solutions in both compartments. The ion concentrations and the membrane potential are uncontrolled, but their readjustment is fully described by Eqns. 2 and 5, respectively.

Note that an alternative approach in which some of these steps are unnecessary will be considered below.

2.2. How to simulate the behavior of the model

The mathematical description will in general lead to a set of nonlinear algebraic equations for steady-state systems or a set of nonlinear differential equations for time-dependent systems (cf. Eqns. 2, 5, and 6 which describe the system shown in Fig. 1A). The following options are then open to the modeller:

(1) Analytical solution of the equations. This is tedious or impracticable except in very simple cases.
(2) Numerical solution of the equations. This can be approached in a number of ways:
  (a) By using an appropriate "package" to solve algebraic equations or to integrate differential equations with a computer[1]. This invariably requires a certain degree of skill and experience in mathematics, particularly in the use of numerical methods.
  (b) By translating the model into a network representation, and using one of the available packages for simulating networks on a computer. This makes it unnecessary to express the uncontrolled parameters as functions of the controlled parameters. Moreover, in the case of time-dependent behavior this approach eliminates the need to express the effects of the flows on the parameters of the system explicitly. In view of these advantages, we shall devote the rest of this chapter to network representations of models and their use for simulations.

2.3. Network representations

2.3.1 Historical survey: King and Altman [3] used a network and graph theoretical representation developed for electrical networks [4] to manipulate and solve the systems of kinetic equations generated by the study of complicated enzymatic reactions in the steady state. Later, Hill [5] independently rediscovered and massively extended this method in the broader context of active transport and related systems. This approach to energy transduction was readily shown to be in harmony with nonequilibrium thermodynamics.

Another approach to the analysis of highly organized dynamic systems was the holistic technique known as "network thermodynamics" [6]. Peusner [7] chose to develop a network representation of the system in terms of simple electrical circuits, rather than the somewhat abstract "bond graphs" used earlier by Oster et al. [6]. However, it should be emphasized that network simulation is not to be confused with the equivalent circuit representation of classical electrophysiology, which cannot deal unambiguously with coupled processes.

It may be asked whether it is clear that the network and graph theoretical methods are indeed the appropriate ones to use in modelling. The answer is that just as calculus is the branch of mathematics suited to the study of rate processes, topology is the branch of mathematics developed for the study of relationships among interconnected elements.

---

[1]A convenient package for use on personal computers is Mlab, obtainable from Civilized Software, Inc., 7735 Old Georgetown Road, Bethesda, MD 20814, USA.

Networks and linear graphs are simply applications of more fundamental topological ideas. The network representation of models combines both types of mathematical operation. Its adaptation to biological modelling by Mikulecky and co-workers [8–11] has opened an entirely new range of possibilities to the biologist.

2.3.2. How to obtain a network: A network representation of the system is obtained by expressing all the processes involved in terms of storage devices and connecting elements. The connecting elements define the flows associated with these processes (e.g. material, heat, charge, or reaction flows) while the storage devices define the storage capacities for whatever it is that happens to be flowing. Thus, for the examples of flows enumerated above the storage capacities would be volume, heat capacity, electrical capacity, and chemical capacity, respectively. The connecting elements constitute the branches of the network; they are hooked together and connected to the storage devices at the nodes of the network. In this way the network also reflects the topology of the system.

2.3.3. Equivalence of network representations: Consider a number of different systems, e.g. hydrodynamic, electrical, mechanical, or chemical. It may be the case that all of these systems have similar network representations. If this is so, the topologies of the various systems are identical, and the mathematical equations governing the behavior of the connecting elements are isomorphous. This means that the equation of a given connecting element in a given system can readily be transformed by parameter substitutions to yield the equations of the corresponding elements in the other systems. In other words, the network representations of the different systems are equivalent.

The important point to note is that once a solution has been found describing the behavior of a given system, it will describe the behavior of all other systems having an equivalent network representation.

2.3.4. Network simulation programs: Any preexisting computer program ("package") which simulates the behavior of a given system in terms of its network representation is a suitable tool for modelling. What is required is a transformation of the network representation of the model into the equivalent representation appropriate for the package. Several such packages are available, for example STELLA uses a network representation in terms of reservoirs as storage devices and pipes as connecting elements. The most common and most elaborate packages deal with electrical networks; examples of these are SPICE[2]

[2]SPICE (Simulation Program with Integral Circuit Emphasis) is available from the Department of Electrical and Computer Sciences, University of California, Berkeley, CA 94720, USA.

and NET2. To the best of our knowledge SPICE is the most versatile for our purposes. While SPICE is designed for implementation on a main-frame computer and is in the public domain, a personal computer version known as PSpice[3] is available commercially which in most respects is identical to SPICE [12]. However, it incorporates a number of additional features that, although intended for the convenience of electrical engineers, may in some cases offer significant extra facilities to those engaged in bioelectrochemical modelling.

For this chapter we have chosen SPICE as the tool for simulations and provide a readily-accessible laboratory manual and reference to the practical use of SPICE.

3. Translation of a model into a form suitable for SPICE

3.1. Outline of the procedure

A prerequisite for translating a model into a form which can be handled by SPICE is that a full mathematical description of the model has been achieved (cf. section 2.1). This comprises a description of the dependence of the flows on the parameters of the system. The parameters may include number of moles, number of charges, concentration, pressure, and electrical or chemical potentials. The flows may be due to diffusion, convection, or other modes of transport of solute, solvent, volume, or electrical charge, as well as to the advancement of chemical reactions.

The procedure to be followed consists in casting the model into a network representation (cf. section 2.3.2) whose storage devices and connecting elements are expressed in terms of an equivalent electrical network by means of suitable parameter substitutions (cf. section 2.3.3). The electrical elements to be used are those offered by SPICE. The operation of each of these elements is well defined and expressed in an equation, called the "constitutive relation", which describes the effect of the element on the electrical parameters of the network. In the network thus obtained, all parameters of the system are represented as electrical potentials and all flows as electrical currents. This does not cause any problems if the general rules to be summarized in section 3.2 are strictly obeyed. Moreover, it is necessary to establish the relationship between the electrical units and the units of the parameters and the flows of the model. This relationship is determined by the chosen parameter substitutions. We refer to this step as the "scaling" of the system (see section 3.3).

The equivalent electrical network designed to represent the model is in a diagrammatic form which cannot be read by a computer. Therefore,

---

[3]Registered trademark of the MicroSim Corporation, 20 Fairbanks, Irvine, CA 92718, USA.

a formal language was created which allows one to describe an electrical network unambiguously and in a format that is "understood" by a computer. The language also includes commands which tell the computer what type of simulations should be performed with the model, and in which form the computer should deliver (or output) the result of the simulation. The rules of this language and the format used to represent the elements of a network are given in Appendix A. The description of a network written in terms of this language consists of a set of statements which we refer to as the "SPICE code" of the model. Obviously this code can only be understood on the basis of the rules of the language. Hence, in order to avoid the possibility that code be mistaken for regular text, all SPICE codes will be printed in a different font.

3.2. General rules

The rules listed below are intended as guide lines for the translation of a model into an electrical network suitable for SPICE. The basic principles are stated first in general terms and then are qualified by applying them to the example introduced in section 1.2 (see also Fig. 1). SPICE elements are mentioned in the rules, but only for the purpose of indicating which element is used in order to implement a given characteristic of a model in the electrical network.

(1) All parameters of the system are associated with nodes. The potentials at these nodes with respect to ground in the electrical network indicate the values of the pertinent parameters. This implies that, e.g., a node is assigned to each species present within each compartment of the system shown in Fig. 1A. Similarly, a node must be assigned to the electrical potential of each compartment. Note that in the latter case the parameter of the system and the actual quantity in the electrical network are formally identical, but nevertheless the units may differ due to the scaling of the model (see section 3.3).

(2) A storage device for a quantity is represented by a capacitor in the electrical network. Each node associated with a quantity subject to storage must be connected to its own storage device even if the actual element providing storage in the system is the same for several quantities. Thus, a capacitor has to be connected to each node that represents a species in a given compartment of the system in Fig. 1A if the simulation is intended to show the effect of storage on the parameters of the system. This is in line with defining the chemical capacity separately for each species in a compartment (Eqn. 1), even if its actual value is the same for all species in a given

compartment (in this case equal to the volume of the compartment).

(3) If a parameter in a system is constant or has a predetermined
behavior in time (e.g. a clamped concentration or an electrical
potential difference which is swept during the experiment) its node
in the electrical network is connected to a SPICE element called an
independent voltage source. Similarly, a flow in a system which is
constant or has a predetermined behavior (e.g. a controlled addition of a reactant to a compartment) can be made so in the
electrical network by means of a SPICE element called an independent current source.

(4) Only nodes whose parameters pertain to the same process or species
in the system may be connected by a connecting element in the
electrical network. This element determines the flow in the branch
and hence the "transfer of a quantity" from one node to another.
Thus, the nodes representing a given species in each of the compartments of the system shown in Fig. 1A may be connected by an
element which determines the flow of the species through the
membrane (cf. Eqn. 6). However, the electric currents associated
with the flows of the ionic species (cf. Eqn. 3) have to be represented by different elements which connect the nodes representing
the electrical potential in each compartment.

(5) The dependence of a flow on the parameters of the system is
modelled in the electrical network by means of a SPICE element
called a controlled current source. Any dependence that can be
expressed in terms of a polynomial can be simulated. The important
feature of a controlled source is that its flow can be dependent on
any parameter represented by nodes anywhere in the whole electrical network without being physically connected to these nodes.
Hence it does not violate rule (4) which requires the nodes and
elements pertaining to a given species in a given process to be
physically separated from all other nodes.

(6) The algebraic manipulations of division, exponentiation, and taking
logarithms of variables can be performed by special independent
circuits (see section 4.3). In these circuits the parameters associated
with the nodes, and the flows through the connecting elements, are
mere representations of variables in the mathematical operation;
they do not have the meaning explained in rule (2). Due to the
feature of controlled sources outlined in rule (5) the result of the
operation can be used in any other part of the electrical network. In
this way the exponential of the membrane potential required in
Eqn. 6 can be calculated. The result, together with the quantities
representing the concentrations of a species in both compartments,
is then used in another circuit to calculate the ratio appearing as the
third factor in Eqn. 6. With this factor the flow expressed in Eqn. 6

Table 1. Examples of correspondence of electrical quantities used in SPICE with quantities for different processes used in models

Electrical quantity	Units	Material flow	Volume flow	Heat flow	Electrical flow
Potential	volts	$\dfrac{\text{number of moles}}{\text{volume}}$	pressure†	temperature†	potential†
Current	amperes	$\dfrac{\text{number of moles}}{\text{time}}$	$\dfrac{\text{volume}}{\text{time}}$	$\dfrac{\text{heat}}{\text{time}}$	$\dfrac{\text{charge}}{\text{time}}$
Charge	coulombs	number of moles	volume	heat	charge
Capacitance	farads	volume	$\dfrac{\text{volume}}{\text{pressure}}$	$\dfrac{\text{heat}}{\text{temperature}}$	$\dfrac{\text{charge}}{\text{potential}}$
Resistance	ohms	$\dfrac{\text{time}}{\text{volume}}$	$\dfrac{\text{pressure} \times \text{time}}{\text{volume}}$	$\dfrac{\text{temperature} \times \text{time}}{\text{heat}}$	$\dfrac{\text{potential}}{\text{current}}$
Time	seconds	time	time	time	time

†The quantities pressure, temperature, and potential in models usually refer to differences rather than absolute values.

is simply the product of two variables and a constant. This can be easily implemented in the polynomial governing the flow through the controlled current source that represents the flow of the species in the electrical network.

3.3. Scaling

The quantities in an electrical network are potential, current, charge, capacitance, resistance, and time. These are described by a self-consistent set of electrical units, namely volts, amperes, coulombs, farads, ohms, and seconds (see Table 1). This set enables SPICE to utilize numerical values for the quantities without further specification of the units. Hence electrical quantities can be substituted for modeller-defined quantities without interfering with SPICE's interpretation of the quantities in electrical terms. Four examples of such substitutions are listed in Table 1. From these the rules for substitution become evident, which will enable the reader to construct additional substitutions if necessary.

A consequence of a substitution is that it becomes the responsibility of the modeller to choose a self-consistent set of units for the substituted quantities. The scheme in Table 1 indicates how self-consistent sets of units corresponding to the electrical units can be constructed for the different processes. In addition, the following hints may be found helpful when scaling the different circuits present in a network representation of a model:

(1) Choose the units to correspond to the magnitude of the system, so that the numbers involved are neither too large nor too small.
(2) Make the selection of units independently for each circuit according to rule (1). However, the time unit must be identical in all circuits.
(3) When parameters from different circuits are to be combined in the polynomial governing a controlled source in a further circuit, use scaling factors to adjust the units appropriately.

4. Elements of SPICE and their use in modelling

4.1. Built-in elements

From among all the elements implemented in SPICE, only those listed in Table 2 are required for simulations of bioelectrochemical systems. The constitutive relation indicates the behavior of each element with respect to its effect on the quantities in the electrical network. An element qualified as "passive" in Table 2 (resistor, capacitor, diode) introduces a dependence of the current I flowing through it on the

Table 2. Elements of SPICE used in electrical networks which represent a model of a bioelectrochemical system

Element	Constitutive relation	Name in a SPICE code	Symbol in a network
Resistor*	$U = RI$	R $\cdots\cdots$	(a)
Capacitor	$Q = CU$ or $I = C(dU/dt)$	C $\cdots\cdots$	(b)
Diode*	$I = I_s[\exp\{\mathscr{F}U/(nRT)\} - 1] + UG_{min}$	D $\cdots\cdots$	(c)
Independent voltage source*	$U = f(t)$	V $\cdots\cdots$	(d)
Independent current source	$I = f(t)$	I $\cdots\cdots$	(e)
Controlled voltage source‡	$U = \text{Poly}(U_1, U_2, \cdots)$ $U = \text{Poly}(I_1, I_2, \cdots)$	E $\cdots\cdots$ H $\cdots\cdots$	(f)
Controlled current source‡	$I = \text{Poly}(I_1, I_2, \cdots)$ $I = \text{Poly}(U_1, U_2, \cdots)$	F $\cdots\cdots$ G $\cdots\cdots$	(g)
Subcircuit	user defined	X $\cdots\cdots$	user defined

Passive elements

(a)   (b)   (c)

Active elements

(d)   (e)   (f)   (g)

The quantities appearing in the constitutive relations are electrical potential difference U, current I, resistance R, capacitance C, charge Q, and time t (see also Table 1). I_s and n denote diode specific parameters, while G_{min} is the minimal conductance adopted by SPICE in a simulation. For functions of time, $f(t)$, implemented in SPICE see section A2.6. Poly indicates a polynomial as specified in Eqns. A6–A9 in Appendix A.
*Element which provides a DC path to ground (see section A1.3).
‡A source is either "voltage controlled" if the arguments in the polynomial of its constitutive relation are potential differences, U_i, or "current controlled" if the arguments are currents, I_i. A "mixed control" is not allowed.

potential difference U between the nodes to which it is connected. In contrast an "active" element determines either the potential difference between the nodes to which it is connected (voltage sources) or the current flowing through it (current sources) with no effect on the other parameter, i.e. current through a voltage source or potential difference across a current source. The last element listed in Table 2, called a subcircuit, is user defined (see section 4.2) and can be a passive or an active element, or a combination of both types.

The symbol used in an electrical network is also shown in Table 2 for each element except subcircuits whose symbols may be defined by the user. Moreover, the letter convention is indicated which is used in a SPICE code for describing a network. According to this convention the first letter of an element's name determines its type as shown in Table 2. An extensive description of all elements, which includes an explanation of the SPICE language, is given in Appendix A.

4.2. User-defined elements: subcircuits

SPICE allows a user to devise his or her own elements. Once defined SPICE handles such elements, called subcircuits, in the very same way as the built-in elements. The design of a subcircuit has to be done by means of built-in elements. However, all elements and nodes of the circuits in a given subcircuit are strictly "local" in the sense that they are "unknown" outside the subcircuit. Therefore particular nodes of the subcircuit have to be declared as "common" in its definition. They serve the purpose of connecting the user-defined element to nodes in the general network. Further explanation, in particular how subcircuits are defined in terms of the SPICE language, can be found in Appendix A.

Subcircuits are very powerful tools since they enable one to condense even complex relations, which frequently recur in a given type of model, into one element which then can be plugged in at any appropriate place in the electrical network representing the model. We will make use of this tool and define several subcircuits. These will serve as building blocks in the models for the examples to be discussed in section 5.

4.3. Algebraic calculations with SPICE

4.3.1. Addition, subtraction, and multiplication of variables: The controlled sources (see Table 2) provide one with the facility of adding, subtracting, and multiplying both constants and variables. If the variables are represented by potential differences between nodes a voltage controlled source is used. If the variables are represented by currents in branches (which are measured by means of ammeters, see section A2.7), a current

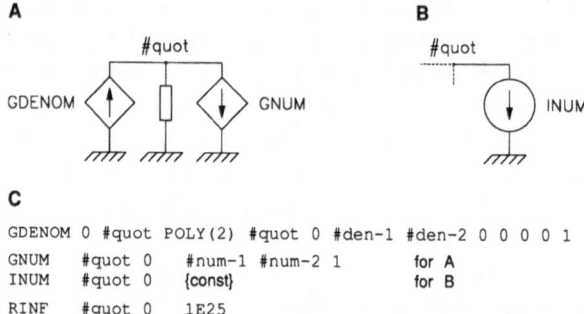

Fig. 2. The division circuit. (A) Circuit for division of a variable by a variable. (B) Replace GNUM at node #quot by INUM to obtain the circuit for division of a constant by a variable. The symbol ⁄⁄⁄ denotes the "electrical ground" which has zero potential. (C) SPICE code. The numerator and the denominator are represented by the potential difference between the node pairs #num-1, #num-2 and #den-1, #den-2, respectively. The symbol {const} has to be replaced by the value of the constant which is to be divided by a variable. The resistor RINF is added to provide the node #quot with a DC path to ground (see section A1.3). For further explanation see text.

controlled source is used. The result can be obtained either as a potential difference between two nodes or a current in a branch by using either a controlled voltage source or a controlled current source, respectively.

Inspection of the formulae for polynomials describing the controlled sources (Eqns. A6–A9 in the Appendix) shows immediately which order of polynomial has to be chosen and which coefficients should be assigned non-vanishing positive or negative values in order to obtain the desired arithmetic operation. For example, suppose the quantity

$$y = (x_1 - x_2)^2 = x_1^2 - 2x_1 x_2 + x_2^2 \tag{7a}$$

is needed. This is readily obtained from POLY(2) (since two variables occur in Eqn. 7a) by using the settings (cf. Eqn. A7 in Appendix A)

$$a_0 = a_1 = a_2 = 0, \quad a_3 = a_5 = 1, \quad a_4 = -2 \tag{7b}$$

No further coefficients need be specified.

4.3.2. Division by a variable: A division is equivalent to forming the ratio (or quotient) of a variable called the numerator to a variable called the denominator. This can be performed by means of two controlled current sources in series (see Fig. 2A) which we call GDENOM and GNUM. The quotient is then represented by the potential of the node denoted by #quot with respect to ground[4]. The specification of

---

[4]Throughout this chapter the symbol " # " will be used as a convenient abbreviation for the word "node" in the node designations used in *describing* the format of a SPICE statement. Thus " #quot" is to be read as "node quot", " #den-1" as "node den-1", and so on. In an actual statement these symbolic designations are replaced by numbers (cf. Appendix A, sections A1.3, A2.1).

the element GDENOM is such that the value of the current flowing through it is equal to the product of the values of the potential difference between the node pair #den-1, #den-2, which represents the denominator, and the potential at node #quot. Provided that the value of RINF is large enough, this current is virtually identical to that flowing through GNUM. The specification of GNUM is such that the value of the current flowing through it is just the value of the potential difference between the node pair #num-1, #num-2 which represents the numerator. Note that GDENOM must be a voltage-controlled current source, while GNUM could in principle be replaced by a current-controlled current source. For division of a constant by a variable the controlled current source GNUM is replaced by the independent current source INUM (see Fig. 2B) whose current is equal to the value of the constant.

4.3.3. Exponentials: An exponential in general terms is a base B raised to the power of a variable (or argument) x;

$$y = B^x = \exp\{x \ln(B)\} \tag{8}$$

The right hand side of Eqn. 8 relates the general exponential to the special exponential with respect to base e; ln denotes the natural logarithm (i.e. with respect to base e). The constitutive relation of a diode (see Table 2) is essentially governed by an exponential with respect to base e provided that the term UG_{min} can be neglected. This condition is easily met if a sufficiently small value for G_{min} is chosen (see section A5.1). Moreover, for reasons of accuracy the potential difference U across the diode should *always be positive*. This element can then serve to calculate an exponential if the diode parameters n and I_s are appropriately chosen.

Let U_x denote the potential difference which represents the variable x, and let U_0 be a bias potential difference which is chosen such that the sum $U = U_0 + U_x$ is positive for all values of x encountered in a given model. Inserting this sum into the constitutive relation yields upon rearranging

$$I + I_s = I_s \exp\{\mathscr{F} U_0/(nRT)\} \exp\{\mathscr{F} U_x/(nRT)\} \tag{9}$$

where I is the current through the diode. Equation 9 shows that, with the following values assigned to the diode parameters

$$n = \mathscr{F}/[RT \ln\{B\}] \tag{10}$$

$$I_s = B^{-U_0} = \exp\{-U_0 \ln(B)\} \tag{11}$$

the exponential of x is represented by $I + I_s$.

Figure 3 presents the circuits which calculate the exponential of a variable. The controlled voltage source in circuit I (Fig. 3A) generates

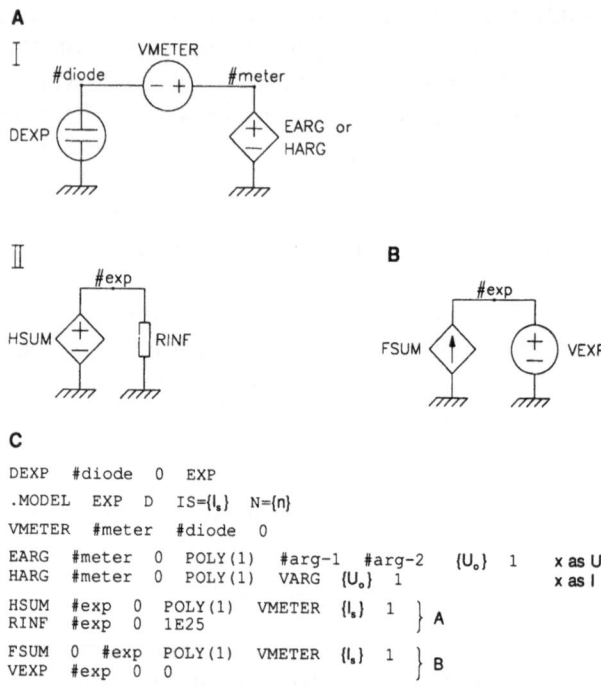

C

```
DEXP    #diode  0   EXP
.MODEL  EXP  D   IS={Iₛ}   N={n}
VMETER  #meter  #diode  0
EARG    #meter  0   POLY(1)   #arg-1   #arg-2   {U₀}   1      x as U
HARG    #meter  0   POLY(1)   VARG   {U₀}   1                x as I
HSUM    #exp  0   POLY(1)   VMETER   {Iₛ}   1   } A
RINF    #exp  0   1E25                              }
FSUM    0   #exp   POLY(1)   VMETER   {Iₛ}   1   } B
VEXP    #exp  0   0                                }
```

Fig. 3. Circuits for calculating exponentials. (A) Circuits for calculating B^x as a potential at the node #exp with respect to ground. The circuits are designated by Roman numerals for reference purposes. Replacing circuit II in (A) by that shown in (B) yields B^x as a current measured by the ammeter VEXP. The symbol ⟂ denotes the "electrical ground" which has zero potential. (C) SPICE code. The argument x is represented either as a potential difference between the node pair #arg-1, #arg-2 ("x as U") or as a current measured by the ammeter VARG ("x as I"). The symbols {n} and {Iₛ} have to be replaced by numbers calculated according to Eqns. 10 and 11, respectively. Similarly, {U₀} has to be replaced by the value chosen for U_0. The resistor RINF is added to provide the node #exp with a DC path to ground. For further explanation see text.

$U_x + U_0$, and the ammeter (see section A2.7) records the current flowing through the diode in response to this potential difference. This current is corrected for I_s in circuit II to yield the exponential of x as the potential of the node #exp with respect to ground. Alternatively, the exponential can be obtained as a current measured by the ammeter VEXP if circuit II is replaced by the circuit shown in Fig. 3B. In scaling the circuits (see section 3.3) it is important to remember that the bias potential difference U_0 compensates for possible negative values of the argument. If it is clear that the argument is always positive, U_0 may be given the value 0. On the other hand, for reasons of accuracy, U_0 must not be so large that I_s falls to a value less than 10^{-33}. Moreover, when using Eqns. 10 and 11, due attention must be paid to the values and units used by SPICE for the parameters R, T, and $\mathscr{F}$ (see section A2.5).

4.3.4. Logarithms: The logarithm with respect to base B applied to a variable (or argument) x reads in general terms

$$y = \log_B\{x\} = \ln\{x\}/\ln\{B\} \qquad (12)$$

The right hand side of Eqn. 12 relates the logarithm with respect to base B to the natural logarithm (base e). Rearranging the constitutive relation of a diode (see Table 2) and taking the natural logarithm yields, provided that the term UG_{min} can be neglected for reasons discussed in the preceding section,

$$U = [\ln\{I + I_s\} - \ln\{I_s\}]nRT/\mathscr{F} \qquad (13)$$

where U is the potential difference across the diode. If the diode parameter n is given the value

$$n = \mathscr{F}/[RT \ln\{B\}] \qquad (10)$$

Eqn. 13 simplifies to

$$U = \ln\{I + I_s\}/\ln\{B\} - U_{log} \qquad (13a)$$

The quantity U_{log} in Eqn. 13a is an abbreviation defined as

$$U_{log} = \ln\{I_s\}/\ln\{B\} \qquad (14)$$

Equation 13a compared to Eqn. 12 shows that a diode can be used to calculate a logarithm. Let I_x denote the current which represents the argument x, then its logarithm is equal to $U + U_{log}$ if the current $I = I_x - I_s$ is forced through the diode.

Figure 4 presents the circuits which calculate the logarithm of a variable. The controlled current source in circuit I (Fig. 4A) generates $I_x - I_s$, and the potential difference across the diode established due to this current is corrected for U_{log} in circuit II to yield the logarithm of x as the potential of the node # log with respect to ground. Alternatively, the logarithm can be obtained as a current measured by the ammeter VLOG if circuit II is replaced by the circuit shown in Fig. 4B. In scaling the circuits (see section 3.3) it is important to remember that the term UG_{min} can only be neglected if $U > 0$ and $I > 0$. Hence the parameter I_s has to be chosen such that its value is less than the lowest value of the argument x to be encountered in a given model. However, for reasons of accuracy, the value of I_s must not be less than 10^{-33}. Moreover, when using Eqn. 10 due attention must be paid to the values and units used by SPICE for the parameters R, T, and $\mathscr{F}$ (see section A2.5).

4.3.5. Subcircuits EXP_FURT, RT_LNC, AH_TO_PH and PH_TO_AH: An exponential frequently encountered in bioelectrochemistry is $\exp\{\mathscr{F}U/(RT)\}$ (see, e.g., Eqn. 6). Inspection of Eqn. 9

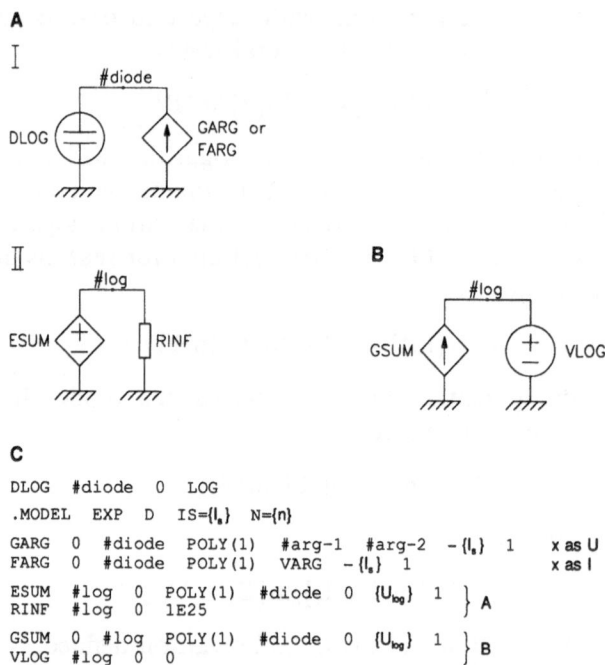

Fig. 4. Circuits for calculating logarithms. (A) Circuits for calculating $\log_B(x)$ as a potential at the node #log with respect to ground. The circuits are designated by Roman numerals for reference purposes. Replacing circuit II in (A) by that shown in (B) yields $\log_B\{x\}$ as a current measured by the ammeter VLOG. The symbol ⁄⁄⁄ denotes the "electrical ground" which has zero potential. (C) SPICE code. The argument x is represented either as a potential difference between the node pair #arg-1, #arg-2 ("x as U") or as a current measured by the ammeter VARG ("x as I"). The symbols {n} and $\{U_{\log}\}$ have to be replaced by numbers calculated according to Eqns. 10 and 14, respectively. Similarly, $\{I_s\}$ has to be replaced by the value chosen for I_s. The resistor RINF is added to provide the node #log with a DC path to ground. For further explanation see text.

shows that this exponential is obtained with the following values for the diode parameters

$$n = 1 \tag{15a}$$

$$I_s = \exp\{-\mathscr{F} U_0/(RT)\} \tag{15b}$$

where U_0 is the bias potential difference introduced in section 4.3.3. Since this exponential is frequently used it is programmed in a subcircuit called EXP_FURT, which then defines a new element with two connections (see Fig. 5A). To one of the connections (node 1) the argument U is supplied as a potential with respect to ground, and the result is obtained at the other connection (node 2) as a potential with respect to ground. The elements inside this subcircuit are identical to those shown in Fig. 3A except for the additional resistor R1 which is used to transmit the argument from the general "outside" to the local "inside" of the subcircuit (cf. section 4.2). In scaling the subcircuit we

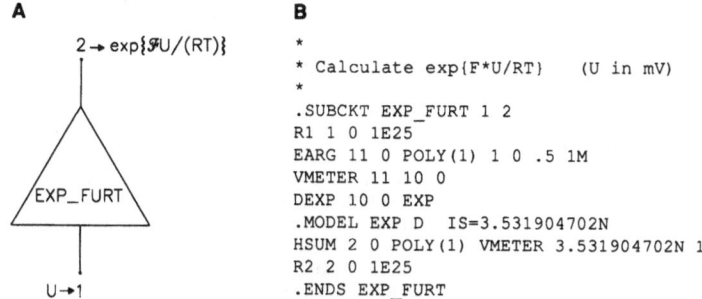

Fig. 5. Subcircuit EXP_FURT. (A) Suggested symbol. The triangular shape of the symbol indicates that the argument U (which is supplied to node 1) is converted to the results (which can be read from node 2) but not vice versa. (B) SPICE code which defines the subcircuit EXP_FURT having two connections at node 1 and 2. The circuits are identical to those of Fig. 3A except for the additional resistor R1. RINF in Fig. 3C is R2, while the node # diode corresponds to node 10, # meter to 11, and # exp to 2, respectively. The two resistors R1 and R2 cause the user-defined element EXP_FURT to provide a DC path to ground. Note that N was omitted in the .MODEL statement because it has the default value of 1.

have chosen mV units for U and $U_0 = 500$ mV, hence the argument cannot be less than -500 mV. The temperature chosen is 25°C, therefore $I_s = 3.531904702 \times 10^{-9}$ (cf. Eqn. 15b). When using the subcircuit the temperature has to be specified in the SPICE code by means of TNOM = 25 in the OPTIONS statement (see section A5.1). If either of of the two conditions are not met I_s has to be adjusted according to Eqn. 15b.

A logarithm frequently encountered in bioelectrochemistry is $RT \ln\{c/1 \text{ M}\}$, where c denotes a concentration. Inspection of Eqn. 13 shows that this logarithm is obtained with the following settings

$$n = \mathscr{F} \tag{16a}$$

$$U_{\log} = RT \ln\{I_s\} \tag{16b}$$

where I_s has to be chosen according to the criteria outlined in section 4.3.4. Again it is advantageous to program this logarithm in a subcircuit called RT_LNC (see Fig. 6). Its elements are identical to those shown in Fig. 4A except for the following modification. The argument is supplied to node 1 as a potential with respect to ground. The result is obtained as a potential difference between the nodes 2 and 3 (node 2 being the positive node, see Fig. 6A) and not as a potential with respect to ground as was the case in the circuits shown in Fig. 4A. Due to this arrangement it is possible to easily simulate an expression such as $RT \ln\{c_{R1}c_{R2}/(c_{R3}1\text{M})\} = RT \ln\{c_{R1}/1\text{M}\} + RT \ln\{c_{R2}/1\text{M}\} - RT \ln\{c_{R3}/1\text{M}\}$ by connecting three subcircuits RT_LNC in proper polarity (for an example see Fig. 20). In scaling the subcircuit we have chosen kJ/mol as units for $RT \ln\{c/1\text{M}\}$, hence n = 96.5 (cf. Eqn. 16a). Moreover, the lower limit for c and the temperature were chosen to be 10^{-14} M and

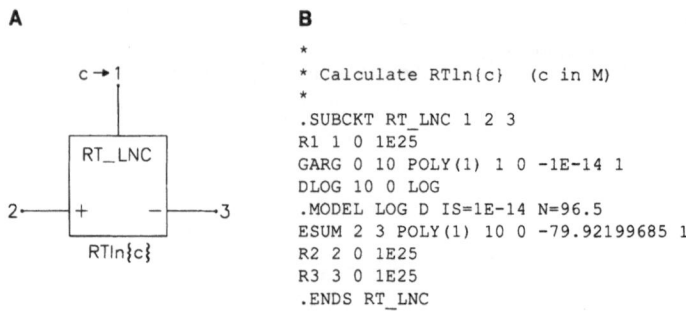

A

c → 1

RT_LNC

2 ——— + — — ——— 3

RTln{c}

B

```
*
* Calculate RTln{c}   (c in M)
*
.SUBCKT RT_LNC 1 2 3
R1 1 0 1E25
GARG 0 10 POLY(1) 1 0 -1E-14 1
DLOG 10 0 LOG
.MODEL LOG D IS=1E-14 N=96.5
ESUM 2 3 POLY(1) 10 0 -79.92199685 1
R2 2 0 1E25
R3 3 0 1E25
.ENDS RT_LNC
```

Fig. 6. Subcircuit RT_LNC. (A) Suggested symbol. The argument c is supplied to node 1, and the result appears as a potential difference between nodes 2 and 3 with polarity as shown by the $+/-$ signs. These signs also indicate that the subcircuit functions as a voltage source at the corresponding node pair which cannot be used as an input for an argument. (B) SPICE code which defines the subcircuit RT_LNC having three connections at nodes 1, 2, and 3. The circuits are identical to those of Fig. 4A except for the additional resistors R1 and R3. RINF in Fig. 4C is R2, while the node # diode corresponds to node 10, and # log to 2, respectively. The three resistors R1, R2, and R3 cause the user-defined element RT_LNC to provide a DC path to ground.

25°C, respectively, which yields $U_{log} = -79.92199685$ (see Eqn. 16b). Note that a SPICE code which uses this subcircuit has to include $TNOM = 25$ in the OPTIONS statement. If the lower limit for the concentration or the temperature are changed U_{log} has to be adjusted according to Eqn. 16b.

Two very common equations in bioelectrochemistry relate the activity of H^+ ions, a_H, in an aqueous solution to the pH value and vice versa

$$pH = -\log_{10}\{a_H/1M\} \quad \text{and} \quad a_H = 1M\ 10^{-pH} \tag{17a}$$

The activity is related to the concentration by (cf. Eqn. 22 in chapter 1)

$$a_H = y_H c_H \tag{17b}$$

where y_H is the activity coefficient. The logarithm and exponential with respect to base 10 in Eqn. 17a are also programmed in subcircuits called AH_TO_PH and PH_TO_AH, respectively (see Fig. 7). The same design as used for the subcircuit EXP_FURT is applied, i.e. argument and result are represented by a potential with respect to ground at node 1 and node 2, respectively. The diode parameters are according to Eqn. 10, Eqn. 14 with $I_s = 10^{-14}$, and Eqn. 11 with $U_0 = 14$ (which compensates for the lowest value of $-pH$); moreover $B = 10$. Note that the value of n is valid only for a temperature of 25°C. Hence, a SPICE code which uses these subcircuits has to include $TNOM = 25$ in the OPTIONS statement, or else the value of n has to be adjusted to the actual temperature (see Eqn. 10).

4.3.6. Integration and differentiation with respect to time: The time integral of a variable is obtained by feeding a flow representing the

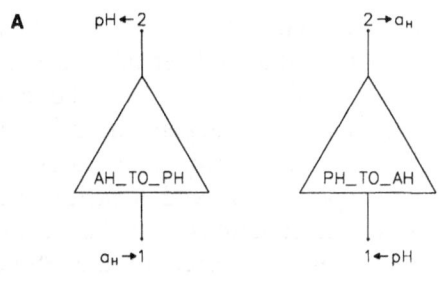

B

```
*
* pH = -log{a(H)}
*
.SUBCKT AH_TO_PH 1 2
R1 1 0 1E25
GARG 0 10 POLY(1) 1 0 -1E-14 1
DLOG 10 0 LOG
.MODEL LOG D IS=1E-14 N=16.90443067
ESUM 2 0 POLY(1) 10 0 14 -1
R2 2 0 1E25
.ENDS AH_TO_PH
```

C

```
*
* a(H) = 10^(-pH)
*
.SUBCKT PH_TO_AH 1 2
R1 1 0 1E25
EARG 11 0 POLY(1) 1 0 14 -1
VTENTO 11 10 0
DEXP 10 0 TENTO
.MODEL TENTO D IS=1E-14 N=16.9044306
HSUM 2 0 POLY(1) VTENTO 1E-14 1
R2 2 0 1E25
.ENDS PH_TO_AH
```

Fig. 7. Subcircuits AH_TO_PH and PH_TO_AH. (A) Suggested symbols. The triangular shape of the symbols indicates that the argument supplied to node 1 (a_H for AH_TO_PH, pH for PH_TO_AH) is converted to the result presented at node 2 (pH for AH_TO_PH, a_H for PH_TO_AH), but not vice versa. (B), (C) SPICE codes which define the subcircuits AH_TO_PH (B) and PH_TO AH (C). Both have two connections, one at node 1 and one at node 2. The circuits are identical to those of Figs. 4A and 3A, respectively, except for the additional resistor R1. The two resistors R1 and R2 cause the user-defined elements AH_TO_PH and PH_TO_AH to provide a DC path to ground.

variable to be integrated into a unit capacitor and reading the potential difference across the capacitor.

The time derivative of a variable is obtained by imposing a potential difference representing the variable to be differentiated on a unit capacitor and reading the flow to the capacitor with an ammeter.

If a non-unitary capacitor is chosen, the integrated or differentiated variable is divided or multiplied, respectively, by the value of the capacitor. Thus, the capacitances listed in Table 1 convert the integrated extensive quantities into intensive quantities, e.g. number of moles into concentrations.

4.3.7. Calculations with PSpice: the Analog Behavioral Modeling option: PSpice allows one to use named parameters and algebraic expressions in place of most numeric values in the circuit description. They cannot be used in the definition of controlled voltage and current sources, but with the addition of the Analog Behavioral Modeling option expressions containing any mixture of potential differences and currents can be used directly in the description of controlled voltage and current sources without having to resort to polynomials. The description can be in the form of a formula or in the form of a tabulated set of values. As a consequence the algebraic circuits described above may be considerably simplified, and rather complex time-independent relationships between processes may readily be introduced. The library of built-in functions includes, inter alia, exponentials and logarithms as well as the trigonometric functions. Other functions may be defined as required. For further details the reader is referred to the PSpice manual.

4.4. Programming hints

4.4.1. Switching: In this section we describe a protocol for simulations of time-dependent behavior which we have found to be extremely valuable, and which in a sense mimics the protocol of an actual experiment. According to this protocol, the system is allowed to undergo an initial period of "pre-incubation" during which it relaxes to a time-independent state. Only then are active processes initiated, for example by exposure to light, addition of reactants, imposition of potential difference or current clamps, or stretching of a fiber. By means of independent sources programmed to perform a step function (see "piece-wise linear" and "pulse" in section A2.6) any parameter in the model can be "switched" on or off. This is done by programming the parameter as a controlled source subject to control by the switch. The most convenient approach is to establish initial conditions for the parameters of state (e.g. concentrations, thermodynamic forces, lengths, and volumes) and to perform the switching operation on kinetic parameters (e.g. permeabilities, rate constants, conductances, etc.). However, the inverse is also possible and for some systems may be preferable.

The advantage of this procedure from the point of view of SPICE is two-fold. Firstly, the initial phase of the simulation is essentially time-independent, which greatly facilitates or in some cases makes possible the calculation of the starting conditions. Secondly, the steep increase of the step function causes a proper integration time step to be chosen for the onset of the active processes. Later a larger time step is used so that the total job time does not become inordinately long. Initially fast processes which become slower with time are often encountered, and should be treated this way whenever possible.

A further advantage of switching is that it can be used to delay the onset of the process under study until other processes have been allowed to reach equilibrium or have been reset in some appropriate fashion. This is accomplished by switching on independent pathways for the other processes to relax which subsequently are switched off before the process under study is initiated.

4.4.2. The use of "building blocks" in constructing complex models: It is often found that the topological map of a system gives rise to rather complicated circuitry. In such cases it is preferable not to attempt to write a SPICE code for the entire model all at once, but rather to identify those parts of the model which can be isolated and tested independently. Bioelectrochemical systems almost invariably comprise a set of such "stand-alone" components, e.g. pumps, permeabilities, channels, carriers, etc. These can be modelled separately and tested by means of auxiliary circuits which impose reasonable values of the parameters. Once tested and found to be working properly, the circuits corresponding to these components can readily be used as building blocks in constructing the complete model. One convenient way of doing this is to define them as subcircuits (see section A2.9). We have used this approach in constructing the redox-driven proton pump (see section 5.4).

While implementing this procedure we have come across an unexpected characteristic of SPICE which is useful to keep in mind when convergence problems arise in circuits involving sets of interconnected controlled sources. Convergence may often be achieved by including an extra variable in one of the polynomials (i.e. taking a polynomial of order $n + 1$ in place of n, see section A2.8). This variable is used as a unitary multiplier of the n variables in the original polynomial (see the polynomial describing the slip transition $2 \rightleftarrows 5$ in section 5.3, the multiplier being the variable sl).

5. Illustrative examples

In this section we present a series of examples of increasing complexity designed to cover all aspects of modelling. For each example we will follow the procedure outlined in the previous sections, arriving eventually at the SPICE code appropriate for the simulation of the behavior of that particular system. We will highlight a number of salient points relating to modelling procedures and will attempt to fit some of these problems into a broader biological and/or electrochemical context. The examples to be considered are (i) the diffusion of salt through a membrane, (ii) a chemical reaction at pseudo-equilibrium and buffering, (iii) an enzyme-catalyzed redox reaction coupled to the movement of H^+ ions, i.e. a redox-driven proton pump, and (iv) the redox-driven proton pump in situ which combines all elements of the first three examples. Note that the pertinent SPICE codes are too long to be included in the text and therefore are collected in Appendix C.

5.1. Permeation of salt through a membrane

This is the system used to illustrate mathematical modelling (see section 1.2). Figure 1A shows a sketch of the elements present in the system, i.e. two compartments which are separated by a membrane, and constitutes the topological map of the elements (cf. steps (1) and (2) in section 2.1). The elemental processes occurring in the system are the flows of ions and the electric current carried by the ions (step (3) in section 2.1), and the mathematical description is set up by Eqns. 2, 5, and 6 (step (4) in section 2.1). We assume that the volume of compartment 1 is much larger than that of compartment 2 and therefore expect the concentrations of ions in compartment 1 to be essentially constant. We want to simulate the time course of the ion flows, the concentrations, and the membrane potential (cf. steps (5) and (6) in section 2.1).

5.1.1. Subcircuit IONPERM:
The flow of a species through a membrane is an elemental process that frequently recurs and hence is worth programming in a subcircuit. This subcircuit will be called IONPERM (see Fig. 8). As is evident from Eqns. 3 and 6 the parameters governing the flow J_i of the ith species, and the current I_i associated with it, are the concentrations $c_{i,1}$ and $c_{i,2}$ as well as the membrane potential $\Delta\phi_m$. These quantities are supplied to the subcircuit at node 10, node 11, and between nodes 1 and 2, respectively. The charge number z_i and the product membrane area times permeability, AP_i, are fed to the subcircuit by means of the nodes 20 and 21, respectively (Fig. 8B).

Circuit I multiplies $\Delta\phi_m$ (the potential difference between nodes 1 and 2) by the charge number of the ion read from node 20 (Fig. 8A). This product is applied to the subcircuit EXP_FURT (see section 4.3.5) which yields the term $\exp\{z_i\mathscr{F}\Delta\phi_m/(RT)\}$ at node 13. Circuit II generates the numerator of the last term of Eqn. 6, while circuit III is a division circuit (see section 4.3.2) which completes the computation of Eqn. 6 using the value of AP_i of the ion read from node 21. In order to avoid the indeterminacy which arises as $\Delta\phi_m$ tends to zero, a small quantity (10^{-5}) is added to the denominator and to the factor $z_i\mathscr{F}\Delta\phi_m/(RT)$. The potential at node 15 representing J_i is fed to the voltage-controlled current sources connecting nodes 10 and 11, and nodes 1 and 2. In the latter case the flow is multiplied by $z_i\mathscr{F}$ to convert it into electric current according to Eqn. 3. Note that the two flows J_i and I_i are circuit-wise totally separated as required by rule (4) in section 3.2. This is also indicated by the two separate arrows in the suggested symbol for the subcircuit (see Fig. 8B).

In scaling the subcircuit the conditions set by the subcircuit EXP_FURT (see section 4.3.5) have to be taken into account, i.e. electrical potentials have to be supplied in millivolts, and the membrane potential cannot be less than -500 mV. If another scaling of a model is chosen or another lower limit is required the pertinent parameters in

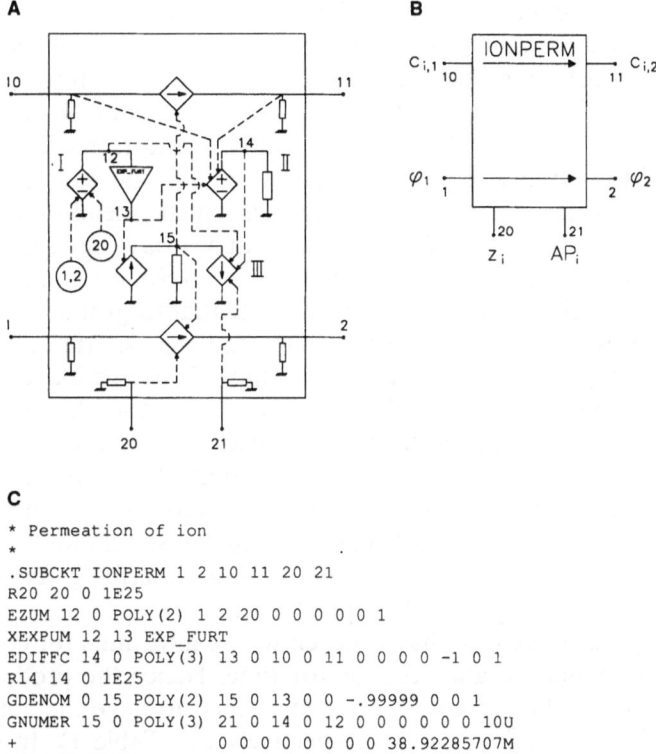

C

```
* Permeation of ion
*
.SUBCKT IONPERM 1 2 10 11 20 21
R20 20 0 1E25
EZUM 12 0 POLY(2) 1 2 20 0 0 0 0 0 1
XEXPUM 12 13 EXP_FURT
EDIFFC 14 0 POLY(3) 13 0 10 0 11 0 0 0 0 -1 0 1
R14 14 0 1E25
GDENOM 0 15 POLY(2) 15 0 13 0 0 -.99999 0 0 1
GNUMER 15 0 POLY(3) 21 0 14 0 12 0 0 0 0 0 0 10U
+                   0 0 0 0 0 0 0 0 38.92285707M
R15 15 0 1E25
R21 21 0 1E25
R10 10 0 1E25
GIONFLOW 10 11 15 0 1
R11 11 0 1E25
R1 1 0 1E25
GCURRENT 1 2 POLY(2) 15 0 20 0 0 0 0 0 96.5K
R2 2 0 1E25
.ENDS IONPERM
```

Fig. 8. Subcircuit IONPERM. Circuits (A), suggested symbol (B), and SPICE code (C). In (A) individual circuits are designated by Roman numerals for reference purposes. Element names are omitted due to lack of space. The frame indicates the "boundary" of the subcircuit which encloses the "local" elements and nodes. The node pairs 1, 2 and 10, 11 represent the connections which transfer the electric current and the flow of the species, respectively, to the "outside" (see B). Nodes 20 and 21 transfer the values for z_i and AP_i of the ionic species to the subcircuit. All resistors are added to provide a DC path to ground. Controlling pathways are indicated by broken arrows. The symbol ⊓⊓⊓ denotes the "electrical ground" which has zero potential. For further explanation see text.

EXP_FURT have to be adjusted. The Faraday constant in the description of GCURRENT is expressed in C/mol, hence the electric current has the same scaling factor as the ion flow (e.g. μA if the ion flow is given in μmol/s).

5.1.2. Electrical network representation of the model: By means of the newly defined element IONPERM we can easily map the experimental system onto the circuit shown in Fig. 1B. The constant concentrations of ionic species 1 and 2 in compartment 1 are determined by the constant voltage sources at nodes 101 and 102. The time-dependent concentrations in compartment 2 are obtained at nodes 201 and 202 by integrating the ion flows by means of the capacitors representing the volume of the compartment (cf. section 4.3.6). The electric currents associated with the ion flows are integrated on the membrane capacity to yield the membrane potential. Note that we have set the electrical potential ϕ_1 in compartment 1 arbitrarily to zero by connecting the node of the subcircuits pertinent to this potential to ground. Ammeters are included which monitor the ion flows and the electric current. The constant voltage sources at nodes 11 and 12 determine the charge numbers of the ions, while those at nodes 21 and 22 provide the respective values of AP_i. Note that the latter sources are programmed piece-wise linear (see section A2.6) which after a brief "pre-experimental" period increases the AP_i values from zero to their operational values in 1 μs. The advantage of commencing the simulation in this way is discussed in section 4.4.1. The SPICE code describing this network can be found in section C1 of Appendix C.

In scaling this system we have chosen to use mM units for concentrations, ml for volumes, and seconds for time. Hence the products AP_i, the flows J_i, and the currents I_i are constrained to have the dimensions ml/s, μmol/s, and μC/s ($=\mu$A), respectively (cf. Table 1). In order to obtain the membrane potential in mV, we can express the membrane capacity in mF.

5.1.3. Simulating the behavior of the system: To simulate the behavior of the system we use $z_1 = 1$, $z_2 = -1$, $c_{1,1} = c_{2,1} = 10$ mM, and the initial values $c_{1,2} = c_{2,2} = 0$. The initial membrane potential is assumed to be zero. Accordingly, the initial conditions for the capacitors at nodes 1, 201, and 202 are omitted since they are equal to the default values. The volume of compartment 2 is taken to be 2 ml. The membrane capacity is 1 μF = 0.001 mF; $AP_1 = 5 \times 10^{-6}$ ml/s, and $AP_2 = 5 \times 10^{-7}$ ml/s. An output generated by SPICE is shown in Fig. 9. Note that V(0, 1) represents the membrane potential since $\Delta\phi_m = \phi_1 - \phi_2$ (see section 1.2). The simulation is very instructive in that it shows how the flows of the two ionic species equalize as a result of the building up of the membrane potential over about 80 ms. Concomitantly the electric current, which charges the membrane capacity, goes to zero. Furthermore, it is seen that a difference of 3.07×10^{-7} mM between the concentrations of cation and anion persists. This finite but tiny deviation from electroneutrality represents exactly the amount of charge needed to account for the membrane potential.

The use of a subcircuit to simulate ion permeation allows parameters to be readily changed. As an example, suppose we would like to

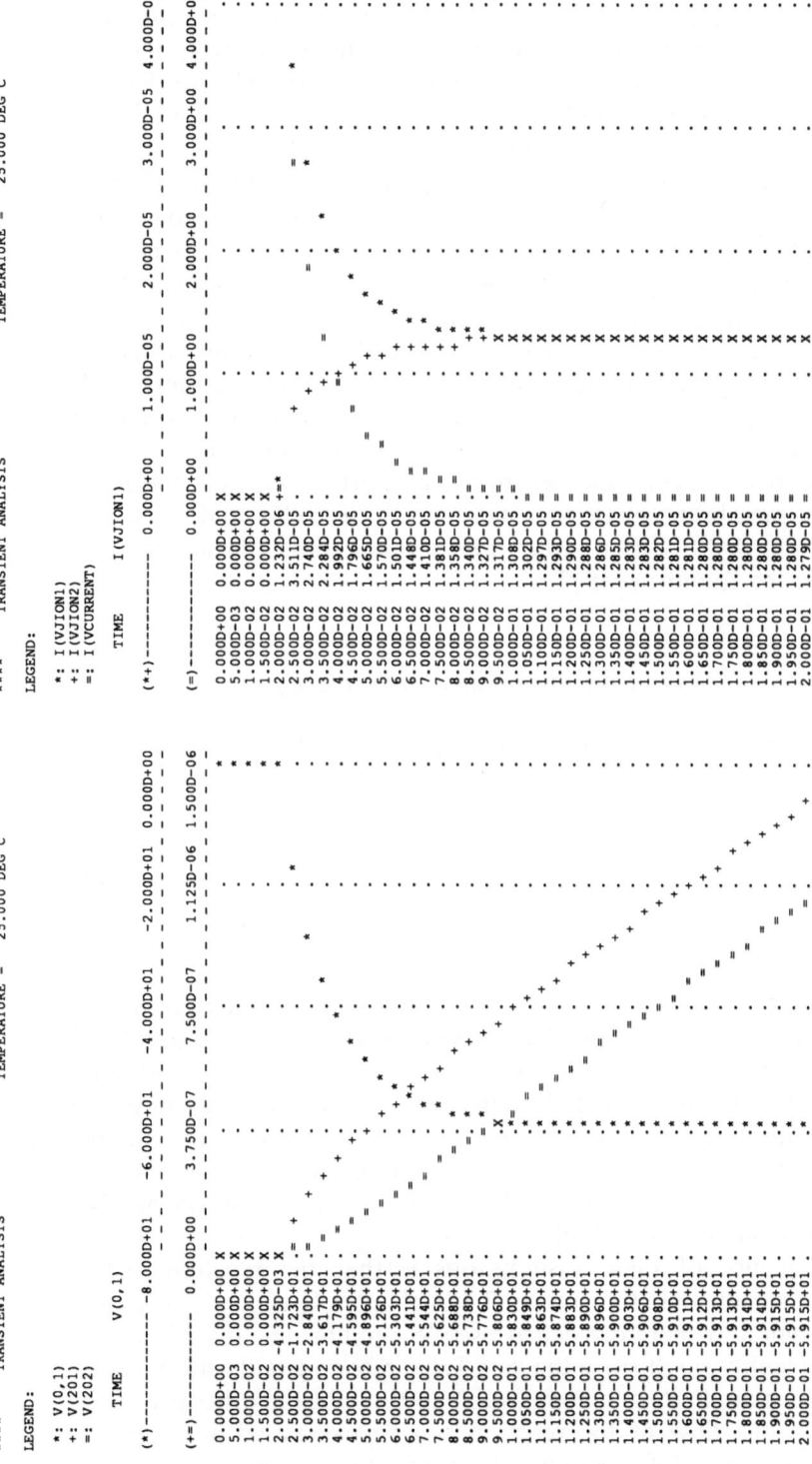

Fig. 9. SPICE output for the permeation of salt through a membrane. $V(0,1)$, $V(201)$, $V(202)$, $I(VJION1)$, $I(VJION2)$, and $I(VCURRENT)$ represent, respectively, $\Delta\varphi_m$ (in mV), $c_{1,2}$, $c_{2,2}$ (in mM units), J_1, J_2 (in μmol/s), and I (in μA). Time is given in seconds. For parameter values see text. For the format of the plots see section A4. Note that the scaling option has been used to ensure the most convenient presentation of the plots. The listing of the SPICE code also appearing in an output is omitted.

simulate the permeation of a salt consisting of a divalent cation and a univalent anion. All we have to do is to change the value of VZION1 to 2 and to assign a value to VION1IN1 which is half of that for VION2IN1. Moreover, IONPERM enables us to easily build more complex topologies. Thus, in order to simulate a system with three compartments in series, we only have to connect two additional subcircuits between the existing nodes 1, 201 and 202 in Fig. 1B, and the new nodes 2, 301, and 302. The latter would then be supplemented by capacitors representing, respectively, the membrane capacity of the second membrane and the volume of the third compartment.

In the above simulation we have not considered the ions which are "dissolved" in the membrane. This allowed us to integrate the Nernst-Planck equation using the constant field approximation (see section 4.3.1 in chapter 1). Moreover, we have neglected that the ions first partition into the membrane before a transport occurs, i.e. the relaxation of the membrane phase to a steady state (see ref. 1 and section 3.4 in chapter 1). A SPICE program which takes these phenomena into account can be found in ref. 2.

5.2. Chemical reaction at pseudo-equilibrium: buffering

Consider a chemical reaction in which a reactant R_1 is split into two fragments R_2 and R_3:

$$R_1 \rightleftarrows R_2 + R_3 \tag{18}$$

The rate constants for the forward and backward reactions are k_f and k_b (cf. section 4.1 in chapter 1). At equilibrium

$$K_c = k_f/k_b = [c_{R2}c_{R3}/(c_{R1}1M)]_{eq} \tag{19}$$

K_c is the equilibrium constant of the reaction which is related to the rate constants according to detailed balancing (see sections 3.5.1 and 5.1.1 in chapter 1). The concentration of reactant R_i is c_{Ri} (activity coefficients are assumed to be unity) and the subscript eq indicates concentrations at equilibrium. Simulate the behavior of the reaction in Eqn. 18 under the assumption that k_f and k_b are very large compared to the rate constants of any other process which may occur in the system. K_c, however, is assumed to be finite.

A consequence of these assumptions is that the reaction relaxes rapidly to a steady state (cf. section 3.4 in chapter 1) which is very close to equilibrium and therefore called pseudo-equilibrium. Its affinity can become negligibly small since the very large rate constants still give rise to flows which are finite and comparable to those of the other processes. In principle, the flow of the reaction can be calculated according to the usual rate law (cf. section 4.1 in chapter 1) and simulated in

SPICE in a way similar to that we shall use for simulating the transition flows in an enzyme cycle (see section 5.3). However, the very large rate constants may cause problems of convergence for SPICE (typical error message: INTERNAL TIME STEP TOO SMALL) which frequently cannot be remedied by the switching technique described in section 4.4.1. We therefore adopt a different procedure which is based on the equilibrium condition in Eqn. 19.

It is easily shown from mass balance considerations and Eqn. 19 that

$$c_{R2} = c_{tot} K_c / (K_c + c_{R3}/1M), \quad c_{R1} = c_{tot} - c_{R2} \tag{20}$$

where c_{tot} denotes the total concentration of R_1 and R_2. Equation 20 shows that the parameter c_{tot} is sufficient to cope with R_1 and R_2. Any change in the concentration of one of these species due to a flow from another process in which it is involved will be immediately translated into concomitant changes in c_{tot} and the concentration of the other species. The same behavior arises with a change in c_{R3}. In either event, a flow for R_3 also occurs which amounts to

$$J_{R3} = V(dc_{R2}/dt) - J_{R2} \tag{21}$$

Here V is the volume of the phase or compartment in which the reaction takes place, and J_{R2} is the flow of R_2 referred to above. Equation 21 takes care of the mass balance for R_3.

5.2.1. Subcircuit REQUIL12: The generality of the relations in Eqns. 20 and 21 makes it convenient to program them in a subcircuit which will be called REQUIL12 (see Fig. 10). This subcircuit yields c_{R1}, c_{R2}, and c_{R3} as potentials with respect to ground at nodes 1, 2, and 3, respectively, together with c_{tot} on node 4. The equilibrium constant K_c and the volume V are applied to the additional nodes 5 and 6, respectively (see Fig. 10B).

The ammeters in circuits I and II monitor the flows of R_1 and R_2 which are summed in circuit III to yield the total flow (Fig. 10A). Circuit IV is a division circuit (see section 4.3.2) which yields c_{R2} at node 40 using c_{R3} and K_c read from nodes 3 and 5, respectively (cf. Eqn. 20). The potential at node 40 and the potential difference between nodes 4 and 40 are fed to the controlled voltage sources in circuits II and I, respectively. Circuit V is a differentiator (see section 4.3.6) which yields dc_{R2}/dt as the current through the ammeter. This current is multiplied by the volume and fed, together with J_{R2} read by the ammeter in circuit II, to the controlled current source in circuit VI (cf. Eqn. 21). Circuit VII simply converts the volume, which is applied as a potential to node 6, to a current[5].

---

[5]In principle, the volume could also be applied by a current source and directly read by an ammeter in the subcircuit. This would result in every subcircuit REQUIL12 in a SPICE code requiring its own current source even if the volume is always the same. Connecting more than one ammeter to a current source is not allowed and causes the error message INDUCTOR/VOLTAGE SOURCE LOOP FOUND CONTAINING $\cdots$

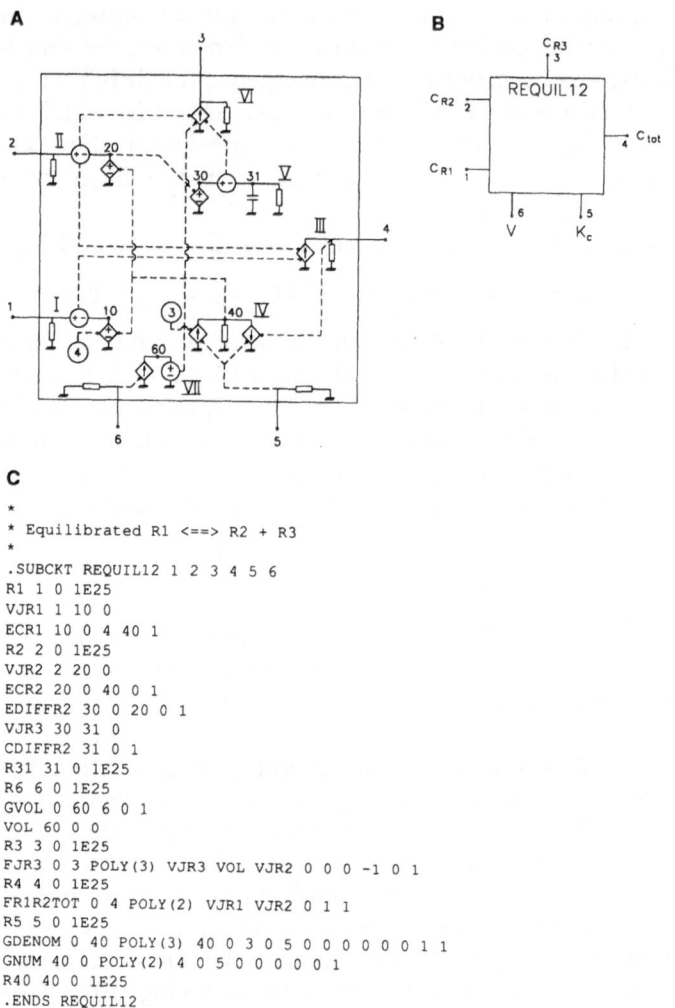

A

B

C

```
*
* Equilibrated R1 <==> R2 + R3
*
.SUBCKT REQUIL12 1 2 3 4 5 6
R1 1 0 1E25
VJR1 1 10 0
ECR1 10 0 4 40 1
R2 2 0 1E25
VJR2 2 20 0
ECR2 20 0 40 0 1
EDIFFR2 30 0 20 0 1
VJR3 30 31 0
CDIFFR2 31 0 1
R31 31 0 1E25
R6 6 0 1E25
GVOL 0 60 6 0 1
VOL 60 0 0
R3 3 0 1E25
FJR3 0 3 POLY(3) VJR3 VOL VJR2 0 0 0 -1 0 1
R4 4 0 1E25
FR1R2TOT 0 4 POLY(2) VJR1 VJR2 0 1 1
R5 5 0 1E25
GDENOM 0 40 POLY(3) 40 0 3 0 5 0 0 0 0 0 0 1 1
GNUM 40 0 POLY(2) 4 0 5 0 0 0 0 0 1
R40 40 0 1E25
.ENDS REQUIL12
```

Fig. 10. Subcircuit REQUIL12. Circuits (A), suggested symbol (B), and SPICE code (C). In (A) individual circuits are designated by Roman numerals for reference purposes. Element names are omitted due to lack of space. The frame indicates the "boundary" of the subcircuit which encloses the "local" elements and nodes. Nodes 1 to 6 represent the connections to the "outside". All resistors are added to provide a DC path to ground. Controlling pathways are indicated by broken arrows. The symbol $\overline{\tau\tau\tau}$ denotes the "electrical ground" which has zero potential. For further explanation see text.

5.2.2. Behavior of reactions at pseudo-equlibrium: The behavior of such reactions as are represented by the subcircuit REQUIL12 can be tested using a small network (see Fig. 11A) which simulates a flow of R_1 by means of an independent current source at node 1 programmed as a pulse (cf. section A2.6). The flow J_{R3} is monitored by an ammeter connected between nodes 3 and 30. the capacitors at nodes 4 and 30 represent the volume, while the potentials at these nodes indicate c_{tot}

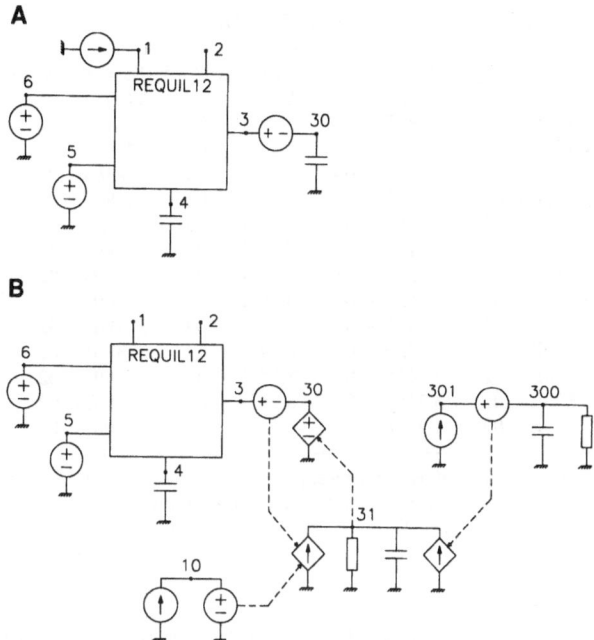

Fig. 11. Circuits for testing subcircuit REQUIL12 (A) and for the simulation of buffering (B). Element names are omitted due to lack of space, but they can be found in the listings of the pertinent SPICE codes in sections C2.1 and C2.2. Horizontally oriented independent voltage sources are ammeters, while vertically oriented sources determine parameter values. Broken arrows indicate control pathways. The symbol $\overline{7\!\!7\!\!7}$ denotes the "electrical ground" which has zero potential. For further explanation see text.

and c_{R3}, respectively. In scaling this model we have chosen mol/l for the concentrations, l for the volume, and seconds for the time. According to Table 1, the unit for the flows is then mol/s. A listing of the pertinent SPICE code is given in section C2.1 for the following parameter values: $K_c = 0.01$, $V = 2$ ml, initial values $c_{tot} = 20$ mM, $c_{R3} = 5$ mM and a pulsed flow of 10 mmol/s for 2 ms. The output created by SPICE is presented in Table 3. It is seen that the adaptation of the reaction to the initial conditions as well as to the pulsed flow of R_1 immediately creates a flow of R_3, which in the case of the adaptation can be rather large.

5.2.3. Buffering: A very important application of the subcircuit REQUIL12 pertains to buffering. The chemical capacity of a given species is increased by means of a buffering substance which can bind or release the species in a fast reaction [1]. In this case R_3 is the species to be buffered, while R_1 and R_2 represent the buffer with and without bound species, respectively. The effect of buffering is demonstrated rather vividly in the following system. Suppose there is a process which, after a certain lag time, starts to periodically produce and consume the species R_3. In order to prevent strong fluctuations of the concentration

Table 3. SPICE output of the program testing REQUIL12

****	TRANSIENT ANALYSIS		TEMPERATURE = 27.000 DEG C		
TIME	V(1)	V(2)	V(3)	V(4)	I(VJR3)
0.000E + 00	1.761E − 02	2.385E − 03	7.385E − 03	2.000E − 02	1.459E + 02
5.000E − 04	1.761E − 02	2.385E − 03	7.385E − 03	2.000E − 02	−1.260E − 14
1.000E − 03	1.761E − 02	2.385E − 03	7.385E − 03	2.000E − 02	−8.314E − 15
1.500E − 03	1.761E − 02	2.385E − 03	7.385E − 03	2.000E − 02	−4.027E − 15
2.000E − 03	1.762E − 02	2.386E − 03	7.386E − 03	2.001E − 02	7.733E − 05
2.500E − 03	1.989E − 02	2.613E − 03	7.613E − 03	2.250E − 02	8.910E − 04
3.000E − 03	2.217E − 02	2.831E − 03	7.831E − 03	2.500E − 02	8.578E − 04
3.500E − 03	2.446E − 02	3.042E − 03	8.042E − 03	2.750E − 02	8.280E − 04
4.000E − 03	2.676E − 02	3.245E − 03	8.245E − 03	3.000E − 02	8.001E − 04
4.500E − 03	2.676E − 02	3.245E − 03	8.245E − 03	3.000E − 02	−6.023E − 18
5.000E − 03	2.676E − 02	3.245E − 03	8.245E − 03	3.000E − 02	−5.120E − 18
5.500E − 03	2.676E − 02	3.245E − 03	8.245E − 03	3.000E − 02	−5.421E − 18
6.000E − 03	2.676E − 02	3.245E − 03	8.245E − 03	3.000E − 02	2.168E − 17

V(1), V(2), V(3), V(4), and I(VJR3) represent c_{R1}, c_{R2}, c_{R3}, c_{tot} (in M units), and J_{R3} (in mol/s), respectively. Time is given in seconds. For parameter values see text. For the format of the output see section A4. The listing of the SPICE code also appearing in an output is omitted.

of the species we add a buffer and compare the resulting time course of c_{R3} with that which would occur without buffer.

When simulating this system we can essentially use the circuit devised to test the behavior of subcircuit REQUIL12 (see Fig. 11A). All we have to do is to disconnect the pulsed current source from node 1 and add a current source at node 30. However, as we have seen above, the addition of a buffer causes the initial value of c_{R3} to be changed. We would observe the same phenomenon in an actual experiment and would adjust the system to the desired initial conditions by an appropriate addition of R_3. We can simulate this adjustment by the following modification of the circuit (see Fig. 11B). The capacitor at node 30 is replaced by a controlled voltage source whose value is equal to the potential at node 31. The capacitor representing the volume for R_3 is now connected to node 31 together with a controlled current source whose value is equal to the current monitored by the ammeter between nodes 3 and 30. It is easily verified that these circuits behave as if the capacitor were still connected to node 30, but with one important difference. We can use a switching circuit programmed in such a way that the current monitored by the ammeter is only transferred to the controlled source after the initial adjustment of the buffer has taken place. As a consequence the potential difference at node 31 (and also at node 30) initially does not change and remains at the desired initial condition for c_{R3}. Since the time course of the unbuffered system is also to be simulated, a third circuit is added consisting of a capacitor, which again represents the volume for species R_3, connected via an ammeter to an independent current source. The latter produces a sinusoidal current after a given lag time and simulates the periodical process. The current monitored by the ammeter is transferred to a controlled current source connected to node 31, which thus feeds the same sinusoidal current to the buffered system.

In the SPICE code describing this network, which is listed in section C2.2, we have used the same parameters as before except for the total concentration of buffer, which was increased to 200 mM. The sinusoidal current, which sets in after a lag time of 2 s, has an amplitude of 10 μmol/s and a frequency of 0.1 Hz. The output created by SPICE (see Fig. 12) clearly shows that c_{R3} in the unbuffered system varies between the initial value of 5 mM and more than 20 mM in response to the sinusoidal current. In the buffered system, however, c_{R3} hardly changes. Most of the R_3 produced or consumed is bound to or released from the buffer system, as is evident from the pertinent flow monitored by the ammeter between nodes 3 and 30.

5.2.4. Buffer capacity for H^+ ions: A species which almost invariably has to be buffered in bioelectrochemical systems is the H^+ ion in an aqueous phase. Buffering for this species is provided by any acid/base pair with a suitable dissociation constant. The latter is usually determined by titration of the buffer with acid or base. Such an experiment can also be simulated by means of the subcircuit REQUIL12. When doing so, it has to be taken into account that water itself is an acid/base pair according to the reaction

$$H_2O \rightleftarrows H^+ + OH^- \tag{22}$$

with a dissociation constant of 1.8×10^{-16}. In the following we will assume unit activity coefficients for all species so that activities become equal to concentrations.

The circuits simulating the titration of two buffers in an aqueous solution with alkali are depicted in Fig. 13A. Three subcircuits REQUIL12 are connected in parallel to two nodes common to all of them: node 3 whose potential represents c_{R3}, which in this case is the H^+ concentration c_H, and node 6 which supplies the value of the volume to all subcircuits. The independent voltage sources at nodes 4, 14, and 24 determine the total concentration of each acid/base pair. The concentrations of the protonated and dissociated species (corresponding to species R_1 and R_2, respectively) are represented by the potentials at nodes 1 and 2 for water, nodes 11 and 12 for the first buffer, and nodes 21 and 22 for the second buffer, respectively. The dissociation constant for each acid/base pair is supplied by the independent voltage sources at nodes 5, 15, and 25. The elements connected to nodes 30, 300, and 50 constitute the switched proton flow integrator as described in the previous example, which assures the proper initial condition for c_H. In addition, the subcircuit AH_TO_PH (see section 4.3.5) is connected to

```
****    · TRANSIENT ANALYSIS                   TEMPERATURE =   27.000 DEG C

LEGEND:

  *: V(3)
  +: V(300)
  =: I(VJR3)
  $: I(VJR3SIN)

     TIME        V(3)

(*+)------------ 0.000D+00   1.000D-02    2.000D-02    3.000D-02  4.000D-02
                - - - - - - - - - - - - - - - - - - - - - - - - - - - - - -

(=$)------------ -3.000D-05   -2.000D-05   -1.000D-05   0.000D+00  1.000D-05
                - - - - - - - - - - - - - - - - - - - - - - - - - - - - - -
5.000D-01  5.000D-03 .   X       .            .              X         .
1.000D+00  5.000D-03 .   X       .            .              X         .
1.500D+00  5.000D-03 .   X       .            .              X         .
2.000D+00  5.000D-03 .   X       .            .              X         .
2.500D+00  5.043D-03 .   *+      .            .       =      .    $    .
3.000D+00  5.156D-03 .   *  +    .            .    =         .      $  .
3.500D+00  5.341D-03 .   *     + .        .   =             .       $ .
4.000D+00  5.575D-03 .    *     .  +      .  =              .        $.
4.500D+00  5.842D-03 .     *    .     +   . =               .         $
5.000D+00  6.117D-03 .     *    .      +  . =               .        $.
5.500D+00  6.372D-03 .     *    .       + .  =              .       $ .
6.000D+00  6.581D-03 .     *    .       +. =               .     $    .
6.500D+00  6.721D-03 .     *    .        .+     =           .  $       .
7.000D+00  6.763D-03 .     *    .        .+           X     .          .
7.500D+00  6.715D-03 .     *    .        .+      $     .  =            .
8.000D+00  6.583D-03 .     *    .       +.     $       .      =        .
8.500D+00  6.374D-03 .     *    .      + .   $         .            =  .
9.000D+00  6.117D-03 .     *    .     +   .$           .             = .
9.500D+00  5.842D-03 .     *    .     +   $            .             = .
1.000D+01  5.573D-03 .    *     .  +      .$           .             = .
1.050D+01  5.341D-03 .    *    + .        .  $         .            =  .
1.100D+01  5.160D-03 .   *  +   .        .     $       .         =     .
1.150D+01  5.042D-03 .   *+     .        .       $     .  =            .
1.200D+01  5.003D-03 .   X      .        .            X.               .
1.250D+01  5.045D-03 .   *+     .        .        =    .  $            .
1.300D+01  5.159D-03 .   *  +   .        .      =      .      $        .
1.350D+01  5.337D-03 .   *    + .        .   =         .       $       .
1.400D+01  5.574D-03 .    *     .  +     .   =         .        $.
1.450D+01  5.842D-03 .     *    .     +  .  =          .         $
1.500D+01  6.118D-03 .     *    .      + .  =          .        $.
1.550D+01  6.374D-03 .     *    .       +.   =         .       $ .
1.600D+01  6.580D-03 .     *    .       +.     =       .     $    .
1.650D+01  6.716D-03 .     *    .        .+     =      .  $       .
1.700D+01  6.769D-03 .     *    .        .+           X.          .
1.750D+01  6.717D-03 .     *    .        .+      $     .  =       .
1.800D+01  6.579D-03 .     *    .       +.     $       .      =   .
1.850D+01  6.373D-03 .     *    .      + .   $         .            = .
1.900D+01  6.118D-03 .     *    .     +   .$           .             = .
1.950D+01  5.842D-03 .    *     .     +   $            .             = .
2.000D+01  ·5.572D-03 .    *     .  +      .$           .             = .
                - - - - - - - - - - - - - - - - - - - - - - - - - - - - - -
```

Fig. 12. SPICE output of the program simulating buffering. V(3) and V(300) represent c_{R_3} (in M units) in the buffered and unbuffered system, respectively. I(VJR3SIN) is the flow generated by a process which periodically produces and consumes R_3, while I(VJR3) indicates the flow of R_3 out of the buffer system (units of flows are mol/s). Time is given in seconds. For parameter values see text. For the format of plot see section A4. The listing of the SPICE code also appearing in an output is omitted.

node 300. It converts $a_H = c_H$ into pH according to Eqn. 17a. Since the dissociated species of the acid/base pair water is the OH^- ion (see Eqn. 22) we can simulate the addition of alkali by an independent current source connected to node 2. This source is programmed "piece-wise linear" (see section A2.6), which mimics the behavior of a good experimenter who waits a certain time to let the system equilibrate before commencing the actual titration.

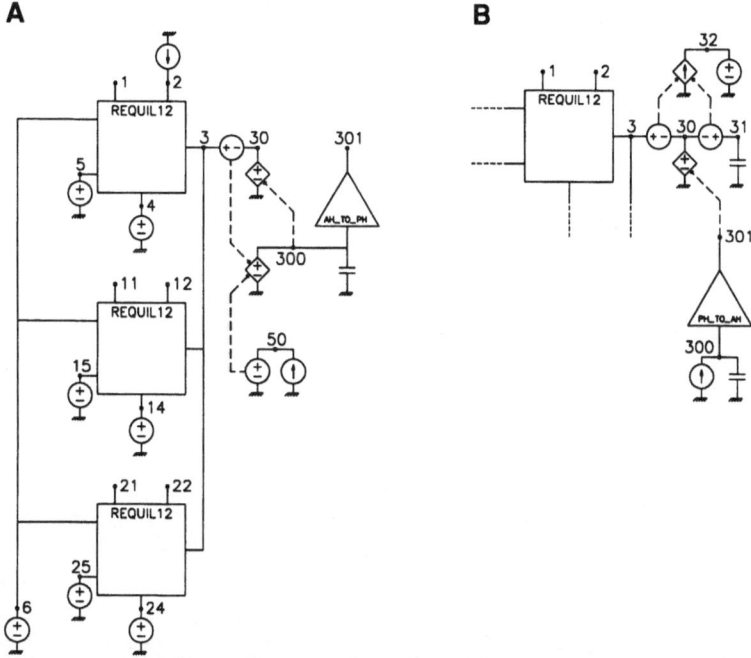

Fig. 13. Circuits for the simulation of a titration (A) and the buffer capacity (B) of two acid/base pairs in a aqueous medium. Element names are omitted due to lack of space, but they can be found in the listings of the pertinent SPICE codes in sections C2.3 and C2.4. When simulating the buffer capacity the elements in A are replaced by those shown in B. The triangular symbol represents the subcircuit AH_TO_PH in A and PH_TO_AH in B, respectively. Broken arrows indicate control pathways. The symbol ⏛ denotes the "electrical ground" which has zero potential. For further explanation see text.

In scaling the system we have chosen molar for concentrations, as required by the conversion of a_H to pH. The units for volume and time are l and min, respectively, which yields mol/min for the flow of added alkali. Figure 14 shows an output generated by the SPICE code listed in section C2.3, using the following parameter values for water, the first and the second buffer: $c_{tot} = 55.5555$ M (equal to 1000 g/18 Da), 40 mM and 20 mM; $K_c = 1.8 \times 10^{-16}$, 10^{-5} and 5×10^{-9}. The volume is 20 ml and the initial H^+ concentration 0.005 M (pH = 2.301). Alkali is added at a constant rate of 0.2 mmol/min, hence adjacent points in the plot after the initial 1 min period correspond to an increment of 0.2 mmol of OH^- ions added. Accordingly, the first, second, and third rise of pH seen in the plot on the left hand of Fig. 14 at 1.5 min, 5.5 min, and 7.5 min indicate the neutralization of 0.1 mmol of H^+ ions initially present, 0.8 mmol of the first buffer, and 0.4 mmol of the second buffer, respectively. In the remaining 1.5 min, 0.3 mmol of OH^- ions in excess are added which yields a final $c_{OH} = 0.015$ M, or pH = 12.18. This interpretation can be verified by inspection of the plot

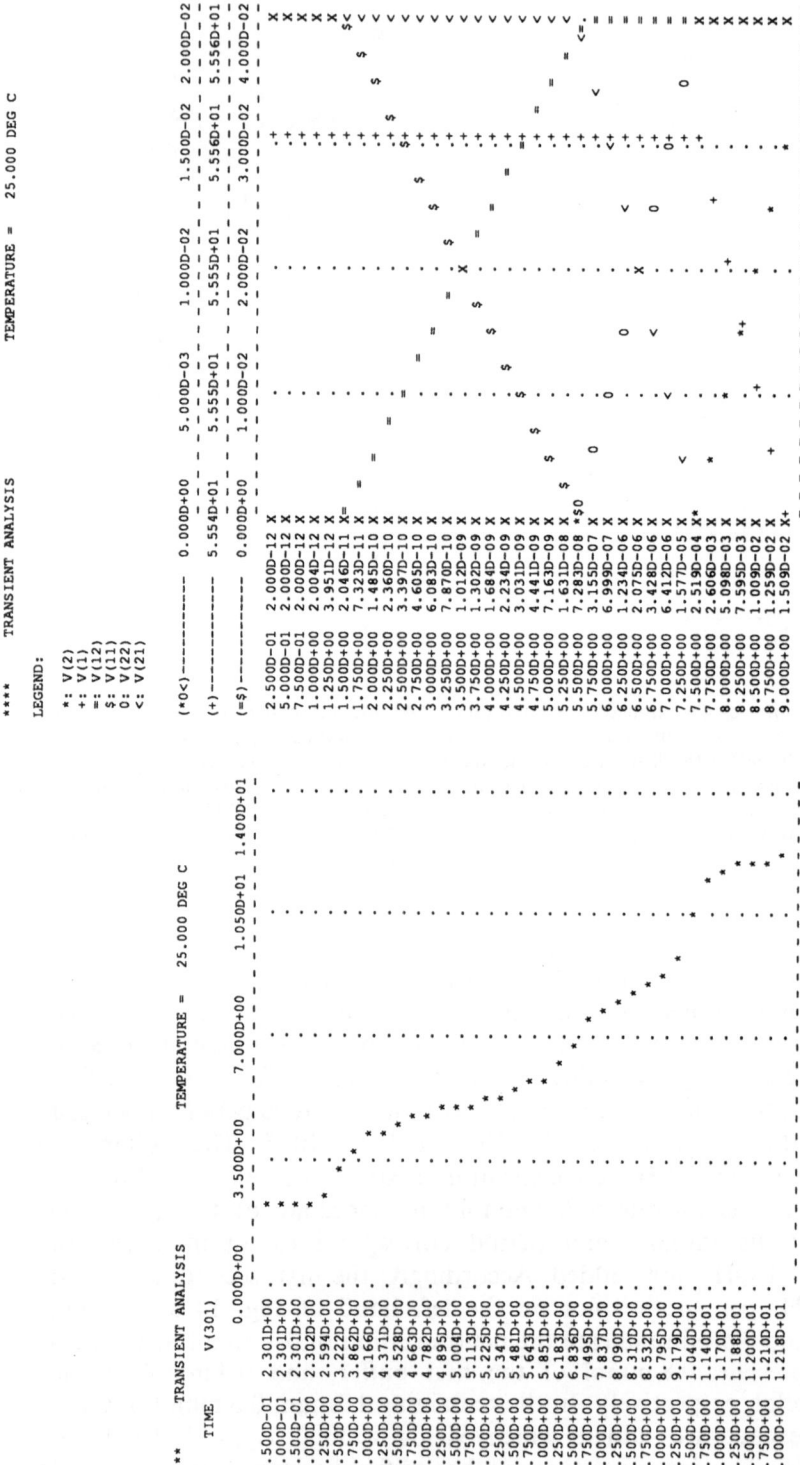

Fig. 14. SPICE output of the program simulating a pH-titration of two acid/base pairs in an aqueous medium. V(301) indicates the pH value. V(2) and V(1) represent the concentrations (in M units) of the dissociated and protonated species of water, respectively. V(12) and V(11), or V(22) and V(21) are the corresponding quantities for the first buffer, or the second buffer, respectively. Time is given in minutes. For parameter values see text. For the format of the plots see section A4. The listing of the SPICE code also appearing in an output is omitted.

on the right of Fig. 14 which presents the change in the concentrations of the protonated and dissociated species of each acid/base pair in the course of the titration.

The less steep parts of the pH-curve in Fig. 14 indicate that buffering of c_H occurs in the corresponding pH ranges. The effectiveness of buffering can be expressed by the buffer capacity β, which is defined as

$$\beta = -(dn_H/dpH)/V = -J_H/[V(dpH/dt)] \tag{23}$$

The second part of Eqn. 23 expresses β in terms of a proton flow and a "pH flow". The capacities of the two buffers in an aqueous medium can be simulated by the circuit representing the acid/base pairs and the modifications shown in Fig. 13B. A pH flow is generated and integrated by the independent current source and the capacitor, respectively, both connected to node 300. The subcircuit PH_TO_AH (see section 4.3.5) converts pH at node 300 to c_H at node 301, which in turn is fed to the controlled voltage source at node 30. The proton flow, which reflects the response of the acid/base pairs to changes in the variable c_H, is monitored by the ammeter between nodes 3 and 30. The unbound protons are represented by the capacitor at node 31, and the corresponding flow is monitored by the ammeter between nodes 31 and 30. Both flows are summed by the controlled current source at node 32, and the resulting flow is monitored by the ammeter connected to the same node. Note that the switching procedure cannot be applied in this case. Since the large flows arising from the adaptation of the system to the initial conditions are meaningless, we will start the simulation at a lower pH than the desired initial value and will omit the first point in the data output.

The scaling of the system is identical to that of the previous example, except for the newly-added pH flow which is in terms of pH units/min. By using the factor $1/[V(dpH/dt)]$ when summing the two proton flows the resulting flow becomes equal to the buffer capacity. Note that the orientation of the ammeters takes care of the minus sign in Eqn. 23. An output generated by the SPICE code listed in section C2.4 with the same parameter values as in the previous example is shown in Fig. 15. Since the dependence of pH on time is linear, the time scale can be easily converted into a pH scale which starts at pH 2 and increases in increments of 0.25 pH units. The points representing the buffer capacity display the bell-shaped curves typical of β of a buffer. The steep rises below pH 3 and above pH 11 are due to the increasing concentrations of H^+ ions and the buffering of the acid/base pair water, respectively. The analytical expression for β reads (see, e.g., ref. 1)

$$\beta = \ln 10 \left\{ 10^{-pH} + 10^{pH}/K_w + \sum_k c_{tot,k} \frac{K_{c,k} 10^{pH}}{(K_{c,k} 10^{pH} + 1)^2} \right\} \tag{24}$$

where $K_w = K_c c_{tot}/1M$ for water, and Σ indicates the sum of the contributions of all buffers present. The three characteristics of the curve in Fig. 15 discussed above, viz. the steep rises at low and high pH

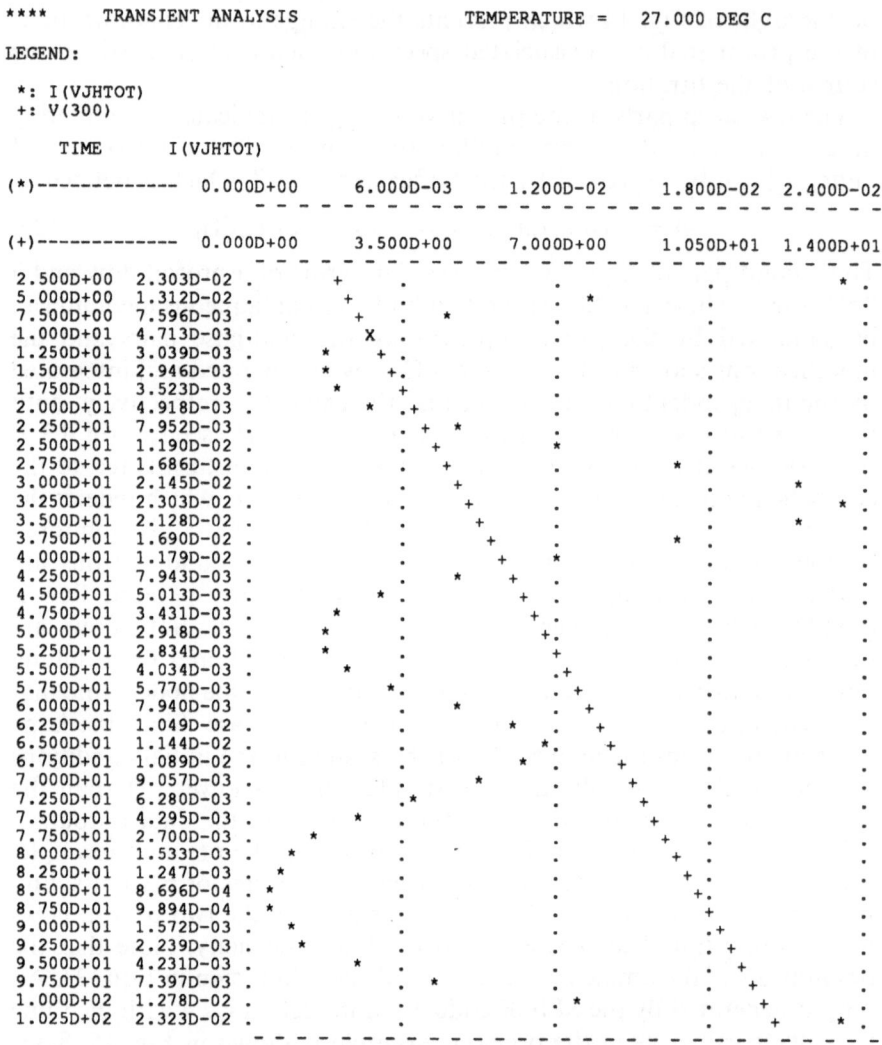

```
****     TRANSIENT ANALYSIS                    TEMPERATURE =   27.000 DEG C

LEGEND:

 *: I(VJHTOT)
 +: V(300)

     TIME      I(VJHTOT)

(*)-------------  0.000D+00     6.000D-03     1.200D-02     1.800D-02  2.400D-02
                  - - - - - - - - - - - - - - - - - - - - - - - - - - - - - - -
(+)-------------  0.000D+00     3.500D+00     7.000D+00     1.050D+01  1.400D+01
                  - - - - - - - - - - - - - - - - - - - - - - - - - - - - - - -
2.500D+00  2.303D-02 .        +       .              .            .       * .
5.000D+00  1.312D-02 .       +        .              .    *       .         .
7.500D+00  7.596D-03 .       +      * .              .            .         .
1.000D+01  4.713D-03 .          X    .              .            .         .
1.250D+01  3.039D-03 .     *    +   .               .            .         .
1.500D+01  2.946D-03 .     *    +.  .               .            .         .
1.750D+01  3.523D-03 .      *    +   .              .            .         .
2.000D+01  4.918D-03 .       *  .+   .              .            .         .
2.250D+01  7.952D-03 .          . +  *              .            .         .
2.500D+01  1.190D-02 .          . +       *         .            .         .
2.750D+01  1.686D-02 .          . +       .      *  .            .         .
3.000D+01  2.145D-02 .          .  +      .         .        *   .         .
3.250D+01  2.303D-02 .          .   +     .         .            .      * .
3.500D+01  2.128D-02 .          .   +     .         .        *   .         .
3.750D+01  1.690D-02 .          .   +     .      *  .            .         .
4.000D+01  1.179D-02 .          .    +    *         .            .         .
4.250D+01  7.943D-03 .          .   *  +  .         .            .         .
4.500D+01  5.013D-03 .       *  .     +   .         .            .         .
4.750D+01  3.431D-03 .     *    .      + .          .            .         .
5.000D+01  2.918D-03 .    *     .      +.           .            .         .
5.250D+01  2.834D-03 .    *     .      +            .            .         .
5.500D+01  4.034D-03 .      *   .     .+            .            .         .
5.750D+01  5.770D-03 .        * .    .  +           .            .         .
6.000D+01  7.940D-03 .          . *   .    +        .            .         .
6.250D+01  1.049D-02 .          .    *.      +      .            .         .
6.500D+01  1.144D-02 .          .   *.       +      .            .         .
6.750D+01  1.089D-02 .          .   * .       +     .            .         .
7.000D+01  9.057D-03 .          .  * .        +     .            .         .
7.250D+01  6.280D-03 .          .*   .         +    .            .         .
7.500D+01  4.295D-03 .       *  .    .         +    .            .         .
7.750D+01  2.700D-03 .     *    .    .          +   .            .         .
8.000D+01  1.533D-03 .   *      .    .           +  .            .         .
8.250D+01  1.247D-03 . *        .    .           +  .            .         .
8.500D+01  8.696D-04 . *        .    .            +.            .         .
8.750D+01  9.894D-04 . *        .    .            + .            .         .
9.000D+01  1.572D-03 .  *       .    .             .+           .         .
9.250D+01  2.239D-03 .    *     .    .             . +          .         .
9.500D+01  4.231D-03 .     *    .    .             .  +         .         .
9.750D+01  7.397D-03 .          .  * .             .   +        .         .
1.000D+02  1.278D-02 .          .    .         *   .      +     .         .
1.025D+02  2.323D-02 .          .    .             .         +  .      * .
                  - - - - - - - - - - - - - - - - - - - - - - - - - - - - - - -
```

Fig. 15. SPICE output of the program simulating the buffer capacity of two acid/base pairs in an aqueous medium. I(VJHTOT) indicates the proton flow in response to a "pH-flow" and is equal to the buffer capacity as a consequence of appropriate scaling (cf. Eqn. 23). V(300) represents the pH value. Time is given in minutes. For parameter values see text. For the format of the plot see section A4. The listing of the SPICE code also appearing in an output is omitted.

values and the bell-shaped regions, can be identified with the three terms in curly brackets of Eqn. 24.

5.3. Enzyme-catalyzed reaction in terms of Hill cycles

Consider the two redox couples (cf. Eqn. 13 in chapter 1)

$$D_{ox} + e^- + H^+ \rightleftarrows HD_{rd} \tag{25a}$$

$$A_{ox} + e^- + H^+ \rightleftarrows HA_{rd} \tag{25b}$$

By choosing the notation D and A for electron donor and electron acceptor, respectively, we have introduced the positive direction for the redox reaction (see section 3.3.2 in chapter 1). The dissociation constants for H^+-binding to the reduced species are assumed to be so small that the unprotonated reduced species can be neglected. Due to kinetic restrictions electron exchange between the couples does not occur without catalysis by an enzyme. Let us assume that this enzyme is embedded in a membrane, and that it exchanges electrons with D_{ox}/HD_{rd} only on side 1 of the membrane, and with A_{ox}/HA_{rd} only on side 2 (cf. section 4.2.5 in chapter 1). When doing so it also exchanges H^+ ions with the redox couples. In order to perform this task, the enzyme has to adopt different conformations (see section 4.2.1 in chapter 1). If we assume that the transitions between these conformations are only possible after the binding or release of two additional H^+ ions on either side of the membrane, we end up with the kinetic cycle of the enzyme shown in Fig. 16A which is generally called a redox-driven proton pump. Note that this scheme also includes a transition between the reduced states of the enzyme without additionally bound H^+ ions on either side of the membrane, which introduces a slip in the cycle (see, e.g., ref. 1). The scheme comprises all elements of the system and presents them in a topological map. Moreover, it contains all elemental processes and thus satisfies steps (1)–(3) listed in section 2.1.

Following Hill [5] (see also section 4.2.1 in chapter 1) the six different species of enzyme shown in Fig. 16A are considered to be six different conformational states of the same macromolecule. Each of these states has a certain probability, denoted by P_i for the ith state, such that for all states

$$\sum_i P_i = 1 \qquad (26)$$

The flow from the ith to the jth state is given by

$$J_{i,j} = \alpha_{i,j} P_i - \alpha_{j,i} P_j \qquad (27)$$

where $\alpha_{i,j}$ and $\alpha_{j,i}$ represent the first-order or pseudo-first-order transition probabilities for the forward and backward transitions. Pseudo-first-order transition probabilities can be expressed as the product of a concentration c and a second-order transition probability $\alpha_{i,j}^0$, in particular

$$\begin{aligned}
\alpha_{1,2} &= c_{HDrd}\,\alpha_{1,2}^0, & \alpha_{2,1} &= c_{Dox}\,\alpha_{2,1}^0, \\
\alpha_{2,3} &= c_{H,1}^2\,\alpha_{2,3}^0, & \alpha_{5,4} &= c_{H,2}^2\,\alpha_{5,4}^0, \\
\alpha_{5,6} &= c_{Aox}\,\alpha_{5,6}^0, & \alpha_{6,5} &= c_{HArd}\,\alpha_{6,5}^0
\end{aligned} \qquad (28)$$

Moreover, some of the transition probabilities will depend on the membrane potential. We assume that the oxidized enzyme is uncharged.

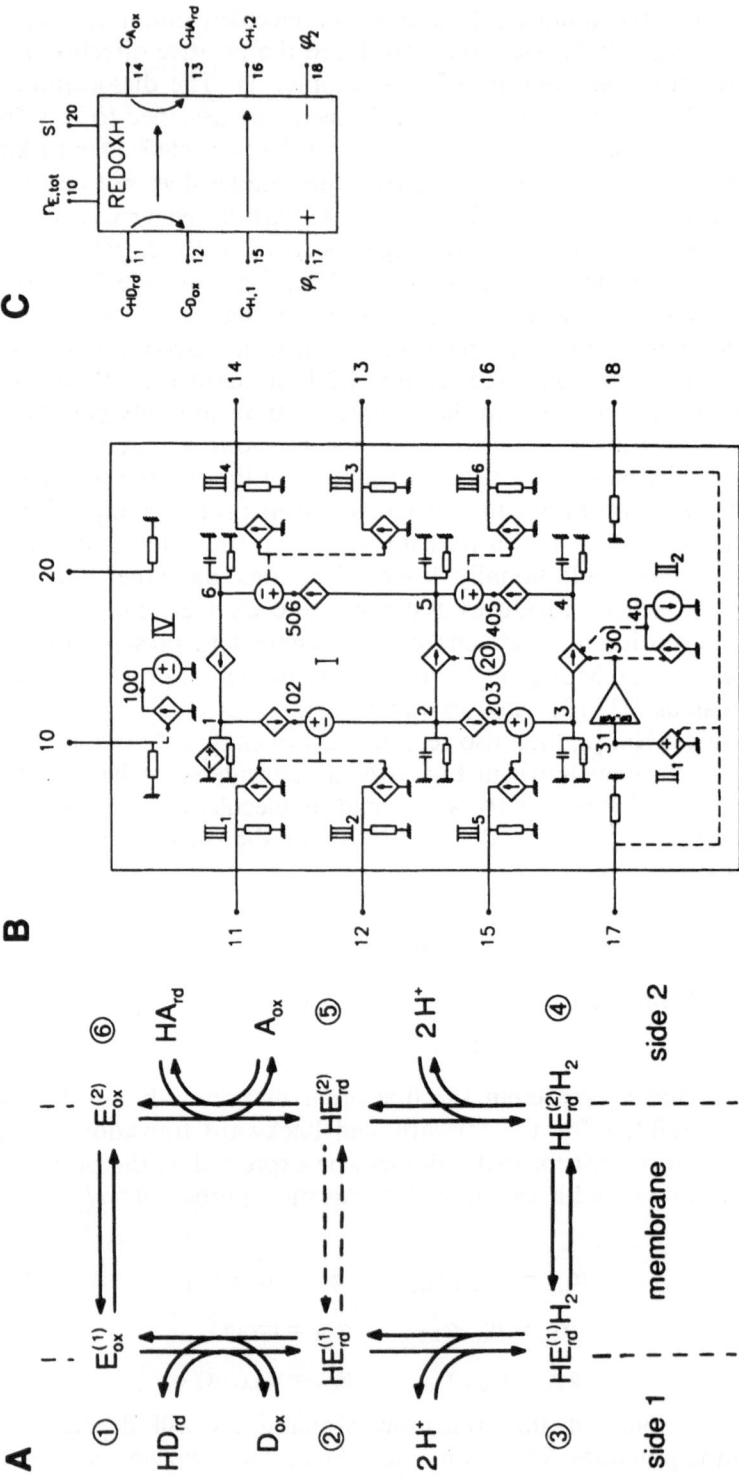

Hence, the reduced enzyme is also uncharged but becomes positively charged upon binding of two H^+ ions. This gives rise to the effective charge numbers $\bar{z}_{3,4} = -\bar{z}_{4,3} = 1$ for the transition $3 \rightleftarrows 4$, accordingly, (cf. Eqns. 49 and 50 in chapter 1)

$$\alpha_{3,4}^* = \alpha_{3,4} \exp\{\mathscr{F} \varDelta\phi_m/(RT)\}, \qquad \alpha_{4,3}^* = \alpha_{4,3} \exp\{-\mathscr{F} \varDelta\phi_m/(RT)\} \qquad (29)$$

In order to calculate the flows of reactants (i.e. the redox species and H^+ ions on both sides of the membrane) the pertinent transition flows must be multiplied by the total mole number of enzyme, $n_{E,tot}$, and the stoichiometric coefficients of reactant binding:

$$J_{Dox} = -J_{HDrd} = n_{E,tot} J_{12},$$

$$J_{HArd} = -J_{Aox} = n_{E,tot} J_{56}, \qquad (30)$$

$$J_{H,1} = -2n_{E,tot} J_{23}, \qquad J_{H,2} = 2n_{E,tot} J_{45}$$

As is evident from Eqns. 25 and Fig. 16A the reduced redox species and the reduced enzyme are present only in the protonated form due to very small dissociation constants for H^+ binding. As a consequence, a $1:1$ cotransport of electrons and H^+ ions occurs. This is equivalent to a transport of hydrogen atoms which are abbreviated by [eH].

Simulate the dependence of the flows on the force generated by the pumped protons (i.e. $\varDelta\tilde{\mu}_H$, cf. section 3.3.2 in chapter 1) at constant concentrations of the redox species, with the enzyme cycle in a steady state. In particular, investigate if the two components of $\varDelta\tilde{\mu}_H$ are kinetically equivalent (cf. section 5.2.2 in chapter 1). This statement, together with Eqns. 26–30, corresponds to steps (4)–(6) as listed in section 2.1.

5.3.1. Subcircuit REDOXH: The network representation of the redox-driven proton pump is fairly complex. Hence according to the policy stated in section 4.4.2, and also for convenience, it is incorporated into a subcircuit called REDOXH (see Fig. 16B,C). In Fig. 16B the enzyme cycle is represented by circuit I. The potentials at nodes 1 to 6 of this circuit indicate the probabilities of the states. Unit capacitors are

Fig. 16. Kinetic scheme of a redox-driven proton pump (A), its implementation in the subcircuit REDOXH (B), and suggested symbol (C). In (A) the membrane-bound enzyme, which is denoted by E, adopts different states (or conformations) in the course of the cycle, as indicated by the subscripts and superscripts. These states are numbered consecutively for reference purposes. For further explanation see text. In (B) element names are omitted due to lack of space, but they can be found in the pertinent SPICE code listed in section C3.1. Individual circuits are designated by Roman numerals for reference purposes. The frame indicates the "boundary" of the subcircuit which encloses the "local" elements and nodes. Nodes 11, 12, 13, and 14 represent the connections which transfer the flows of HD_{rd}, D_{ox}, HA_{rd}, and A_{ox}, respectively, to the "outside". Nodes 15 and 16 are the corresponding connections for the proton flows. Nodes 10 and 20 transfer the values of $n_{E,tot}$ and sl to the subcircuit, while the node pair 17, 18 transfers the membrane potential. All resistors are added to provide a DC path to ground. Controlling pathways are indicated by broken arrows.

connected to nodes 2 to 6 in order to integrate the transition flows in the adjacent branches over time (cf. section 4.3.6). At node 1 a controlled source is used to satisfy the condition in Eqn. 26. The transition flows in each branch are simulated by means of voltage-controlled current sources according to Eqns. 27–29. The terms $\exp\{\pm\mathscr{F}\Delta\phi_m/(RT)\}$ required for the probabilities of the transition $3 \rightleftarrows 4$ are generated as potentials at nodes 30 and 40 in circuits II_1 and II_2, respectively, with the membrane potential being read from nodes 17 and 18. Ammeters are introduced in those branches of the circuit which involve binding of reactants in the cycle of Fig. 16A. The flows monitored by these ammeters are multiplied by $n_{E,tot}$ and the pertinent stoichiometric co-efficients, and fed to the current-controlled current sources in circuits III_1 to III_6, thus yielding the flows of reactants according to Eqns. 30. Note that $n_{E,tot}$ is not included as a parameter value in the descriptions of the controlled sources. Instead it is supplied as a voltage to node 10 which is then converted into a current in circuit IV. In this way the value for $n_{E,tot}$ can be easily changed and even supplied to several pumps in case they have the same value for $n_{E,tot}$ (cf. footnote 5). Similarly, a parameter called sl which determines the slip in the cycle is applied as a potential difference to node 20 (in the absence of slip sl is zero).

5.3.2. Flow-force relations: The flow of H^+ ions is positive from side 1 to side 2 (cf. Eqns. 30) according to the sign convention already introduced by the electron transfer between the redox couples across the membrane. Hence the conjugate force is (cf. Eqn. 17 in section 3.3.2 in chapter 1), in view of Eqns. 17 with $y_H = 1$,

$$\Delta\tilde{\mu}_H = \tilde{\mu}_{H,1} - \tilde{\mu}_{H,2} = RT \ln 10(pH_2 - pH_1) + \mathscr{F}\Delta\phi_m \qquad (31)$$

where $\Delta\phi_m = \phi_1 - \phi_2$ is the membrane potential. We choose side 1 of the membrane to be the reference point, i.e. $pH_1 = $ const and $\phi_1 = 0$. It then follows that

$$pH_2 = pH_1 + \Delta\tilde{\mu}_H/(RT \ln 10) \quad \text{for } \Delta\phi_m = 0 \qquad (32a)$$

and

$$\Delta\phi_m = \Delta\tilde{\mu}_H/\mathscr{F} \quad \text{for } pH_1 = pH_2 \qquad (32b)$$

Figure 17 shows the circuits which simulate the flow-force relations for *both cases in Eqns. 32 in one simulation.* The independent voltage source at node 1 is used to simulate $\Delta\tilde{\mu}_H$ (its value will sweep through the desired range in the DC analysis). Two subcircuits REDOXH, one for each case according to Eqns. 32, are connected to the common nodes 10, 20, 11, 12, 13, and 14 whose potentials are determined by the independent voltage sources representing $n_{E,tot}$, sl, c_{HDrd}, c_{Dox}, c_{HArd}, and c_{Aox}, respectively. Ammeters are interspersed between the subcircuits and node 13, which monitor the flow of HA_{rd} and thus $J_{[eH]}$. Node

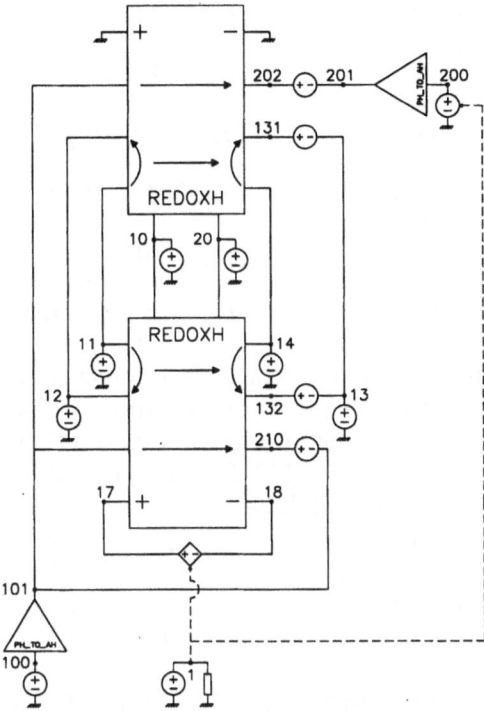

Fig. 17. Circuit for simulating flow-force relations of a redox-driven proton pump. Element names are omitted due to lack of space, but they can be found in the listing of the pertinent SPICE code in section C3.2. Horizontally oriented independent voltage sources are ammeters, while vertically oriented sources determine parameter values. The triangular symbol represents the subcircuit PH_TO_AH. The upper part of the main circuit containing the subcircuit REDOXH simulates the case of a vanishing membrane potential, while the lower part simulates the case of equal pH values on both sides of the membrane. Controlling pathways are indicated by broken arrows. The symbol ⊓⊓ denotes the "electrical ground" which has zero potential. For further explanation see text.

101, which is also common to both subcircuits, represents $c_{H,1}$ as determined by the constant pH_1 at node 100 and the subcircuit PH_TO_AH. In the case of $pH_2 = pH_1$ node 210 is also connected to node 101 via an ammeter which monitors J_H. The voltage source between nodes 17 and 18 is controlled by the potential at node 1 according to Eqn. 32b and yields the membrane potential. The variable pH_2 in the case of $\Delta\phi_m = 0$ (see Eqn. 32a) is produced by the controlled voltage source at node 200. It is converted by the subcircuit PH_TO_AH to $c_{H,2}$ at node 201 and fed to the subcircuit REDOXH via the ammeter (which monitors J_H) between nodes 202 and 201.

In scaling the system we have chosen to use molar for concentrations and seconds for time. The values for the transition probabilities are as follows:

$$\alpha_{1,2}^0 = 2 \times 10^5 \, \text{M}^{-1}\,\text{s}^{-1} \qquad \alpha_{2,1}^0 = 4 \times 10^6 \, \text{M}^{-1}\,\text{s}^{-1}$$

$$\alpha_{2,3}^0 = 2 \times 10^{16} \text{ M}^{-2} \text{ s}^{-1} \quad \alpha_{3,2} = 200 \text{ s}^{-1}$$

$$\alpha_{3,4} = 2000 \text{ s}^{-1} \quad \alpha_{4,3} = 200 \text{ s}^{-1}$$

$$\alpha_{4,5} = 200 \text{ s}^{-1} \quad \alpha_{5,4}^0 = 2 \times 10^{13} \text{ M}^{-2} \text{ s}^{-1}$$

$$\alpha_{5,6}^0 = 2 \times 10^6 \text{ M}^{-1} \text{ s}^{-1} \quad \alpha_{6,5}^0 = 2 \times 10^3 \text{ M}^{-1} \text{ s}^{-1}$$

$$\alpha_{6,1} = 200 \text{ s}^{-1} \quad \alpha_{1,6} = 200 \text{ s}^{-1}$$

$$\alpha_{2,5} = \text{sl} \times 0.5 \text{ s}^{-1} \quad \alpha_{5,2} = \text{sl} \times 5 \times 10^{-5} \text{ s}^{-1}$$

where sl ≥ 0 is the parameter which determines how much slipping occurs in the pump cycle. This set of transition probabilities yields, in view of thermokinetic balancing (see section 5.1.2 in chapter 1), an equilibrium constant $K_c = 5 \times 10^5$ for the overall reaction and the hydrogen atom transport, while $K_c = 1$ for H^+ ion transport. The constant concentrations of redox species are $c_{\text{HDrd}} = 10$ mM, $c_{\text{Dox}} = 20$ μM, $c_{\text{HArd}} = 100$ mM, and $c_{\text{Aox}} = 1$ mM. The thermodynamic force for the hydrogen atom transport across the membrane constituted by the redox couples is (cf. Eqn. 78b in chapter 1)

$$\Delta \tilde{\mu}_{[\text{eH}]} = RT \ln\{K_c \, c_{\text{HDrd},1} c_{\text{Aox},2}/(c_{\text{Dox},1} c_{\text{HArd},2})\} \tag{33}$$

and amounts to 36.52 kJ/mol. The constant pH_1 is 7.5, and the total mole number of enzyme is 1 (without units) which means that the flows are turnover numbers.

An output generated by SPICE with the code listed in section C3.2 is shown in Fig. 18 for sl = 1. The result is very instructive in several ways. We notice that the two terms of $\Delta \tilde{\mu}_H$ are far from being equivalent. In the case of a variable chemical potential the flows display the hyperbolic tangent type of dependence discussed in section 5.2 in chapter 1. Static head of proton pumping (i.e. $J_H = 0$) occurs at $\Delta \tilde{\mu}_H = -18.11$ kJ/mol, a value which is slightly lower than expected from the force set up by the redox couples (i.e. $\frac{1}{2} \times 36.52$ kJ/mol $= 18.26$ kJ/mol) because of the slip in the cycle[6]. For the same reason $J_{[\text{eH}]}$ does not vanish but has a finite value of 0.225 s^{-1}. The flows in the case of a variable membrane potential are dramatically different. The probabilities for the transition $3 \rightleftarrows 4$ are well-adjusted for $\Delta \phi_m \approx 0$, but adopt increasingly unfavorable values with respect to pumping if $\Delta \phi_m < 0$ (e.g. $\alpha_{3,4}^* = 60.2$ s^{-1} and $\alpha_{4,3}^* = 6643$ s^{-1} for $\Delta \phi_m = -90$ mV). As a consequence, the proton flow becomes very low for $\Delta \tilde{\mu}_H \leq -9$ kJ/mol. Static head also occurs at $\Delta \tilde{\mu}_H = -18.11$ kJ/mol, but $J_{[\text{eH}]}$ at this state amounts to 0.421 s^{-1}.

---

[6]The value of $\Delta \tilde{\mu}_H$ at static head can be found by running a simulation with the PLOT statement replaced by a PRINT statement, and a type of analysis statement with the following specifications: .DC VD_MUH -18.1 -18.15 .01

```
****       DC TRANSFER CURVES                   TEMPERATURE =    25.000 DEG C

LEGEND:

  *: I(VJE_PH)
  +: I(VJH_PH)
  =: I(VJE_PHI)
  $: I(VJH_PHI)

     VD_MUH          I(VJE_PH)

 (*+=$)---------- -1.000D+01    0.000D+00     1.000D+01     2.000D+01   3.000D+01
         - - - - - - - - - - - - - - - - - - - - - - - - - - - - - - - - -
  0.000D+00  1.315D+01 .              .              X        .        X      .
 -1.000D+00  1.315D+01 .              .            . = *      .      $ . +     .
 -2.000D+00  1.315D+01 .              .          =. *      $ .        +     .
 -3.000D+00  1.315D+01 .              .       =      X        .        +      .
 -4.000D+00  1.314D+01 .              .   =    $ .    *      *  .        +      .
 -5.000D+00  1.314D+01 .            . = $       .      *        .        +      .
 -6.000D+00  1.314D+01 .          . =$          .      *        .        +      .
 -7.000D+00  1.313D+01 .          .=$           .      *        .        +      .
 -8.000D+00  1.312D+01 .          .X            .      *        .        +      .
 -9.000D+00  1.308D+01 .          $=            .      *        .       +       .
 -1.000D+01  1.300D+01 .          $=            .      *        .      +        .
 -1.100D+01  1.281D+01 .          $=            .      *        .     +         .
 -1.200D+01  1.242D+01 .          $=            .    *          .    +          .
 -1.300D+01  1.161D+01 .          $=            .   *           .  +            .
 -1.400D+01  1.011D+01 .          $=            .  *          +. .              .
 -1.500D+01  7.723D+00 .          $=       .  *        +        .              .
 -1.600D+01  4.806D+00 .          $=  * *       +.              .              .
 -1.700D+01  2.171D+00 .          $= * +         .              .              .
 -1.800D+01  3.700D-01 .          XX            .       .    .                 .
 -1.900D+01 -6.397D-01 .       +*$=             .              .              .
 -2.000D+01 -1.144D+00 .     +  *  $=           .              .              .
 -2.100D+01 -1.381D+00 .     +  *  $=           .              .              .
 -2.200D+01 -1.489D+00 .   +    *  $=           .              .              .
 -2.300D+01 -1.538D+00 .   +    *  $=           .              .              .
 -2.400D+01 -1.560D+00 .   +    *  $=           .              .              .
 -2.500D+01 -1.570D+00 .   +    *  $=           .              .              .
 -2.600D+01 -1.574D+00 .   +    *  $=           .              .              .
 -2.700D+01 -1.576D+00 .   +    *  $=           .              .              .
 -2.800D+01 -1.577D+00 .   +    *  $=           .              .              .
 -2.900D+01 -1.577D+00 .   +    *  $=           .              .              .
 -3.000D+01 -1.577D+00 .   +    *  $=           .              .              .
         - - - - - - - - - - - - - - - - - - - - - - - - - - - - - - - - -
```

Fig. 18. SPICE output of the program simulating flow-force relations for a redox-driven proton pump. VD_MUH indicates $\Delta\tilde{\mu}_H$ in kJ/mol. I(VJE_PH) and I(VJH_PH) represent the turnover numbers (in s^{-1}) for hydrogen atoms and protons, respectively, in the case of a vanishing membrane potential (cf. Eqn. 32a). I(VJE_PHI) and I(VJH_PHI) are the corresponding quantities in the case of equal pH values on both sides of the membrane (cf. Eqn. 32b). For parameter values see text. For the format of the plot see section A4. The listing of the SPICE code also appearing in an output is omitted.

5.4. Redox-driven proton pump

In the preceding section we have investigated the redox-driven proton pump in its "isolated state", i.e. with all parameters on both sides of the membrane controlled. Such an investigation would be very difficult (if not impossible) to do in practice, which underscores the advantages offered by modelling. In this section we explore the behavior of the pump in situ, i.e. in a vesicular system that does not provide direct access to the compartment on side 2 of the membrane (see Fig. 19). In addition to considering the pump, we have to take into account the fact that the membrane is permeable to A_{ox} and HA_{rd}, but not to D_{ox} and HD_{rd}. Moreover, it may be permeable to protons (usually called the

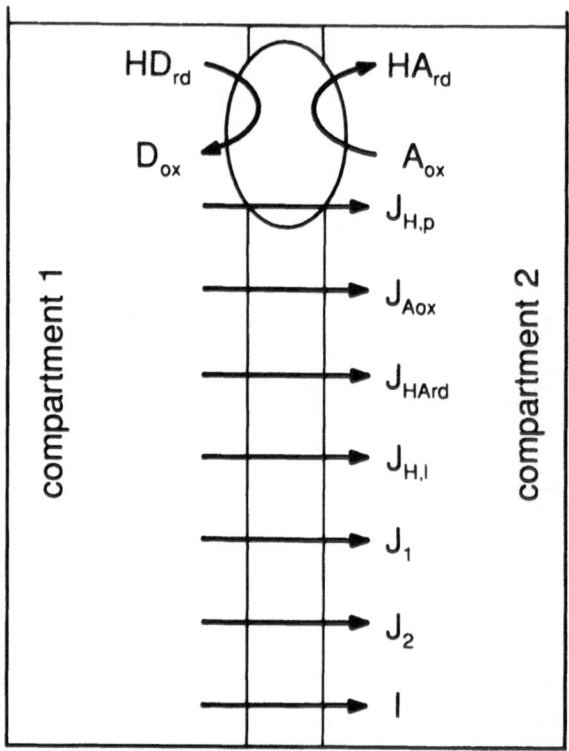

Fig. 19. Topological map of a redox-driven proton pump in situ. The system consists of a membrane which separates compartment 1 from the inaccessible compartment 2. The membrane contains the redox-driven proton pump which creates the flow of pumped protons, $J_{H,p}$, when transferring hydrogen atoms from the donating to the accepting redox couple. The membrane is permeable to protons which gives rise to the leak flow $J_{H,l}$. In addition, the membrane is permeable to the redox species of the electron-accepting couple and to the ions of a uni-univalent salt present. The corresponding flows are indicated by arrows. The arrow marked by I represents the electric current associated with the flows of charged species (cf. Fig. 1A).

proton leak) as well as to the cations and anions of salts present. We start from an equilibrium state in the presence of a uni-univalent salt and all reactants except HD_{rd}. We want to simulate the approach of the system to static head which evolves after the addition of HD_{rd}.

The circuits used to simulate this system comprise all the elements whose behavior has been discussed extensively in sections 5.1 to 5.3. We therefore leave it to the reader as an exercise to interpret the circuitry presented in Fig. 20. The following hints may prove helpful (see also the legend to Fig. 20). The concentrations of H^+ ions, D_{ox}, HD_{rd}, HA_{rd}, A_{ox}, cation, and anion in compartment 1 are represented by the potentials at nodes 100, 101, 102, 103, 104, 105, and 106, respectively. The corresponding concentrations in compartment 2 (except for D_{ox} and HD_{rd}) are the potentials at nodes 200 to 206. The buffer capacity for H^+ ions in compartment 1 is assumed to be constant due to added

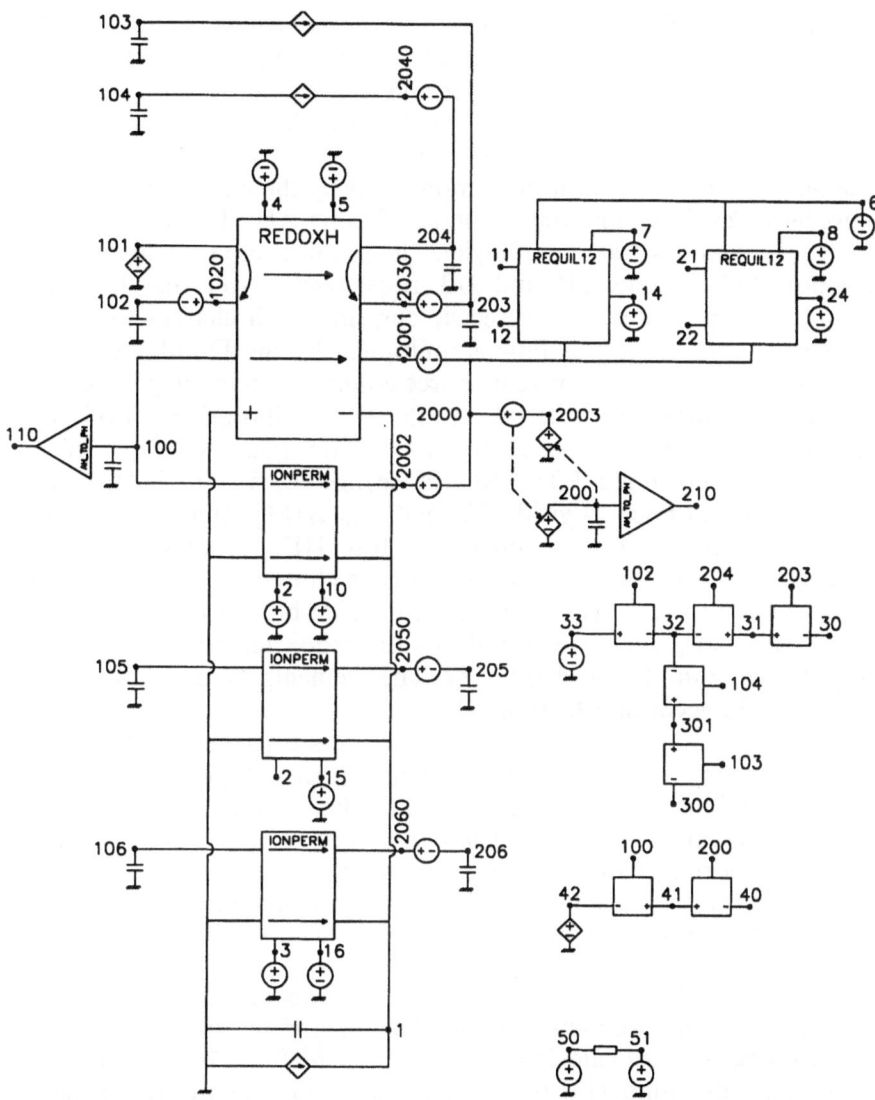

Fig. 20. Circuits for simulating a redox-driven proton pump in situ. Element names are omitted due to lack of space, but they can be found in the listing of the pertinent SPICE code in section C3.3. Horizontally oriented independent voltage sources are ammeters, while vertically oriented sources determine parameter values. The symbol $\overline{\overline{}}$ denotes the "electrical ground" with zero potential. The triangular symbol represents the subcircuit AH_TO_PH. Controlling pathways are indicated by broken arrows. The auxiliary circuits comprising nodes 30 to 33, 300 and 301, and 40 to 42 calculate $\Delta\tilde{\mu}_{[eH]}$, $\mathscr{A}_{[eH]}$, and $\Delta\tilde{\mu}_H$, respectively. The circuit comprising nodes 50 and 51 generates a potential difference at node 50 and a current through the ammeter connected to node 51 which switch from zero during the "pre-experimental" period to 1 at the beginning of the "actual" experiment.

impermeable buffers. This causes the volume for these ions to be formally increased [1]. The subcircuits REQUIL12 connected to node 2000 simulate the buffering by water and a membrane-bound buffer on side 2, respectively. The pre-experimental equilibration period is again simulated by the switching device discussed in section 5.2.3 which operates on the total proton flow monitored by the ammeter connected between nodes 2000 and 2003. The subcircuits AH_TO_PH (see section 4.3.5) at nodes 100 and 200 yield the pH in compartments 1 and 2 as a potential at nodes 110 and 210, respectively. Note that J_H of the pump also contributes to the electric current which charges the membrane capacity. Hence this flow is multiplied by the Faraday constant and fed to the controlled current source connected between ground and node 1. This current is switched on after the equilibration period. The elements connected to nodes 30 to 33 calculate $\Delta\tilde{\mu}_{[eH]}$ (see Eqn. 33) by means of the subcircuits RT_LNC (cf. section 4.3.5). The independent voltage source at node 33 yields RT $\ln\{K_c c_{HDrd}/1M\}$ after the equilibration period. During this period (i.e. before HD_{rd} is added) $\Delta\tilde{\mu}_{[eH]}$ is undetermined, and the value of the source is chosen such that the force becomes close to zero. The elements in the side branch leading to node 300 calculate $\mathscr{A}_{[eH]}$, i.e. the affinity of the redox reaction in compartment 1 (see Eqn. 15 in chapter 1). The elements at nodes 40 to 42 calculate $\Delta\tilde{\mu}_H$ according to Eqn. 31.

In scaling the system we choose molar units for concentrations and seconds for time. For the other parameters we adopt the procedure common in bioenergetics, i.e. we relate them to a quantity which expresses the amount of membrane present (e.g. the amount of protein for mitochondria, or the mole number of chlorophyll, n_{chl}, for chloroplasts). We use parameter values typical for thylakoids of chloroplasts (see, e.g., ref. 1) and obtain 10^4 and 20 l/(mol chl) for the volumes of compartments 1 and 2, respectively, while $n_{E,tot}/n_{chl} = 0.005$ (i.e. one pump per 200 chlorophylls). The transition probabilities for the pump are as listed in the preceding section. The membrane capacity amounts to 10 kF/(mol chl) as calculated from 1 μF/cm^2 for a biological membrane and 1 m^2/(μmol chl) for the specific surface of thylakoid membranes. This capacity is given in kF units in order to obtain the membrane potential in millivolts. The dissociation constant and the concentration of the membrane-bound buffer are 10^{-5} and 50 mM, respectively. The impermeable buffers in compartment 1 are assumed to yield a formal volume of 2×10^7 l(mol chl)$^{-1}$ for H$^+$ ions. The concentration of the uni-univalent salt present is 0.1 M. The unit for the flows in this scaling becomes mol (mol chl)$^{-1}$ s^{-1}. The unit for the product of membrane area and permeability (see section 5.1) is 1 (mol chl)$^{-1}$ s^{-1}, and its value was chosen to be 200 for H$^+$ ions, 0.4 for cations, 0.2 for anions, 20 for A_{ox}, and 1000 for HA_{rd}, respectively.

A listing of the SPICE code can be found in section C3.3. Figure 21 presents, by way of example, the time course of some relevant parame-

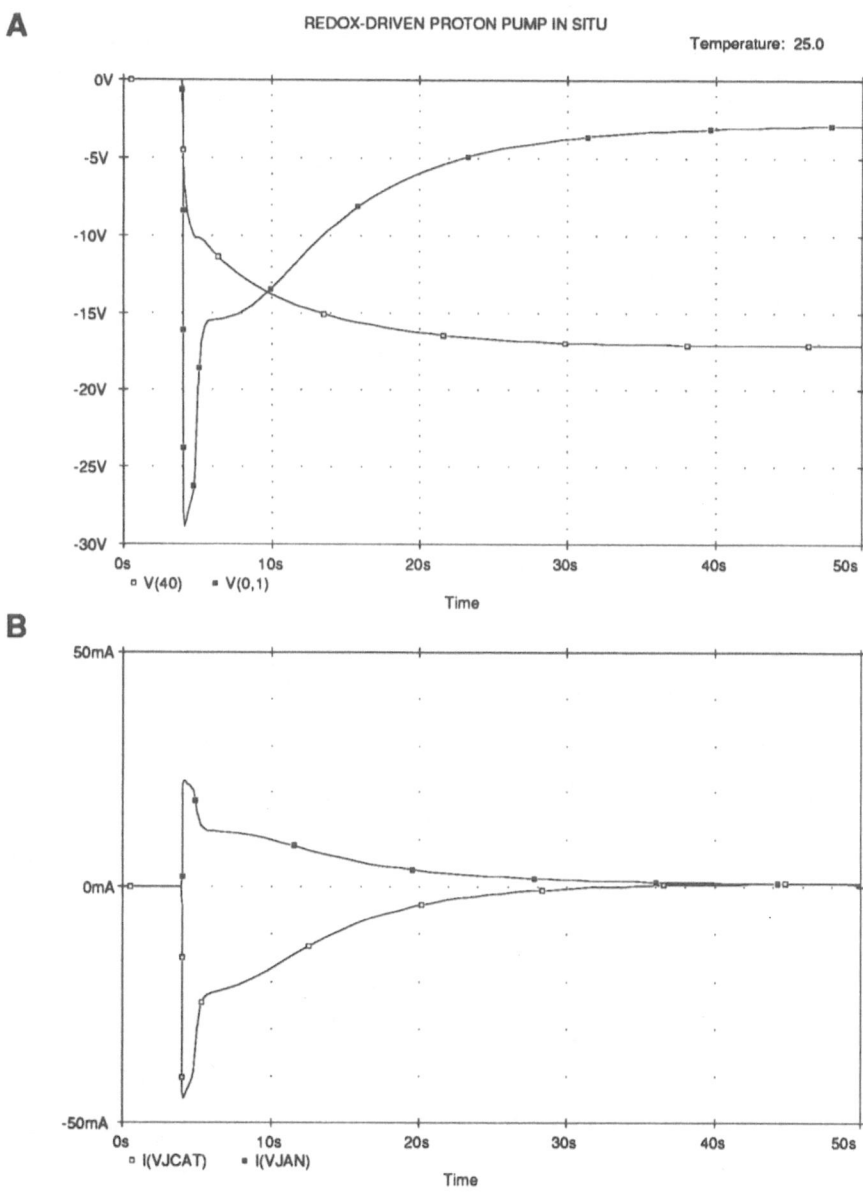

Fig. 21(A)–(B).

ters of the system for initial conditions listed in the legend. Such graphs can be obtained by running PROBE in PSpice. It is beyond the scope of this chapter to discuss the result in detail. However, some instructive points are worth mentioning. (1) The absolute value of the membrane potential initially rises very fast and then decreases to a small value

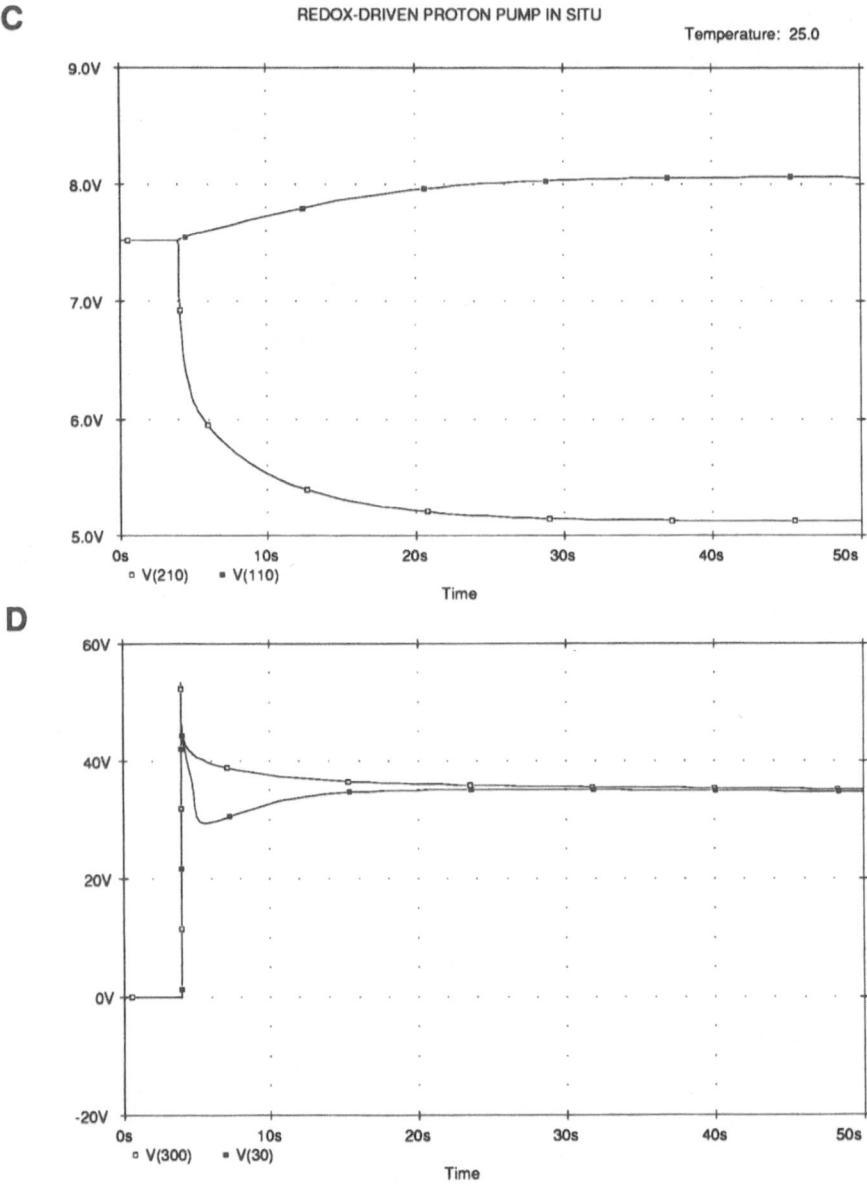

Fig. 21(C)–(D).

(Fig. 21A). This is due to the fact that chemical capacities and electrical capacity are substantially different, and that permeation of ions starts (Fig. 21B) as soon as the membrane potential deviates from zero. (2) The initial drop of pH in compartment 2 slows down (Fig. 21C) because of the thermodynamic control exerted by the increasing $\Delta\tilde{\mu}_H$ on $J_{[eH]}$ [1]

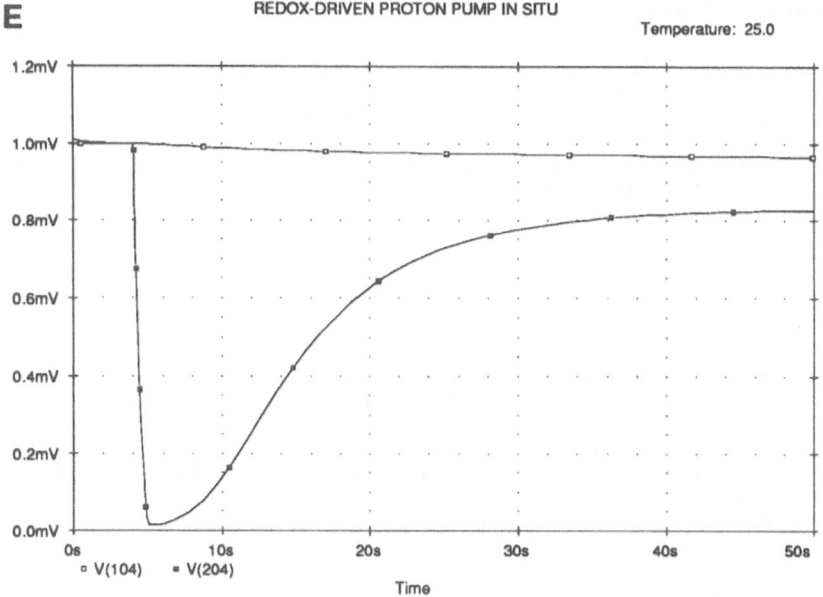

Fig. 21(E).

Fig. 21. PSpice output of the program simulating a redox-driven proton pump in situ. The graphs were obtained using the PROBE option of PSpice. V(0, 1) and V(40) in (A) represent the membrane potential (in mV) and $\Delta\tilde{\mu}_H$ (in kJ/mol), respectively. I(VJCAT) and I(VJAN) in (B) are the flows of univalent cation and anion (in mol/s). V(110) and V(210) in (C) indicate the pH in compartment 1 and 2, respectively. V(30) and V(300) in (D) represent the forces (in kJ/mol) $\Delta\tilde{\mu}_{[eH]}$ and $\mathscr{A}_{[eH]}$, respectively, while V(104) and V(204) in (E) indicate the concentration of A_{ox} (in M units) in compartment 1 and 2, respectively. The initial concentrations of redox species are $c_{Dox,1} = 10^{-10}$ M, $c_{Aox,1} = c_{Aox,2} = 1$ mM, $c_{HArd,1} = c_{HArd,2} = 0.1$ M; $c_{HDrd,1}$ is switched from zero to 10 mM after the pre-experimental period. The initial membrane potential is zero. For other parameter values see text.

but also because the pH in this compartment enters the range where the membrane-bound buffer becomes effective. (3) The affinity of the redox reaction as determined by the concentrations of the redox species in compartment 1 can substantially deviate from the actual driving force of $J_{[eH]}$, viz. $\Delta\tilde{\mu}_{[eH}$ (Fig. 21D). Due to the relatively small permeability for A_{ox}, concentration of this species in compartment 2 rapidly drops to almost zero and approaches the value in compartment 1 only when the system reaches the steady state (Fig. 21E).

The reader, who by now has very likely become acquainted to some extent with SPICE modelling (providing he or she has the appropriate facilities), is invited to explore further the behavior of the redox-driven proton pump (or any other model from his or her own field of interest). He or she will then experience the full power of modelling, since every parameter of the system, and what is even more important the interplay of the parameters, can easily be followed. The experimenter can only dream of such a situation.

Acknowledgements

We are indebted to Dr. I. Yuli for carefully reading the manuscript and many valuable comments and suggestions.

D. W. acknowledges financial support by the Swiss National Science Foundation and by the Julius Bär Foundation, Zürich. S.R.C. was supported by the Basic Research Foundation administered by the Israel Academy of Sciences and Humanities. D.R.L.S. was supported by grants from the South African Medical Research Council and the Foundation for Research and Development, South Africa.

References

1. D Walz, Biochim. Biophys. Acta 1019 (1990) 171–224.
2. J Horno, F González-Caballero and CF González-Fernández, Eur. Biophys. J. 17 (1990) 307–313.
3. EL King and C Altman, J. Phys. Chem. 60 (1956) 1375–1378.
4. SJ Mason and HJ Zimmerman, *Electronic Circuits, Signals, and Systems*, M.I.T. Press, Cambridge, MA, 1970.
5. TL Hill, *Free Energy Transduction in Biology*, Academic Press, New York, 1977.
6. GF Oster, A Perelson and A Katchalsky, Q. Rev. Biophys. 6 (1973) 1–134.
7. L Peusner, *Studies in Network Thermodynamics*, Elsevier, Amsterdam, 1986.
8. DC Mikulecky, Biophys. J. 25 (1979) 323–339.
9. DC Mikulecky, in *Membrane Biophysics II: Physical Methods in the Study of Epithelia*, M Dinno, AB Calahan and TC Rozzell (eds), Alan R. Liss, New York, 1983, pp. 257–282.
10 DC Mikulecky, EG Huf and SR Thomas, Biophys. J. 25 (1979) 87–105.
11 JC White and DC Mikulecky, Pharmacol. Ther. 15 (1981) 251–291.
12 PW Tuinenga, *SPICE: A Guide to Circuit Simulation and Analysis Using PSpice*, Prentice Hall, Englewood Cliffs, NJ, 1988.
13 D Walz, Biochim. Biophys. Acta 505 (1979) 279–353.
14 A Vladimirescu, K Zhang, AR Newton, DO Pederson and A Sangiovanni-Vincentelli, *SPICE Version 2G.6 User's Guide*, University of California, Berkeley, 1981.

Appendix A. Lexicon of SPICE

This is an adaptation of the SPICE manual [14] directed at bioelectrochemical simulations. It deals with the conventions and program statements of SPICE based on version 2G.6. Older versions will not necessarily have all of the options or elements listed here, while more recent versions may have useful additional features.

A1. Format of instructions to SPICE

A1.1. General rules

(1) All instructions to SPICE except titles and comments must be typed in capital letters. Individual items are separated by one or more blank spaces. (PSpice allows the use of upper and lower case letters for all instructions.)

(2) Lines must contain no more than 80 characters. A statement that requires more than 80 characters can be continued on subsequent lines, each of which must begin with a plus sign in the first space $(+\cdots)$.

(3) The first line, which may be blank, is always interpreted by SPICE as a title. This title appears in the heading of each SPICE output, and may not exceed 1 line.

(4) The order of the statements in a set of instructions is irrelevant, except that (a) continuation lines must obviously be in sequence, and (b) the ALTER statement(s) (see section A5.3) must precede the END statement.

(5) Comments may be inserted at any point. Each line of comment must commence with an asterisk $(*\cdots)$, which need not be in the first space of the line. Any characters may be used following the asterisk, including lower-case letters.

(6) The set of instructions must be terminated with the END statement which has the following format: .END

A1.2. Numbers

Numbers in SPICE may be in one of three formats:

integer (e.g. -32, 120)
floating point (e.g. 0.45, .32, 320.6789)
exponential (e.g. $3.57E - 8$ where "E" represents "ten to the power")

Certain important powers of ten in the exponential format can be abbreviated by the use of suffixes as listed in Table 4. Thus $3.56E - 12$, 3.56P, and 3560F all represent 3.56×10^{-12}. Note that a digit must precede the "E" in the exponential format, e.g. $1E - 3$. Furthermore, no blank spaces may appear within a number (in particular, a minus or plus sign must not be separated by a blank from the number it qualifies). The range of numbers accepted by SPICE is $1E - 35$ to $1E + 35$. (PSpice cannot handle voltages or currents exceeding $1E + 10$.)

Table 4. Suffixes used to express powers of ten

Power of ten	Suffix	Greek derivation	Power of ten	Suffix	Greek derivation
$E - 15$	F	femto			
$E - 12$	P	pico	$E + 12$	T	tera
$E - 9$	N	nano	$E + 9$	G	giga
$E - 6$	U	micro	$E + 6$	MEG	mega
$E - 3$	M	milli	$E + 3$	K	kilo

A1.3. Nodes

Nodes in SPICE are the imaginary points between which the elements are connected. They are represented by positive integers. These numbers need not be sequential, but for clarity a systematic choice is best. For example, all nodes of a section of a circuit pertinent to a particular species may be numbered from 100 up, while those pertinent to a second species are numbered from 200 up, and so forth.

Node number 0 has a special meaning. In electrical network terms it represents the ground which has zero potential. For model simulations it will represent the zero point from which any given quantity (concentration, pressure, chemical potential, electrical potential, etc.) is measured. All nodes require the following:

(1) at least two connections, and
(2) a "DC path to ground".

The second requirement stipulates that each node must be connected to ground via at least one pathway containing one or more of the following elements only: resistors, independent voltage sources, and diodes. This requirement can always by fulfilled by adding resistors with appropriately chosen values.

A2. Element specifications

A2.1. General format

The format for specifying any element is

name # -1 # -2 ⟨ # -3 ⋯ ⟩ description ⟨IC = initial value⟩

The items between the signs ⟨ ⟩ apply only to certain elements. Each element must be given its own name. The name of an element consists of up to eight characters, chosen from the set of capital letters A to Z and numbers 0 to 9. The first character determines the element type according to the following list (see also Table 2):

C   capacitor
D   diode
E   voltage-controlled voltage source
F   current-controlled current source
G   voltage-controlled current source
H   current-controlled voltage source
I   independent current source
R   resistor
V   independent voltage source
X   subcircuit

The remaining characters of a name (up to seven) may be freely chosen according to taste. The symbols # -1, # -2, # -3 etc. indicate the nodes to which the element is connected (cf. footnote 4). They are replaced in an actual statement by the corresponding numbers of the nodes in a circuit. The description either comprises the parameters in the constitutive relation of the element or else refers to another statement in which they are specified. The IC (initial condition) option applies only to capacitors, diodes, and controlled sources and will be discussed in the sections dealing with these elements. Each element will be considered individually below, including typical examples and specifying the constitutive relation. This relation correlates potential difference U, current I, resistance R, capacitance C, charge Q, and time t. The symbols suggested for the elements are given in Table 2. The items between the signs ⟨ ⟩ are optional.

A2.2. Conventions

A2.2.1. Sign convention: For all elements except subcircuits the following sign convention for potential difference and current holds. A potential difference is counted positive if the node # -1 has a higher potential than the node # -2, and a current is counted positive if it flows from the node # -1 through the element to the node # -2. (Note that this convention does not indicate which direction these parameters attain in an actual simulation.) Subcircuits comprise more than one element, hence their nodes are usually not subject to the above sign convention. However, there are cases where a node pair of a subcircuit is equivalent to the terminals of an element. It is then advantageous to indicate the polarity according to the above sign convention by means of $+/-$ signs or an arrow, as shown in Figs. 6A, 8B, and 16C.

A2.2.2. Units: The units for potential difference, current, resistance, capacitance, and time are not specified in SPICE, but taken to be the volt, the ampere, the ohm, the farad, and the second, respectively. These electrical units can be translated into the relevant units of a model as explained in section 3.3.

A2.3. Resistor

R · · · · · · ·   # -1   # -2   value
Example: RPERM12   101 102 1K
Constitutive relation: $U = RI$ (A1)

The description of a resistor is the value of its resistance, R.

A2.4. Capacitor

$$C \cdots \cdots \quad \#\text{-}1 \quad \#\text{-}2 \quad \text{value} \quad \langle \text{IC} = \text{initial value}\rangle$$
Example: CVOLCA1   201   0   1M   IC=20U
Constitutive relation: $Q = CU$ or $I = C(dU/dt)$ \hfill (A2)

The description of a capacitor is the value of its capacitance, C. The initial value refers to the potential difference across the capacitor at time zero and will be set to zero if not specified.

A2.5. Diode

$$D \cdots \cdots \quad \#\text{-}1 \quad \#\text{-}2 \quad \text{modname} \quad \langle \text{IC} = \text{initial value}\rangle$$
Example: DMUNA   311   0   LOG
Constitutive relation: $I = I_s[\exp\{\mathscr{F} U/(nRT)\} - 1] + UG_{min}$ \hfill (A3)

where $\mathscr{F}$, R, and T are the Faraday constant, the gas constant, and the absolute temperature, respectively. I_s and n are diode-specific parameters called "saturation current" and "emission coefficient", respectively. G_{min} is a minimal conductance which dominates the diode current at large negative values of U. The initial value refers to the potential difference across the diode, and need not be specified in most cases of biological interest. The description of a diode consists of a model name only, which refers to a separate model statement having the following format:

$$.\text{MODEL} \quad \text{modname} \quad D \quad \langle \text{IS} = \text{value } N = \text{value}\rangle$$
Example: .MODEL LOG   D   IS=1E-12   N=16.90398213

The model statement must be included, but one statement is sufficient if several diodes have the same specification. If IS and N (i.e., I_s and n) are not specified, they will automatically be given the values $1E - 14$ and 1, respectively. For the specification of G_{min} see section A5.1.

SPICE uses the following values for the physical constants, (the values of R and $\mathscr{F}$ are derived from the Boltzmann constant and the elementary charge, respectively):

$$R = 8.3154889 \text{ J mol}^{-1} \text{ K}^{-1}$$

$$\mathscr{F} = 96\,500 \text{ C mol}^{-1}$$

$$0 \text{ K} = -273.15°\text{C}$$

Note that SPICE assumes a temperature of 27°C (T = 300.15 K) unless specified otherwise using TNOM in the OPTIONS statement (see section A5.1).

A2.6. Independent voltage and current sources

These two elements have formally identical descriptions.

Voltage source

$$V \cdots \cdots \quad \# \text{-}1 \quad \# \text{-}2 \quad \text{description}$$

Example: VCNAEX   401   0   10

Constitutive relation: $U = f(t)$ (A4)

Current source

$$I \cdots \cdots \quad \# \text{-}1 \quad \# \text{-}2 \quad \text{description}$$

Example: IJSUBSTR   0   9   .02

Constitutive relation: $I = f(t)$ (A5)

The description in each case is simply the definition of the function $f(t)$ given in the constitutive relation. The following functions are available:

(a) Constant

The constant value of the voltage or current is specified.

(b) Piece-wise linear

The function varies linearly with time over specified intervals as shown in Fig. 22A.

Format: $\text{PWL}(t_1 \, f(t_1) \cdots t_i \, f(t_i) \cdots t_n \, f(t_n))$

Note that all t_i must be positive and that at least t_1 must be smaller than timestop which is the total time of simulation (see section A3.2). SPICE uses $f(\text{timestop}) = f(t_n)$ if $t_n <$ timestop or $f(\text{timestop}) = f(t_i)$ where t_i is the first time specified in PWL which is larger than timestop. The following condition holds for times less than t_1

$$f(t) = f(t_1) + (t - t_1)[f(t_2) - f(t_1)]/(t_2 - t_1)$$

Hence, if only t_1 and $f(t_1)$ are given, $t_2 =$ timestop with $f(t_2) = f(\text{timestop}) = f(t_1)$ and $f(t) = f(t_1)$ for all t, i.e. a constant value is obtained as in (a). (PSpice simply uses $f(t) = f(t_1)$ for times less than t_1.)

(c) Pulse

The function switches back and forth between two levels f_1 and f_2 at specified intervals as shown in Fig. 22B.

Format: $\text{PULSE}(f_1 \quad f_2 \quad t_d \quad t_r \quad t_f \quad t_w \quad t_p)$

where t_d indicates delay time, t_r rise time, t_f fall time, t_w pulse width, and t_p pulse period. Note that the following obvious condition must be satisfied:

$$t_p \geq t_r + t_w + t_f$$

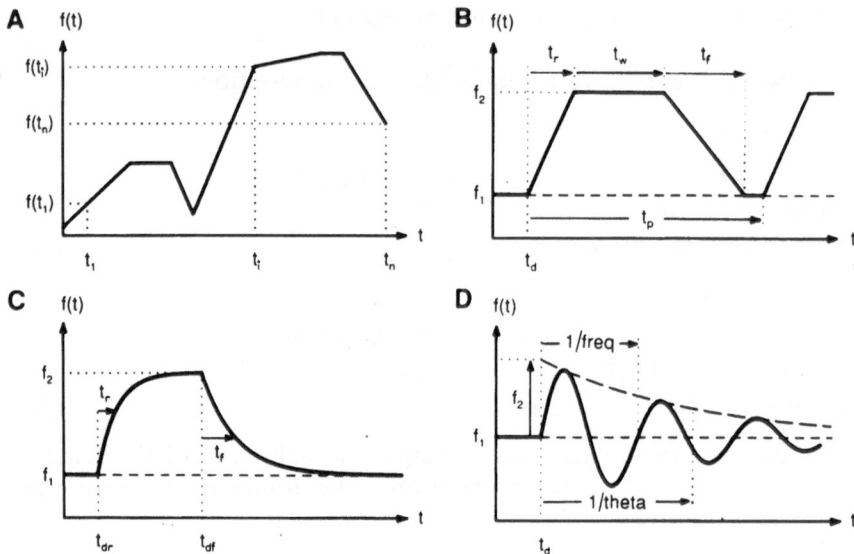

Fig. 22. Examples of different functions f(t) for independent sources. Piece-wise linear (A), pulse (B), exponential (C), and sinusoidal (D). For further explanation see text.

If t_w or t_p are set to zero SPICE uses the default value for these parameters which is timestop, i.e. the total time of simulation (see section A3.2). Hence, a single step function at time t_d with a rise time t_r is obtained if $t_w = 0$, and one pulse as specified is performed if $t_p = 0$. If t_r or t_f are set to zero SPICE uses the default value timestep (see section A3.2). Hence, unexpected results may be obtained depending on the parameters chosen for the transient analysis. However, t_r and t_f can be given very small values, if desired.

(d) Exponential

The function varies exponentially between two levels f_1 and f_2 at specified intervals as shown in Fig. 22C.

Format: EXP(f_1 f_2 t_{dr} t_r t_{df} t_f)

where t_{dr} and t_{df} are delay times, and t_r and t_f time constants, for the rise and fall, respectively.

(e) Sinusoidal

The function varies sinusoidally around a level f_1 with amplitude f_2 and frequency freq (see Fig. 22D).

Format: SIN(f_1 f_2 freq t_d theta)

where t_d is a delay time, and theta a damping factor which gives an exponential decay of the amplitude with time constant 1/theta.

Note that certain of the above parameters (not including f_1 and f_2) will be assigned predetermined values if not specified, e.g. t_d and theta are set to zero.

A2.7. Ammeter

All independent voltage sources can be used to measure current. The current through the source is reported by SPICE in accordance with the sign convention (see section A2.2.1). To determine the current in a branch which does not contain an independent voltage source, or does contain such sources but with inappropriate orientation, a dummy source of zero voltage is introduced.

A2.8. Controlled voltage and current sources

These four elements have formally identical descriptions, which it is convenient to deal with first. The description specifies the dependence of the source value on a set of controlling parameters, and is always given in the form of a polynomial. The controlling parameters are either potential differences, each defined by a pair of nodes which may be situated anywhere in the circuit (voltage-controlled sources), or currents, each measured by an ammeter (current-controlled sources). Note that mixed control by both potential difference and current is not permitted, and that nodes and elements within a subcircuit are not known to the outside circuit (see section A2.9).

Description format for voltage control:

POLY(n)　　# cp1-1　　# cp1-2　　# cp2-1　　# cp2-2 · · ·
　　　　　　# cpn-1　　# cpn-2　　coefficients

Description format for current control:

POLY(n)　　Vam-1　　Vam-2 · · · Vam-n　　coefficients

Here n represents the number of controlling parameters, # cpi-1 and # cpi-2 is the node pair defining the ith controlling potential difference, and Vam-i is the ammeter measuring the ith controlling current. Both potential difference and current follow the sign convention described in section A2.2.1. "Coefficients" represent a list of coefficients of the polynomial, $a_0, a_1, a_2, \ldots$, whose significance varies with the value of n according to the following pattern.

For POLY(1) which is a polynomial of one controlling variable x_1:

$$Poly(x_1) = a_0 + a_1 x_1 + a_2 x_1^2 + a_3 x_1^3 + a_4 x_1^4 + \cdots \qquad (A6)$$

For POLY(2) which is a polynomial of two controlling variables x_1, x_2:

$$Poly(x_1, x_2) = a_0 + a_1 x_1 + a_2 x_2 + a_3 x_1^2 + a_4 x_1 x_2 + a_5 x_2^2 +$$

$$a_6 x_1^3 + a_7 x_1^2 x_2 + a_8 x_1 x_2^2 + a_9 x_2^3 + a_{10} x_1^4 + \cdots \qquad (A7)$$

For POLY(3) which is a polynomial[7] of three controlling variables x_1, x_2, x_3:

$$Poly(x_1, x_2, x_3) = a_0 + a_1 x_1 + a_2 x_2 + a_3 x_3 + a_4 x_1^2 + a_5 x_1 x_2 +$$
$$a_6 x_1 x_3 + a_7 x_2^2 + a_8 x_2 x_3 + a_9 x_3^2 + a_{10} x_1^3 +$$
$$a_{11} x_1^2 x_2 + a_{12} x_1^2 x_3 + a_{13} x_1 x_2^2 + a_{14} x_1 x_2 x_3 +$$
$$a_{15} x_1 x_3^2 + a_{16} x_2^3 + a_{17} x_2^2 x_3 + a_{18} x_2 x_3^2 +$$
$$a_{19} x_3^3 + a_{20} x_1^4 + \cdots \tag{A8}$$

For POLY(4) which is a polynomial of four controlling variables x_1, x_2, x_3, x_4:

$$Poly(x_1, x_2, x_3, x_4) = a_0 + a_1 x_1 + a_2 x_2 + a_3 x_3 + a_4 x_4 + a_5 x_1^2 + a_6 x_1 x_2 +$$
$$a_7 x_1 x_3 + a_8 x_1 x_4 + a_9 x_2^2 + a_{10} x_2 x_3 + a_{11} x_2 x_4 +$$
$$a_{12} x_3^2 + a_{13} x_3 x_4 + a_{14} x_4^2 + a_{15} x_1^3 + a_{16} x_1^2 x_2 +$$
$$a_{17} x_1^2 x_3 + a_{18} x_1^2 x_4 + a_{19} x_1 x_2^2 + a_{20} x_1 x_2 x_3 +$$
$$a_{21} x_1 x_2 x_4 + a_{22} x_1 x_3^2 + a_{23} x_1 x_3 x_4 + a_{24} x_1 x_4^2 +$$
$$a_{25} x_2^3 + a_{26} x_2^2 x_3 + a_{27} x_2^2 x_4 + a_{28} x_2 x_3^2 + a_{29} x_2 x_3 x_4 +$$
$$a_{30} x_3^3 + a_{31} x_3^2 x_4 + a_{32} x_3 x_4^2 + a_{33} x_4^3 + a_{34} x_1^4 + \cdots \tag{A9}$$

The following abbreviated forms of the description format are permitted:

(1) In the case $n = 1$, POLY(1) can be omitted. However, PSpice does require POLY(1) if more than one coefficient is included.
(2) If POLY(1) is omitted, and only one coefficient is given, this coefficient is taken to be a_1.

The general format for the controlled sources is as follows:

Voltage-controlled voltage source

$$\text{E} \cdots \quad \#\text{-}1 \quad \#\text{-}2 \quad \text{description} \quad \langle IC = \cdots \rangle$$
Example: ESUM   5   0   POLY(2) 12 6 18 2 0 1 1
Constitutive relation: $U = Poly(U_1, U_2, \ldots)$ $\qquad$ (A10)

The example given sums the potential differences between nodes 12 and 6 and nodes 18 and 2.

Current-controlled current source

$$\text{F} \cdots \quad \#\text{-}1 \quad \#\text{-}2 \quad \text{description} \langle IC = \cdots \rangle$$
Example: FJHNET   0   7   POLY(2) VREAC1 VREAC2 0 2 -1
Constitutive relation: $I = Poly(I_1, I_2, \ldots)$ $\qquad$ (A11)

---

[7]Note that POLY(3) as given in ref. 12 is not correct since the terms with coefficients a_{13}, a_{14}, a_{15}, and a_{18} are missing.

This example pertains to the net flow of protons in a system in which two protons are produced by reaction 1 and one is consumed by reaction 2.

Voltage-controlled current source

$$G \cdots\cdots \quad \#\text{-}1 \quad \#\text{-}2 \quad \text{description} \quad \langle IC = \cdots \rangle$$

Example: GDENOM 0   8   POLY(2) 8 0 200 0 0 0 0 0 1

Constitutive relation: $I = \text{Poly}(U_1, U_2, \ldots)$ (A12)

The current produced by the source in this example is proportional to the product of the two potential differences specified. This source is a typical element in a division circuit (see section 4.3.2).

Current-controlled voltage source

$$H \cdots\cdots \quad \#\text{-}1 \quad \#\text{-}2 \quad \text{description} \quad \langle IC = \cdots \rangle$$

Example: HCONVERT 11   0   VCURRENT 1

Constitutive relation: $U = \text{Poly}(I_1, I_2, \ldots)$ (A13)

In this example the current through the source VCURRENT is converted to a potential difference of the same numerical value. (Note that the abbreviated forms have been used.)

The initial condition option refers to initial values of the controlling parameters, which in most cases of bioelectrochemical interest need not be specified. If it is desired to specify them, the initial values are given sequentially separated by commas or spaces. The default values are zeros.

A2.9. Subcircuits

These are user-defined elements (see section 4.2) which are called by a statement having the following format:

$$X \cdots\cdots \quad \#\text{-}1 \quad \#\text{-}2 \quad \#\text{-}3 \cdots \#\text{-}n \quad \text{subname}$$

Example: XSYMP12 101   102   201   202   SYMPORT

Constitutive relation: specified by subcircuit

The description "subname" refers to the subcircuit definition statement which has the format:

$$.\text{SUBCKT} \quad \text{subname} \quad \# \text{sc-}1 \quad \# \text{sc-}2 \quad \# \text{sc-}3 \cdots \# \text{sc-}n$$

Example: .SUBCKT   SYMPORT   1   2   3   4

The nodes $\#$ sc-1, $\#$ sc-2, $\#$ sc-3, etc. are used to connect the subcircuit to the main circuit. Note that SPICE automatically connects the subcircuit ground node to the main circuit ground node (node 0 in each case). Hence the subcircuit node list may not include 0. All elements following the SUBCKT statement are considered by SPICE to be elements of that subcircuit. The specification of elements belonging to the subcircuit is concluded with the ENDS statement having the following format:

.ENDS ⟨subname⟩

Example: .ENDS SYMPORT

Repetition of subname in the ends statement is optional, but strongly recommended for clarity and to avoid possible problems.

The following characteristics of subcircuits should be noted:
(1) The number of nodes and their order of presentation must be strictly identical in the subcircuit call statement and in the subcircuit definition statement.
(2) All elements within a subcircuit and all nodes except for those specified in the subcircuit definition statement are unknown outside the subcircuit. Furthermore, the nodes specified in the subcircuit definition are only known outside the subcircuit by the numbers appearing in the subcircuit call statement.
(3) The statements within a subcircuit can only consist of element specifications, including call statements for other subcircuits.
   MODEL statements for diodes and definitions of other subcircuits may but need not be included in the set of subcircuit statements. If included, such statements and definitions are unknown outside the subcircuit.

A3. Types of analysis

SPICE offers several types of analysis, of which only two are generally useful in bioelectrochemical modelling. Direct current (or DC) analysis gives one immediately the steady-state solution for the system. Transient analysis gives the time-dependent behavior of the system.

A3.1. Steady-state analysis

The statement which tells SPICE to perform a steady-state analysis has the general format:

.DC indsource1 startval1 stopval1 incr1
        ⟨indsource2 startval2 stopval2 incr2⟩

Example: .DC VMEMBRAN   −90M   −30M   10M

A maximum of two independent sources may be included, either of which may be a voltage or a current source. SPICE calculates a set of steady-state solutions varying the value of each independent source between its starting value and its stopping value in the incremental steps given. If two sources are included, the first will be swept over its range for each value of the second.

A3.2. Transient analysis

The statement that tells SPICE to perform a time-dependent analysis has the general format:

.TRAN timestep timestop ⟨timestart ⟨timemax⟩⟩ UIC
Example: .TRAN   2   50   0   5M   UIC

The first three time values determine the form of the output data. The stepping, stopping, and starting times are, respectively, the time interval, total time, and initial time used in the output tables or plots. If the starting time is omitted, it is assumed to be zero. However, a value must be given for the starting time if timemax is to be defined. The latter is the maximum step size SPICE will use while calculating the solution, and if not specified will be taken to be either (timestop − timestart)/50 or timestep, whichever is smaller. In certain cases this value may not be small enough, and thus causes inaccurate results. A method of dealing with this problem that is frequently applicable involves "switching" (see section 4.4.1). Note that decreasing timemax increases computing time.

The symbol UIC, whose inclusion is mandatory in bioelectrochemical simulations, tells SPICE to "use initial conditions" as given in the element specifications or by means of an IC statement having the following format:

.IC V(# -1) = value1   ⟨V(# -2) = value2 · · ·⟩
Example: .IC   V(201)=20U   V(202)=5M

Value1, value2, etc. are the *potentials with respect to ground* to be assigned initially to the nodes # -1, # -2, etc. Note that the IC statement differs in this respect from IC given in the element specification which sets the initial *potential difference* across the element. If initial conditions for a given node appear both in a pertinent element specification and in an IC statement, the former overrides the latter. If no initial conditions are specified anywhere, SPICE assumes the potential difference across that element to be initially zero. (Note that omitting the UIC statement does not in general result in the potential differences across all elements being set initially to the values given in the IC statement, in contrast to what is said in the SPICE manual.)

A4. Data output

Output variables in the form of potential differences between any two
nodes [format: V(# -1, # -2)] and currents through independent current sources or ammeters [format: I(V · · ·)] can be obtained in tabular
and/or pseudo-graphic form by means of PRINT and PLOT statements. At least one of these statements must be included in each set of
instructions. The number of such statements that can be included is
limited only by the computer memory available. The statement producing tables has the format:

$$\text{.PRINT} \quad \text{type} \quad \text{ov1} \quad \langle \text{ov2} \cdots \text{ov8} \rangle$$

Example: .PRINT TRAN V(100) V(10,20) I(VNAFLOW)

The statement producing plots has the format:

$$\text{.PLOT} \quad \text{type} \quad \text{ov1} \quad \langle (\text{lolim1,hilim1}) \rangle$$
$$\langle \text{ov2} \langle (\text{lolim2,hilim2}) \rangle \cdots$$
$$\text{ov8} \langle (\text{lolim8,hilim8}) \rangle \rangle$$

Example: .PLOT DC V(201) (-1,2) I(VNA) I(VK) (0,10M)

In the above statements the type of analysis is first identified. Up to
eight output variables are then specified (the ground node need not be
listed). In the PLOT statement the lower and upper limits lolim and hilim
are optional. If specified, all output variables to the left of a pair of plot
limits will be plotted with the same limits. If not specified, SPICE will
automatically determine the minimum and maximum values of all
output variables plotted and scale the plot to fit. More than one scale is
used if necessary. Each output variable is represented by a different
symbol (*, +, =, etc.) whose meaning is indicated in a legend appearing
in the heading of the graph. Overlapping points in a plot are indicated
by the symbol X. The first column in a graph contains the values of the
independent variable (time for a transient analysis, values of the independent source for a steady state analysis) at which the output variables are
plotted. The second column contains the values of the first output
variable specified in the PLOT statement which are printed as well as
plotted. In the remaining columns the symbols of the output variables
appear at the printing position nearest to their actual value according to
the scale given on the top of the graph. The position of the values printed
on the scale are indicated in the plot by columns of dots.

In each output the SPICE code used for the simulation is printed first.
It is followed by the tables and/or pseudo-graphs according to the
PRINT and PLOT statements included in the code. All numbers are
given in exponential format (see section A1.2); however, depending on
the type of computer used the letter E may be replaced by D or even be
missing. The default number of digits printed is 4. This value may be
altered by including NUMDGT = n in the OPTIONS statement (see

section A5.1), where n has a maximum value of 7. Nevertheless, without resetting the error tolerances by also including RELTOL = x and TRTOL = y, the extra digits may not be significant. If 7 digits are required, suggested settings are RELTOL = 1E − 6 and TRTOL = 0.5. These options increase the computing time.

PSpice includes an optional graphics post-processor PROBE which permits interactive viewing of simulation results with high-resolution graphics instead of pseudo-graphs. Functions of the output variables, including derivatives and integrals, are allowed. However, PROBE lacks one extremely useful feature of the pseudo-graphs – the ability to use multiple ordinate scales (which may differ by orders of magnitude) when superimposing several curves on one plot. For details of PROBE the reader is referred to the PSpice manual.

A5. Program options

A5.1. The OPTIONS statement

This statement has the format:

```
        .OPTIONS  opt1  ⟨opt2 ···⟩
Example: .OPTIONS  PIVTOL=1E-35  GMIN=1E-35  NOPAGE
```

To cover all possible eventualities, the inclusion of PIVTOL = 1E − 35 and GMIN = 1E − 35 is required. The major options useful in bioelectrochemical simulations are:

LIST	Lists the elements by element type and gives node connections.
NODE	Lists the nodes and the elements connected to them.
NOMOD	Prevents the listing of the model parameters of all diodes.
NOPAGE	Prevents the commencement of new pages for each item.
LIMPTS = n	Resets the maximum number of points tabulated or plotted. The default value is 201.
TNOM = x	Resets the nominal temperature. The default value is 27°C
NUMDGT = n	Resets the number of digits printed. The default value is 4.
ITL1 = n	Resets the number of iterations allowed for convergence of a steady-state (DC) analysis. The default value is 100.

ITL5 = n	Resets the total number of iterations allowed in a transient analysis. The default value is 5000. ITL5 = 0 allows an unlimited number of iterations.
PIVTOL = x	Resets the minimal value accepted by SPICE in the iterative procedure during a simulation.
GMIN = x	Resets the minimal value accepted for conductances. The default value is $1E - 12$.
RELTOL = x	Resets the relative error tolerance. The default value is 0.0001.
TRTOL = x	Resets the transient error tolerance which is an estimate of the truncation error occurring in computations. The default value is 7.
METHOD = GEAR	The default procedure used by SPICE for integration is the trapezoidal method. This may give errors if the variables simulated have very disparate time constants, which cause some variables to change very rapidly while others change slowly. Problems occur because an integration time step which is small enough to cope with rapidly changing variables may give rounding errors in slowly changing variables. Such a problem can be resolved by activating the optional method of integration specified as METHOD = GEAR. The order of the integration can be predetermined by including MAXORD = n in the OPTIONS statement. The minimum (and default) value of n is 2, the maximum value is 6. However, the use of high-order GEAR integration is not recommended owing to the minimal gain in stability compared to the large increase in time.

A5.2. The WIDTH statement

This statement has the format:

          .WIDTH   IN = characnum   OUT = characnum
Example: .WIDTH   IN=60   OUT=80

Here characnum is the position in the line of the final character read or printed by SPICE in a line of input or output, respectively. The setting takes effect with the next line read. The default values are 80 and 132, respectively. The input width does not exceed 80 characters, while the output width is limited to 80 or 132 characters.

A5.3. The ALTER statement

This statement has the format:

.ALTER
Element specification
.ALTER
Element specification
.
.
.END

The ALTER statement causes SPICE to recalculate with an altered element description. No topological change of the circuit is allowed. There is no limit to the number of ALTER statements permitted, but they must immediately precede the END statement. Each altered element specification remains in force unless altered again.

A5.4. The NODESET statement

This statement has the format:

$$.\text{NODESET} \quad V(\#\text{-}1) = \text{value1} \quad \langle V(\#\text{-}2) = \text{value2} \cdots \rangle$$
Example: `.NODESET   V(201)=20U   V(202)=5M`

Value1, value2, etc. are the potentials with respect to ground to be assigned to the nodes $\#$-1, $\#$-2, etc. Although rarely required, NODESET may be useful occasionally to facilitate or even make possible the calculation of the starting conditions for a simulation. Unlike the IC specification either as an independent statement or as part of an element description, NODESET supplies a list of "initial guesses" of node potentials which guide SPICE to a preliminary solution. Thereafter the NODESET potentials are released and SPICE calculates the proper initial condition.

Appendix B. Error messages in SPICE

The error messages in SPICE are essentially self-explanatory. When searching for the reasons for an error, the following measures are usually worth taking:

(1) Check the program for typing errors.
(2) Check the circuit topology. The options LIST and NODE may be helpful.
(3) Check the scaling.
(4) Check the translation of the model into electric circuits.

The more frequently encountered errors are listed below (in alphabetical order) together with brief comments.

ABOVE LINE ATTEMPTS TO REDEFINE $\cdots$

This occurs when more than one element is present with the same name. (Remember that SPICE reads only the first 8 characters in a name.)

ATTEMPT TO REFERENCE UNDEFINED NODE nn – NODE RESET TO 0

This is a PLOT or PRINT statement error. A table or plot from a nonexistent node has been required.

ELEMENT $\cdots$ HAS BEEN REFERENCED BUT NOT DEFINED

This occurs when a nonexistent ammeter is referenced either in the description of a current-controlled source or in a PLOT or PRINT statement.

ILLEGAL NUMBER – SCAN STOPPED AT COLUMN $\cdots$

This indicates that the corresponding parameter or option value is too high or too low (remember that the range of values acceptable by SPICE is $10^{-35}-10^{35}$). Alternatively a blank space between two numbers is missing which creates a number with two decimal points, or a blank space appears between a sign and a number.

INDUCTOR/VOLTAGE SOURCE LOOP FOUND, CONTAINING $\cdots$

This indicates that two or more voltage sources are connected to a node whose other nodes are on ground (one of these sources is named).

INTERNAL TIME STEP TOO SMALL IN TRANSIENT ANALYSIS

The internal integration time step is less than 10^{-9} of its initial value. This is frequently due to scaling errors (units should be consistent with each other). If the message is printed without giving a solution for time zero (or timestart), it is probable that insufficient iterations have been allowed for determination of the initial starting conditions. Reset ITL1 in the OPTIONS statement (e.g. ITL1 = 500).

LESS THAN TWO CONNECTIONS AT NODE nnn

Either an element (or a TITLE statement) has been omitted, or an incorrect node number has been assigned to an element, or a node number is used in the description of a controlled source which does not exist.

MAXIMUM ENTRY IN THIS COLUMN AT STEP··· IS LESS THAN PIVTOL

This is generally caused by a resistance which exceeds 0.1/PIVTOL. This message can also appear instead of or together with INTERNAL TIME STEP TOO SMALL···(or NO CONVERGENCE IN DC···) and then usually indicates scaling errors.

MEMORY NEEDS EXCEED···

This indicates that SPICE has run out of space to store the data. Remove any unnecessary variables from both PLOT and PRINT statements. If this does not solve the problem, increasing timestep (and timemax if specified) may help.

NO CONVERGENCE IN DC ANALYSIS···

If all other checks fail to disclose the cause, insufficient iterations have been allowed for determination of the steady-state solution. Reset ITL1 in the OPTIONS statement (e.g. ITL1 = 500).

NO CONVERGENCE IN DC TRANSFER CURVES AT···

This indicates an error while sweeping a range of potentials or currents using a DC statement. See "NO CONVERGENCE IN DC ANALYSIS" above.

SUBCIRCUIT NAME MISSING

The name of the subcircuit was omitted in a statement which calls a subcircuit.

UNABLE TO FIND X···

This indicates that a subcircuit is called by the main program which is not included in the SPICE code.

UNKNOWN DATA CARD···

This error indicates that the first character of a statement is not a valid letter which determines the element type (see section A2.1). It also occurs if the letter is valid but typed in lower case.

VALUE IS MISSING OR IS NONPOSITIVE

The value in the description of a resistor or capacitor was omitted, or the value is zero or negative. Note that no error message occurs if the value for an independent voltage or current source is omitted but the value is set to zero.

X ··· HAS DIFFERENT NUMBERS OF NODES THAN ···

The number of nodes included in a statement which calls a subcircuit
(X ···) is different from the number of nodes in the statement which
defines the subcircuit (.SUBCKT ···).

In addition to the above messages, some error messages may be
obtained from the FORTRAN processor of the computer system used.
These cannot be dealt with in detail as they will vary from system to
system.

Appendix C. SPICE codes for the examples

This section presents the listing of the SPICE programs for the illustrative examples discussed in section 5. Note that programs which contain
the comment "* SUBCIRCUITS (incl. ···)" require the addition of the
subcircuits listed in the comment before the codes are executable by
SPICE. In PSpice this can be easily done by including statements ".INC
⟨file name⟩", where ⟨file name⟩ denotes the name of files in which the
pertinent subcircuits are stored. Alternatively, a library of all subcircuits
can be generated and stored in a file named ⟨library name⟩. In this case
".LIB ⟨library name⟩" is sufficient to make all subcircuits available to
PSpice.

C1. Permeation of salt through a membrane

```
PERMEATION OF SALT THROUGH A MEMBRANE BETWEEN TWO COMPARTMENTS
*
* Permeabilities and charge numbers
*
VPION1 21 0 PWL(0 0 19.999M 0 20M 5E-6 1000 5E-6)
VZION1 11 0 1
VPION2 22 0 PWL(0 0 19.999M 0 20M 5E-7 1000 5E-7)
VZION2 12 0 -1
*
* Permeation of ions
*
VION1IN1 101 0 10
XION1 0 10 101 2010 11 21 IONPERM
VJION1 2010 201 0
CION1IN2 201 0 2
VION2IN1 102 0 10
XION2 0 10 102 2020 12 22 IONPERM
VJION2 2020 202 0
```

```
CION2IN2 202 0 2
VCURRENT 10 1 0
CMEM 1 0 1M
*
* SUBCIRCUITS (incl. IONPERM, EXP_FURT)
*
* CONTROL
*
.TRAN 5M .2 UIC
.PLOT TRAN V(0,1) (-80,0) V(201) V(202) (0,1.5E-6)
.PLOT TRAN I(VJION1) I(VJION2) (0,4E-5) I(VCURRENT)
.OPTIONS NOPAGE PIVTOL=1E-35 GMIN=1E-35 TNOM=25
.WIDTH OUT=80
.END
```

C2. Chemical reaction at pseudo-equilibrium: buffering

C2.1. Behavior of reactions at pseudo-equilibrium

```
TEST EQUILIBRATED REACTION
XREACTION 1 2 3 4 5 6 REQUIL12
IJR1 0 1 PULSE(0 10M 1.999M 1U 1U 2M)
VJR3 3 30 0
CR3 30 0 2M
CRTOT 4 0 2M
VKC 5 0 1M
VOLUME 6 0 2M
*
* SUBCIRCUITS (incl. REQUIL12)
*
* CONTROL
*
.TRAN .5M 6M UIC
.OPTIONS NOPAGE PIVTOL=1E-35 GMIN=1E-35
.PRINT TRAN V(1) V(2) V(3) V(4) I(VJR3)
.IC V(30)=5M V(4)=20M
.END
```

C2.2. Buffering

```
BUFFERING OF SPECIES R3
*
* Sinusoidal input flow and unbuffered c(R3)
```

```
*
CR3UNBF 300 0 2M
R300 300 0 1E25
VJR3SIN 301 300 0
IJR3 0 301 SIN(0 10U .1 2)
R31 31 0 1E25
*
* Buffered c(R3)
*
CR3 31 0 2M
FJR3SIN 0 31 VJR3SIN 1
FJR3 0 31 POLY(2) VJR3 VSWI 0 0 0 0 1
XREACTION 1 2 3 4 5 6 REQUIL12
VJR3 3 30 0
ECR3 30 0 31 0 1
CRTOT 4 0 2M
VKC 5 0 10M
VOLUME 6 0 2M
*
* Switch for buffered flow
*
ISW 0 10 PWL(0 0 .999 0 1 1 1000 1)
VSWI 10 0 0
*
* SUBCIRCUITS (incl. REQUIL12)
*
* CONTROL
*
.TRAN .5 20 .5 UIC
.OPTIONS NOPAGE PIVTOL=1E-35 GMIN=1E-35
.PLOT TRAN V(3) V(300) (0,.04) I(VJR3) I(VJR3SIN) (-3E-5,1E-5)
.IC V(4)=.2 V(31)=5M V(300)=5M
.WIDTH OUT=80
.END
```

C2.3. Titration of buffers

```
PH-TITRATION OF TWO ACID/BASE PAIRS IN AN AQUEOUS MEDIUM
*
* c(H) and proton flow
*
.IC V(300)=.005
VJHBF 3 30 0
ECH 30 0 300 0 1
```

```
FJH 0 300 POLY(2) VJHBF VSWI 0 0 0 0 1
CH 300 0 20M
XPH 300 301 AH_TO_PH
*
* Addition of OH ions and switch for buffered proton flow
*
IJOHADD 0 2 PWL(0 0 .999 0 1 .2M 1000 .2M)
ISW 0 50 PWL(0 0 .4999 0 .5 1 1000 1)
VSWI 50 0 0
*
* Buffers
*
VOLUME 6 0 20M
XWATER 1 2 3 4 5 6 REQUIL12
VCWATER 4 0 55.5555
VKWATER 5 0 1.8E-16
XBUFF1 11 12 3 14 15 6 REQUIL12
VCBUFF1 14 0 40M
VKBUFF1 15 0 1E-5
XBUFF2 21 22 3 24 25 6 REQUIL12
VCBUFF2 24 0 20M
VKBUFF2 25 0 5E-9
*
* SUBCIRCUITS (incl. REQUIL12 and AH_TO_PH)
*
* CONTROL
*
.TRAN .25 9 .25 UIC
.OPTIONS NOPAGE PIVTOL=1E-35 GMIN=1E-35 TNOM=25
.PLOT TRAN V(301) (0,14)
.PLOT TRAN V(2) V(1) V(12) V(11) V(22) V(21)
.WIDTH OUT=80
.END
```

C2.4. Buffer capacity

```
BUFFER CAPACITY OF TWO ACID/BASE PAIRS IN AN AQUEOUS MEDIUM
*
* pH flow
*
.IC V(300)=1.75
IPH 0 300 .1
CPH 300 0 1
XCH 300 301 PH_TO_AH
```

```
*
* c(H) and proton flow
*
VJHBF 3 30 0
ECH 30 0 301 0 1
VJHFR 31 30 0
CH 31 0 20M
FJHTOT 0 32 POLY(2) VJHBF VJHFR 0 500 500
VJHTOT 32 0 0
*
* Buffers
*
VOLUME 6 0 20M
XWATER 1 2 3 4 5 6 REQUIL12
VCWATER 4 0 55.5555
VKWATER 5 0 1.8E-16
XBUFF1 11 12 3 14 15 6 REQUIL12
VCBUFF1 14 0 40M
VKBUFF1 15 0 1E-5
XBUFF2 21 22 3 24 25 6 REQUIL12
VCBUFF2 24 0 20M
VKBUFF2 25 0 5E-9
*
* SUBCIRCUITS (incl. REQUIL12 and PH_TO_AH)
*
* CONTROL
*
.TRAN 2.5 102.5 2.5 UIC
.OPTIONS NOPAGE PIVTOL=1E-35 GMIN=1E-35 TNOM=25
.PLOT TRAN I(VJHTOT) (0,.024) V(300) (0,14)
.WIDTH OUT=80
.END
```

C3. Enzyme-catalyzed reactions

C3.1. Subcircuit REDOXH

```
*
* Redox-driven H-pump
*
.SUBCKT REDOXH 10 20 11 12 13 14 15 16 17 18
ESUMP 1 0 POLY(5) 2 0 3 0 4 0 5 0 6 0 1 -1 -1 -1 -1 -1
R1 1 0 1E25
```

```
GJ12 1 102 POLY(4) 1 0 11 0 2 0 12 0 0 0 0 0 0
+                              0 2E5 0 0 0 0 0 0 -4E6
VJD 102 2 0
C2 2 0 1
R2 2 0 1E25
GJ23 2 203 POLY(3) 15 0 2 0 3 0 0 0 0 -200 0 0 0 0 0 0 0 2E16
VJH1 203 3 0
C3 3 0 1
R3 3 0 1E25
GJ34 3 4 POLY(4) 3 0 30 0 4 0 40 0 0 0 0 0 0
+                              0 2000 0 0 0 0 0 -200
C4 4 0 1
R4 4 0 1E25
GJ45 4 405 POLY(3) 16 0 5 0 4 0 0 0 0 200 0 0 0 0 0 0 0 -2E13
VJH2 405 5 0
C5 5 0 1
R5 5 0 1E25
GJ56 5 506 POLY(4) 5 0 14 0 6 0 13 0 0 0 0 0 0
+                              0 2E6 0 0 0 0 0 -2E3
VJA 506 6 0
C6 6 0 1
R6 6 0 1E25
GJ61 6 1 POLY(2) 6 0 1 0 0 200 -200
* Slip
R20 20 0 1E25
GJ25 2 5 POLY(3) 20 0 2 0 5 0 0 0 0 0 0 .5 -5E-5
* Exp{F*mem.pot/RT}
R17 17 0 1E25
R18 18 0 1E25
EUM 31 0 17 18 1
XEXPUM 31 30 EXP_FURT
GDENEXP 0 40 POLY(2) 30 0 40 0 0 0 0 0 1
INUMEXP 40 0 1
R40 40 0 1E25
* Flows
R10 10 0 1E25
GNENZ 0 100 10 0 1
VNENZ 100 0 0
FJDRD 0 11 POLY(2) VJD VNENZ 0 0 0 0 -1
R11 11 0 1E25
FJDOX 0 12 POLY(2) VJD VNENZ 0 0 0 0 1
R12 12 0 1E25
FJARD 0 13 POLY(2) VJA VNENZ 0 0 0 0 1
R13 13 0 1E25
FJAOX 0 14 POLY(2) VJA VNENZ 0 0 0 0 -1
R14 14 0 1E25
```

```
FJH1 0 15 POLY(2) VJH1 VNENZ 0 0 0 0 -2
R15 15 0 1E25
FJH2 0 16 POLY(2) VJH2 VNENZ 0 0 0 0 2
R16 16 0 1E25
.ENDS REDOXH
```

C3.2. Flow-force relations

```
FLOW-FORCE RELATIONS FOR A REDOX-DRIVEN PROTON PUMP
*
* Del-mu(H), mole number and slip
*
VD_MUH 1 0 0
R1 1 0 1E25
VNENZ 10 0 1
VSLIP 20 0 1
*
* Pumps, chem. (_PH) or elect. part (_PHI) of del-mu(H) varied
*
VPH1 100 0 7.5
XCH1 100 101 PH_TO_AH
XRDH_PH 10 20 11 12 131 14 101 202 0 0 REDOXH
VJE_PH 131 13 0
VJH_PH 202 201 0
XCH2_PH 200 201 PH_TO_AH
EPH2_PH 200 0 POLY(1) 1 0 7.5 .175170798
XRDH_PHI 10 20 11 12 132 14 101 210 17 18 REDOXH
VJE_PHI 132 13 0
VJH_PHI 210 101 0
EMEMPOT 17 18 1 0 10.3626943
*
* Concentration of redox species
*
VDRD 11 0 10M
VDOX 12 0 20U
VARD 13 0 100M
VAOX 14 0 1M
*
* SUBCIRCUITS (incl. REDOXH, EXP_FURT, PH_TO_AH)
*
* CONTROL
*
.DC VD_MUH 0 -30 1
```

```
.PLOT DC I(VJE_PH) I(VJH_PH) I(VJE_PHI) I(VJH_PHI) (-10,30)
.OPTIONS NOPAGE NOMOD PIVTOL=1E-35 GMIN=1E-35 TNOM=25
.WIDTH OUT=80
.END
```

C3.3. Redox-driven proton pump

```
REDOX-DRIVEN PROTON PUMP IN SITU
*
* Permeabilities and charge numbers
*
VP_HLEAK 10 0 200
VP_CATION 15 0 .4
VP_ANION 16 0 .2
GP_AOX 104 2040 104 204 20
GP_HARD 103 203 103 203 1000
VZCATION 2 0 1
VZANION 3 0 -1
*
* n(enzyme), slip and initial conditions
*
VNENZ 4 0 5M
VSLIP 5 0 1
.IC V(100)=3E-8 V(200)=3E-8
.IC V(102)=1E-10 V(103)=.1 V(203)=.1 V(104)=1M V(204)=1M
.IC V(105)=.1 V(205)=.1 V(106)=.1 V(206)=.1
*
* Proton flows
*
CH_1 100 0 2E7
XPH_1 100 110 AH_TO_PH
XRDH 4 5 101 1020 2030 204 100 2001 0 1 REDOXH
VJH 2001 2000 0
XHLEAK 0 1 100 2002 2 10 IONPERM
VJHLEAK 2002 2000 0
XWATER 11 12 2000 14 7 6 REQUIL12
VCWATER 14 0 55.5555
VKWATER 7 0 1.8E-16
XMEMBUF 21 22 2000 24 8 6 REQUIL12
VCMEMBUF 24 0 50M
VKMEMBUF 8 0 1E-5
VOLUME_2 6 0 20
VJHTOT 2000 2003 0
ECH_2 2003 0 200 0 1
```

```
CH_2 200 0 20
XPH_2 200 210 AH_TO_PH
FJHTOT 0 200 POLY(2) VJHTOT VSWI 0 0 0 0 1
*
* Membrane potential
*
CMEM 1 0 10
FJHPUMP 0 1 POLY(2) VJH VSWI 0 0 0 0 96.5K
*
* Permeation of ions
*
CCAT_1 105 0 10000
XCAT 0 1 105 2050 2 15 IONPERM
VJCAT 2050 205 0
CCAT_2 205 0 20
CAN_1 106 0 10000
XAN 0 1 106 2060 3 16 IONPERM
VJAN 2060 206 0
CAN_2 206 0 20
*
* Conc. redox species and electron flows
*
EHDRD 101 0 50 0 10M
CDOX 102 0 10000
VJE_1 1020 102 0
CHARD_1 103 0 10000
VJE_2 2030 203 0
CHARD_2 203 0 20
CAOX_1 104 0 10000
VJE 2040 204 0
CAOX_2 204 0 20
*
* Thermodynamic forces
*
* Delta-mu(eH) and A(eH)
VLNKCDR 33 0 PWL (0 -45.665 3.9999 -45.665 4 21.11636 1000 21.116
XLNDOX 102 33 32 RT_LNC
XLNAOX_2 204 31 32 RT_LNC
XLNARD_2 203 31 30 RT_LNC
XLNAOX_1 104 301 32 RT_LNC
XLNARD_1 103 301 300 RT_LNC
* Delta-mu(H)
EFMEMPOT 42 0 0 1 .0965
XLNCH_1 100 41 42 RT_LNC
XLNCH_2 200 41 40 RT_LNC
```

```
*
* Switch
*
VSW 50 0 PWL(0 0 3.9999 0 4 1 1000 1)
RUNITY 50 51 1
VSWI 51 0 0
*
* SUBCIRCUITS (incl. REDOXH, IONPERM, REQUIL12, EXP_FURT
*                     RT_LNC, AH_TO_PH)
*
* CONTROL
*
.TRAN 1 50 UIC
.PLOT TRAN V(0,1) V(110) V(210) V(40)
.PLOT TRAN V(30) V(300) V(104) V(204)
.PLOT TRAN I(VJCAT) I(VJAN)
.OPTIONS NOPAGE NOMOD PIVTOL=1E-35 GMIN=1E-35 TNOM=25
.END
```

Bioelectrochemistry: General Introduction
ed. by S. R. Caplan, I. R. Miller and G. Milazzot
© 1995 Birkhäuser Verlag Basel/Switzerland

CHAPTER 3
Structure of water and ionic hydration

Werner Kunz[1], Marie-Claire Bellissent-Funel[2] and
Patrick Calmettes[2]

[1]*Université de Technologie de Compiègne, Département Génie Chimique, B.P. 649,
COMPIEGNE cedex, France*
[2]*Laboratoire Léon Brillouin, C.E.A.-C.N.R.S., C.E. de Saclay, GIF-SUR-YVETTE cedex,
France*

1. Introduction

One of the fundamental problems in physical chemistry and in biology is an understanding of the structure and the dynamics of water. A detailed knowledge of the microscopic properties of water is the key to a proper description of the interaction between water and other molecules. These interactions are manifestly important in lifeless nature, for example in aqueous solutions such as sea water or in hydrated minerals. Even more important are the interactions between water and biomolecules since water is essential for life.

Pure water is already a very complicated system whose unique behaviour is not yet completely understood. Solutions and mixtures which contain other components in addition to water are more complex. The degree of complexity increases from "simple" aqueous solutions of small ions to biological systems containing numerous macromolecules in contact with water and small ions. While the "simpler" systems can be described better and better by refined models, our actual knowledge of water interactions with biomolecules remains vague and sometimes ambiguous.

The present chapter reflects those different levels of complexity. In section 2 we give a survey of experimental and computational methods used to study pure water. X-ray and neutron scattering techniques appear to be efficient tools to investigate the structure and the dynamics of water at the microscopic level. The results are sufficiently precise to allow a detailed discussion of water properties to be made.

Section 3 is devoted to a general introduction to theoretical and experimental methods allowing the study of aqueous solutions. This section begins with a description of simple thermodynamic models and experimental methods and finishes with more complex computational theories and elaborate techniques.

Finally, section 4 deals with some features of the interaction between water and the two main types of biomolecules, proteins and nucleic acids. Due to the complexity of such interactions the phenomena involved and the experimental results are discussed in a fairly descriptive way.

2. Pure water

2.1 Water molecules and H-bonding

Water is the most abundant liquid on earth. It is well known that water forms a necessary constituent of the cells of all animals and plant tissues and that life cannot exist, even for a limited period, in the absence of water, with the exception of seeds, encysted embryos and a few other examples known up to now.

The structure of a water molecule [1] is shown in Fig. 1a. The oxygen atom has eight electrons and the $1s^2 2s^2 2p_z^2 2p_x^1 2p_y^1$ electronic configuration. Of the six electrons in the outer orbital of the oxygen atom, two are involved in covalent bonds to the hydrogens (σ bonds). The other four exist in non-bonded pairs and form hybridized orbitals. The angle between the σ bonds (between the oxygen and the two hydrogen atoms) is 104.523° for H_2O, which is very close to the tetrahedral angle (109°).

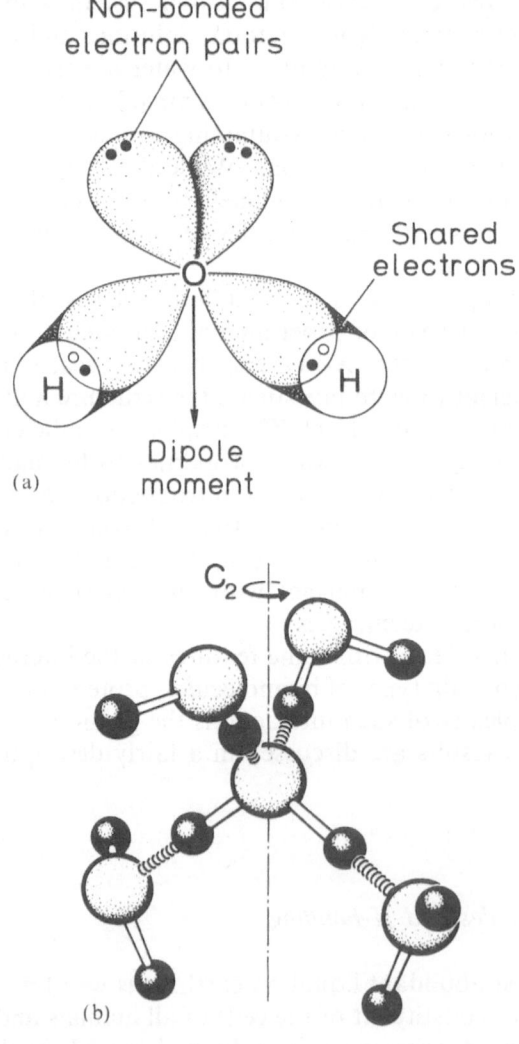

Fig. 1(a). The structure of a water molecule. (b) Hydrogen bonding and local tetrahedral geometry of a water molecule. Small spheres, H atoms; large spheres, O atoms. Disks refer to hydrogen bonds. (Adapted from Ref. [2]).

Because of this geometry and the polarization of electron charge densities, the water molecule has an appreciable dipole moment which can be thought of as a vector bisecting the angle between the O-H bonds.

Another effect of the polarization of charges is to lead to interactions between water molecules, known as "hydrogen bonds". Each water molecule is capable of making four hydrogen bonds. It acts as a donor of hydrogen to two water molecules and as an acceptor of hydrogen from two more. Thus, each water molecule sits at the centre of a tetrahedron of water molecules [2] (Fig. 1b).

Hydrogen bonds are characterized by interaction energies comparable to those of ionic bonds, and by a high directionality. These interaction energies are extremely strong when compared to the intermolecular interaction energies in atomic liquids such as argon. However, if H-bonds are stable at room temperature, they are nevertheless "fragile" as compared with covalent bonds. This explains the importance of H-bonding in biochemical reactions where the energies involved are low. Table 1 summarizes the properties of water.

2.2 Anomalous properties of liquid water

Water molecules form hydrogen bonds whose characteristic energy and directionality are at the origin of the unusual physical and transport properties of liquid water, which are especially anomalous at low temperature, i.e. in the supercooled state. The metastable limit for the liquid phase is $-45°C$ for H_2O ($-40°C$ for D_2O), at which point it demonstrates a behaviour similar to that of a critical point.

2.2.1 Static properties:

- The density [3] d of water exhibits a maximum as a function of the temperature (4°C for H_2O, 11°C for D_2O) from which an increase in molecular volume results, while the density increases for a normal liquid when the temperature decreases (Fig. 2).

Table 1. Physical properties of H_2O and D_2O at 25°C

	H_2O	D_2O
Molecular weight (g/mole)	18.02	20.03
Melting point (°C)	0	4
Boiling point (°C)	+100	+101.42
Heat of vaporization (kJ/mole)	40.71	—
Density (g/cm³)	0.997	1.104
Dielectric constant	78.54	77.93
Viscosity (g/(cm s))	0.890×10^{-2}	—
Surface tension (dyne/cm)	71.97	—

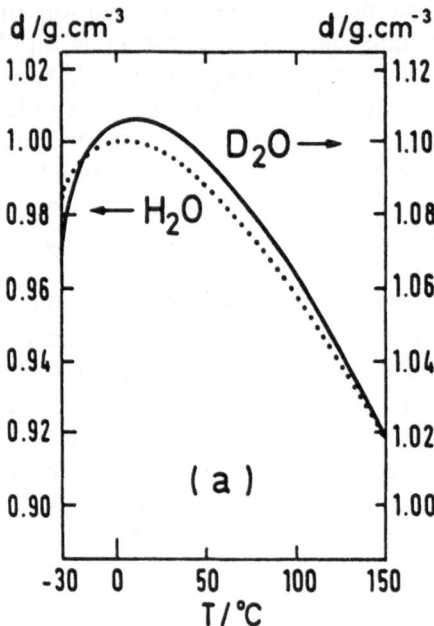

Fig. 2. Densities of H_2O (dotted line) and D_2O (full line) as a function of temperature. (Data taken from Ref. [3]).

- The isothermal compressibility [3] κ_T of water exhibits a minimum (46°C for H_2O, 52°C for D_2O) while for a normal liquid it increases when the temperature increases.
- The thermal expansivity [4] α_p of water becomes negative when the temperature decreases while for a normal liquid it remains almost constant.
- The heat capacity at constant pressure [5] C_p of water has a behaviour as a function of the temperature which is very different from that observed in normal supercooled systems such as alcohols [5]. Below a minimum temperature (T_{min} H_2O: 308 K, D_2O: 373 K) C_p^{liq} of H_2O and D_2O increases with falling temperature, and this increase becomes very steep in the metastable region in the vicinity of the homogeneous nucleation temperature.
- From the anomalies of d, κ_T and C_p discussed above, the speed of ultrasound v in liquid water into the supercooled region should exhibit unusual properties. The experiments performed by several groups lead to different results; comments and references to these studies are given in the review paper by Lang and Lüdemann [5].
- The static dielectric constant ε_0 [6, 7] of supercooled water does not reveal any anomalies, but in this case the measurement averages the dipole orientation in water over dimensions that are large compared to the particle diameter. This is an indication that the peculiarities of

the structure of supercooled water are restricted to rather small domains corresponding to the arrangement of the closest neighbours of a water molecule.

2.2.2 Transport and relaxation properties:

- The self diffusion [8] D and the shear viscosity v [8] exhibit a non-Arrhenius behaviour at low temperatures (Fig. 3). However, the previous properties approximately obey the relations valid for a normal liquid [9]:

$$\frac{D\eta}{T} = 7.73 \times 10^{-10} \, \text{g cm K}^{-1} \, \text{s}^{-2} \tag{1}$$

$$D\tau = 2.11 \times 10^{-16} \, \text{cm}^2 \tag{2}$$

where τ represents some relaxation time and the constants are nearly T-independent.

The hydrogen-bond lifetime τ_{HB} has been evaluated in the frame of the percolation model of Stanley and Teixeira [10] and estimated from depolarized light scattering [11, 12] and dielectric relaxation time measurements [12]. It exhibits Arrhenius behaviour in contrast to the temperature dependences of the dielectric relaxation time τ_d and the orientational relaxation time τ_r (see Fig. 12) and its value is about 1 ps at $-20°C$. The activation energy associated with τ_{HB} has been found to be equal to 1.85 kcal/mole and it represents the energy of the hydrogen bond.

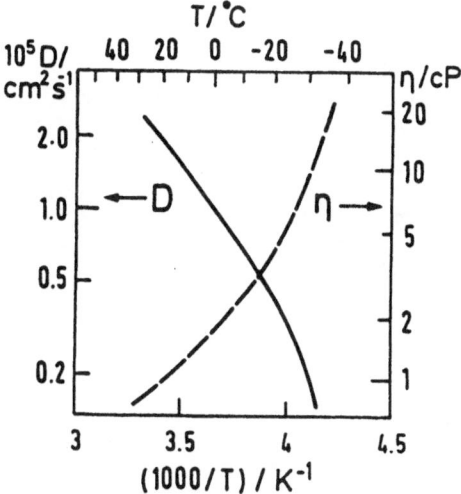

Fig. 3. Self-diffusion constant (full line) and shear viscosity (dashed line) of H_2O as a function of temperature (Data taken from Ref. [8]).

The next sections deal with the structure and dynamics of water. There are different scattering experiments such as light, X-ray, and neutron scattering which are used to characterize the structure and dynamics of water at the microscopic level.

2.3 The structure of liquid water

The detailed structural arrangement of molecules in liquid water has been of fundamental interest for a considerable time. A comprehensive survey [14–19] is presented in many reviews. The methods typically used to study the structure of liquid water are X-ray scattering and neutron scattering. The structure of a liquid is defined in terms of the static pair correlation function g(r) which involves the probability of finding an atom at a distance r from a reference atom.

In a scattering experiment the intensity scattered by the sample is a function of the diffraction angle 2Θ and the wavelength λ ($\lambda \simeq 1$ Å). These quantities are related to the so-called wave-number transfer Q (Fig. 4),

$$Q = \frac{4\pi \sin \Theta}{\lambda}. \tag{3}$$

From the recorded intensity I(Q), it is possible to evaluate the structure factor S(Q) from which the Fourier transform gives the pair correlation function g(r).

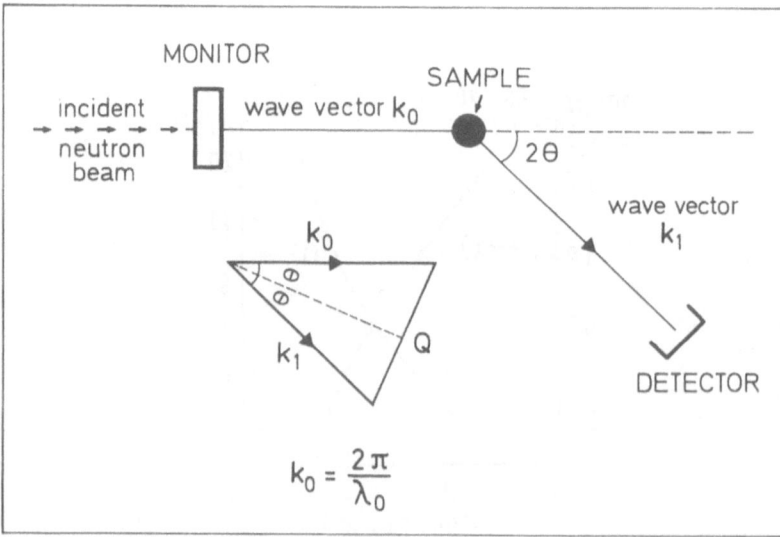

Fig. 4. General scheme of neutron scattering studies. λ_0 is the wavelength of the incident neutron beam. Q is the momentum transfer.

While X-ray scattering is mainly sensitive to the oxygen position, neutron scattering provides the pair correlation between the relative positions of deuterium and oxygen atoms and thus gives information on the orientational correlations of water molecules.

As seen in section 2.2, liquid water has the unique property of showing a maximum in the density versus temperature curve, and this feature allows experiments to be performed at two different temperatures, but at the same density. This allows us to define the isochoric temperature derivative (ITD) functions for a pair of temperatures T_1 and T_2 on opposite sides of the density maximum for which the density d is the same and $\Delta T = T_1 - T_2 > 0$ and to follow the evolution of the structure of liquid water as a function of the temperature.

2.3.1 X-ray diffraction: The X-ray diffraction technique gives access to the electronic spatial distribution. The main contribution to the intensity scattered by H_2O is due to the O-O correlations (64%); the other contributions relative to the O-H and H-H correlations represent respectively only 32% and 4%. The measured X-ray structure factor $S_X(Q)$ can be approximately related to the partial structure factor $S_{OO}(Q)$ by

$$S_X(Q) = \langle F^2(Q) \rangle S_{OO}(Q) \qquad (4)$$

where $F(Q)$ is the molecular form factor for X-rays. In Eqn. 4 it is assumed that the quantity $\langle F^2(Q) \rangle$ is equal to $\langle F(Q) \rangle^2$. This can be done because the electronic distribution in a water molecule is very nearly spherical [20]. Figure 5 represents the evolution of the pair correlation function $g_{OO}(r)$ versus the temperature [20]; $g_{OO}(r)$ gives approximately the distribution of the molecular centers. The area of the first peak of $g_{OO}(r)$ located at 2.9 Å gives approximately the number of the first nearest neighbours at the distance $r = 2.9$ Å from a central molecule. The number of first-nearest neighbours is approximately equal to 4.4 and is slightly temperature-dependent. The broad second maximum represents the distribution of the second-nearest neighbours in the O-O-O configuration.

The ITD of the X-ray structure factor is defined as

$$\Delta S_{OO}(Q, \rho, \Delta T) = \frac{S_X(Q, \rho, T_1) - S_X(Q, \rho, T_2)}{\Delta T} \qquad (5)$$

from which it is possible to evaluate the temperature variation of the number of neighbours up to a distance R from a central molecule:

$$\Delta N(R, \Delta T) = \rho \int_0^R \Delta g_{OO}(r, \Delta T) 4\pi r^2 \, dr \qquad (6)$$

where ρ is the number density of water. The variation of $\Delta N(R, \Delta T)$ as

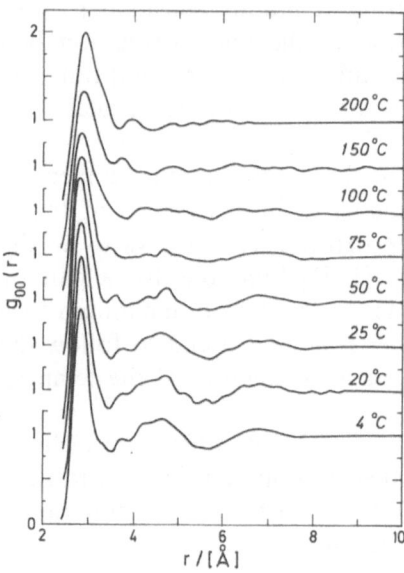

Fig. 5. Pair correlation functions $g_{OO}(r)$ for liquid water. (Adapted from Ref. [20]).

a function of R [21] led to interesting structural information which is given in the conclusion.

The conclusions of the X-ray scattering studies are summarized below. The number of first-nearest neighbours is not significantly affected when the temperature is varied; and the important arrangement of the oxygen atoms occurs at the second-nearest neighbour shell. When the temperature is lowered, there is a tendency towards a more regular tetrahedral coordination. In section 2.2.1. we mentioned the anomalous increase of the molecular volume with temperature. A way to explain this effect is found when considering the positions of the second neighbours which essentially depend on the O-O-O angle.

The O-O-O angular correlation is connected with a finite correlation length, evidence of which has been given both by small angle X-ray [22] and neutron scattering [23] measurements. The overall conclusion drawn from the X-ray scattering is that the completion of the tetrahedral structure is due to the opening of the O-O-O angle with decreasing temperature.

2.3.2 Neutron scattering: Neutrons are scattered by oxygen, hydrogen, and deuterium nuclei in different ways. The scattering from an isolated atom gives spherical scattering, which consists of coherent and incoherent parts that depend on nuclear parameters. The incoherent part results from averages over nuclear spin and isotopic compositions of otherwise identical scatterers. The coherent wave can exhibit interference effects due to scattering from neighbouring centres. For a diatomic molecule like the water molecule, the coherent scattering is dependent on all the positions and orientations to give a molecular form factor $f_1(Q)$; and there are additional coherent contributions from the neighbouring molecules. Thus the total coherent scattering involves a summation over all the molecules. Because the incoherent scattering cross-section of hydrogen atoms is very large when compared with the coherent one, scattering experiments are performed in general on D_2O samples (cf. Table 2).

For water (D_2O), there are three pair partial correlation functions $g_{OO}(r)$, $g_{OD}(r)$ [$\equiv g_{DO}(r)$], and $g_{DD}(r)$ which are required to define the structure. Thus, in the case where pair partial functions must be inferred, H_2O samples as well as D_2O/H_2O mixtures are used. Details about the different methods employed to extract the pair partial functions and the comparison with computer molecular dynamics simulations are reviewed in Refs. [18] and [19]. Mixtures with small amounts of H_2O (from 0 to 20%) seem the more appropriate ones. They give a better accuracy on the partial functions because the effects of H/D non-equivalence, inelastic corrections and incoherent scattering are minimized.

This section will focus mainly on the variation of the structure of liquid water as a function of temperature.

Theory. The liquid structure factor $S_M(Q)$ can be written in terms of a molecular form factor $f_1(Q)$ and a difference function $D_M(Q)$ which contains all the inter-molecular contributions, i.e. [14–19]

$$S_M(Q) = f_1(Q) + D_M(Q) \tag{7}$$

For D_2O molecules, $f_1(Q)$ is given by

$$f_1(Q) = [b_O^2 + 2b_D^2 + 4b_O b_D j_0(Qr_{OD}) \exp(-\gamma_{OD}Q^2)$$
$$+ 2b_D^2 j_0(Qr_{DD}) \exp(-\gamma_{DD}Q^2)]/(b_O + 2b_D)^2 \tag{8}$$

where r_{OD} and r_{DD} are respectively the intramolecular oxygen-deuterium and deuterium-deuterium distances, and $\gamma_{ij} = 1/2\langle u_{ij}^2 \rangle$ where $\langle u_{ij}^2 \rangle$ is the mean-square amplitude of displacement from equilibrium position due to the normal mode vibration of the molecule. The quantities b_O and b_D are the coherent scattering lengths of oxygen and deuterium atoms ($b_O = 0.5805 \times 10^{-12}$ cm, $b_D = 0.6674 \times 10^{-12}$ cm).

Table 2. Neutron scattering cross-sections σ in barn ($1b = 10^{-24}$ cm^2)

Atom	$\sigma_{coherent}$	$\sigma_{incoherent}$
H	1.8	79.9
D	5.6	2.0
O	4.2	0.0

The total pair correlation function $g(r)$ is related to the Fourier transform of $S_M(Q)$ by

$$4\pi r \rho_M [g(r) - 1] = \frac{2}{\pi} \int_0^\infty Q[S_M(Q) - S_M(\infty)] \sin Qr \, dQ \qquad (9)$$

where $S_M(\infty) = (b_O^2 + 2b_D^2)/(b_O + 2b_D)^2$ is the asymptotic value of $f_1(Q)$ at large Q and ρ_M is the molecular number density.

The function $g(r)$ is a combination of the separate partial correlation functions and includes sharp peaks due to the intramolecular distances. It is more convenient to remove the intramolecular terms by subtracting the molecular form factor from $S_M(Q)$ to give $D_M(Q)$, which may be transformed by

$$d_L(r) = 4\pi r \rho_M [g_L(r) - 1] = \frac{2}{\pi} \int_0^\infty Q D_M(Q) \sin Qr \, dQ \qquad (10)$$

to give the pair correlation functions $d_L(r)$ and $g_L(r)$ for the intermolecular terms only.

The composite function $g_L(r)$ (or $d_L(r)$) consists of a weighted sum of partial terms which for neutron diffraction of D_2O is expressed by

$$g_L(r) = 0.489 g_{DD}(r) + 0.421 g_{OD}(r) + 0.090 g_{OO}(r) \qquad (11)$$

where the dominant contributions concern the DD and OD spatial correlations.

Measurements at different temperatures allow one to follow the evolution of the structure of liquid water, either from a temperature of reference (for instance the temperature for which the density is maximum) or from the isochoric temperature derivative function [19, 24, 25].

For measurements at different temperatures, the function $\Delta D_M(Q, \Delta T)$ is defined as

$$\Delta D_M(Q, \Delta T) = S_M(Q, T_1) - S_M(Q, T_{ref}) = \Delta S_M(Q, \Delta T) \qquad (12)$$

with $\Delta T = T_1 - T_{ref}$ and where it is assumed that the molecular conformation remains unchanged. Fourier transform then gives the change in the real-space correlation function

$$\Delta d_L(r, \Delta T) = \frac{2}{\pi} \int_0^{Q_M} Q \Delta D_M(Q, \Delta T) \sin Qr \, dQ. \qquad (13)$$

The defintion of the ITD function is the same as for X-rays. An example of this treatment is given in Fig. 6. The two curves 6a and 6b can be superimposed except for the first peak near 2 Å. This suggests that there is a universal ITD curve which defines the structural change over the 2–10 Å range. There is a smaller change in the composite peak at 2 Å for the data corresponding to the higher ΔT value. It is possible to interpret this as a saturation effect in the local hydrogen-bonding whose

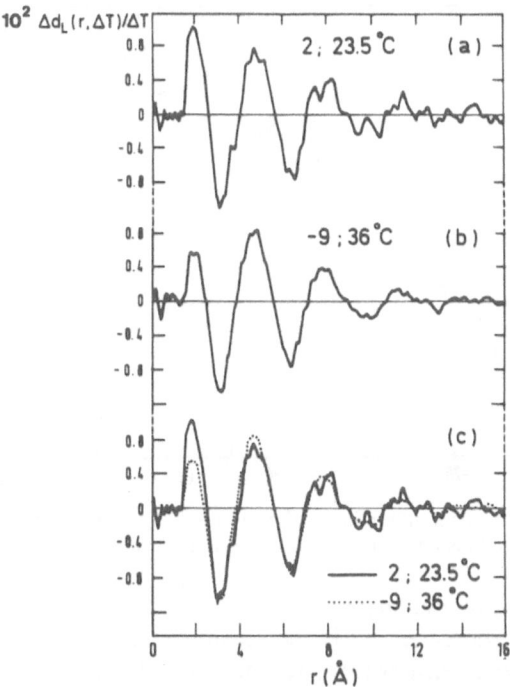

Fig. 6. Real-space distribution functions $\Delta d_L(r, \Delta T)/\Delta T$ for isochoric temperature derivative ITD functions. a) between 2°C and 23.5°C. b) between −9°C and 36°C. The overlapping data are shown below (c). (From Ref. [19]).

contribution at 1.8 Å is predominant (hydrogen-bonded OD distance ≈ 1.8 Å). This interpretation is in agreement with the spectroscopic work of Walrafen [2] and the fourfold coordination occurring in the crystalline and amorphous forms of ice. The mean effect of the other changes can be understood in terms of increased orientational correlation over larger distances in the H-bond network.

Temperature variation studies. The most recent results have been obtained on the 7C2 spectrometer of the Orphée reactor, at the Laboratoire Léon Brillouin. The experimental procedure and data analysis have already reported elsewhere [19, 26, 27].

a) Structure factors $S_M(Q)$

A striking feature of the neutron structure factor is its extremely strong temperature dependence, characterized by a displacement of the main diffraction peak position Q_0 towards smaller Q values with decreasing temperature, as already observed in previous experiments mentioned in Refs. [19] and [27].

The evolution of Q_0 as a function of temperature is given in Fig. 7. It appears that the observed Q_0 tends towards the value 1.70 Å^{-1} which is

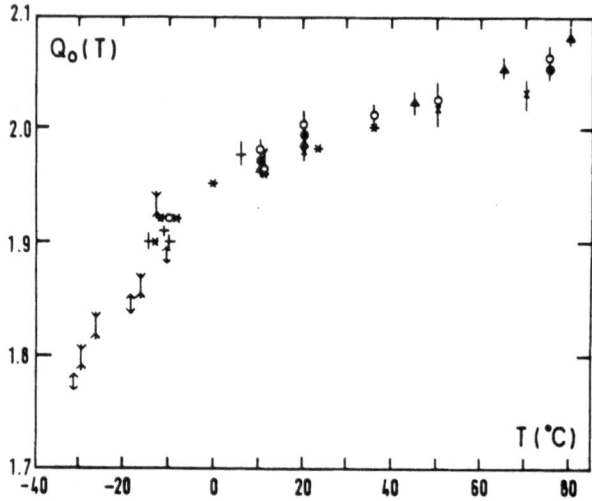

Fig. 7. Evolution of the main diffraction peak $Q_0(T)$ of $S_M(Q)$ versus the temperature. (From Ref. [26]). For the symbols and further details see Refs. [26] and [27].

characteristic of low-density amorphous ice. Moreover the slope of the $Q_0(T)$ function increases with decreasing temperature. The slope of $Q_0(T)$ in the lower temperature regions correlates with the decrease in bulk density and emphasizes the link between the microscopic and macroscopic features.

Some important structural information results from the determination of $S_M(Q)$ with high precision in the low Q range [23] (Fig. 8). The inset shows in more detail the low Q region and gives the calculated values of $S_M(Q)$. One observes clearly the increase of $S_M(Q)$ with decreasing Q, the anomaly being more pronounced at 5°C with a minimum occurring clearly around 0.5 Å^{-1}. The absolute scale enables the determination of a small correlation length of the order of 6 Å with the same hypothesis as used before for the treatment of the "normal" contribution to compressibility [22]. This anomalous behaviour already obtained by X-ray scattering [22] has just been confirmed by neutron scattering [23] and indicates the presence of density fluctuations with a correlation length ξ that increases up to about 8 Å as T is decreased to 253 K. This result supports the theoretical description [10] and the simulation of the hydrogen-bond network of water [30, 31]. Tiny "patches" of this network, perhaps those in which all bonds are intact, are of lower density than the rest of the network and hence give rise to anomalous density fluctuations.

b) Pair correlation functions $d_L(r)$ [19, 27]

In Fig. 9, the pair correlation function of low-density amorphous ice is plotted together with that of water at 27°C and that of supercooled water at −10.5°C and −31.5°C.

In the small-r range, when the temperature is lowered, some features characteristic of amorphous ice show up: for example, two peaks appear

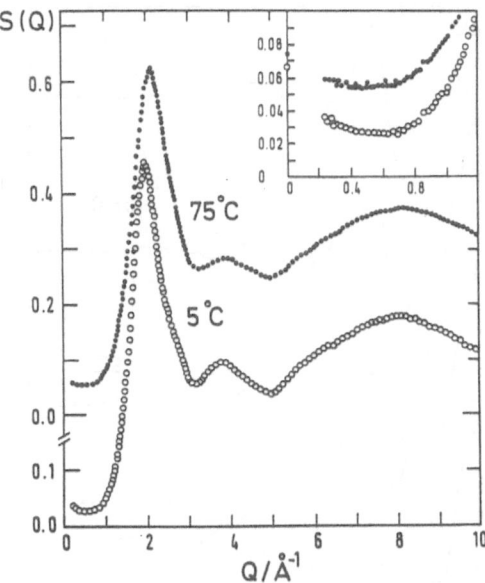

Fig. 8. Structure factors $S_M(Q)$ of D_2O obtained respectively at 5°C (circles) and 75°C (full dots). Inset: Note the decrease of $S_M(Q)$ with increasing Q, at low temperature. (From Ref. [23]).

at 1.75 Å and 2.30 Å; the first corresponds to the hydrogen-bond distance and the second to the intermolecular D-D distance. In the larger r range one observes additional small oscillations, which are not present at -10.5°C, and an out-of-phase behaviour for the broad oscillations (r > 10 Å). These features are similar to those observed in low-density amorphous ice.

Neutron diffraction is sensitive to both position and angular correlations and $d_L(r)$ is dominated by the partial functions $g_{OD}(r)$ and $g_{DD}(r)$ (see Eqn. 10). Because X-ray data show that oxygen-oxygen distances are not strongly temperature-dependent [24], the observed strong temperature evolution of $g(r)$ is mainly due to the temperature-dependence of the $g_{OD}(r)$ and $g_{DD}(r)$ pair correlation functions. The increased effect of hydrogen-bonding therefore produces enhanced orientational correlations between neighbouring molecules and, as the hydrogen-bonded clusters increase in size, there is a greater spatial correlation between second- and third-bonded molecules. However, the assembly still possesses liquid characteristics so that the correlations of the transient hydrogen-bonded assemblies are less developed than in the static structure of amorphous ice.

c) Comparison with amorphous ice

In order to interpret the behaviour of the pair correlation function of water at -31.5°C, we recall that an appropriate description of the structure of amorphous ice has been given in terms of a continuous

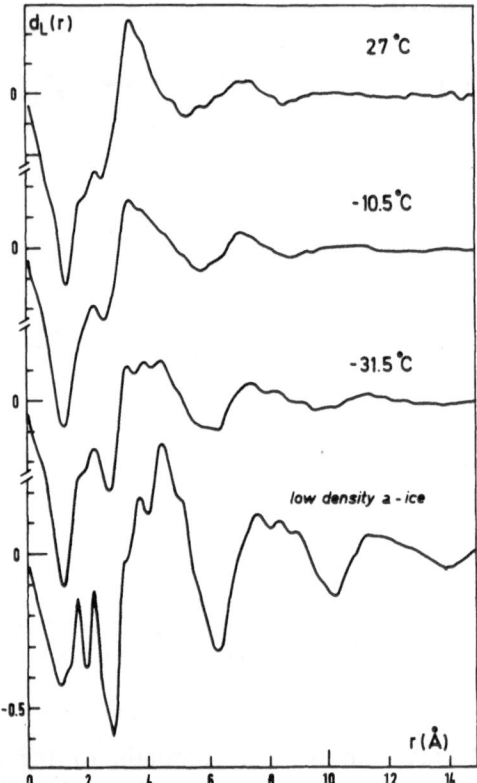

Fig. 9. Pair correlation function $d_L(r)$ of D_2O obtained respectively at 27°C, -10.5°C, -31.5°C. (From Ref. [27]). The function $d_L(r)$ of low density amorphous ice is given for comparison. (From Ref. [27]).

random network model (CRN) [32]. In this model, the basic building unit is a water pentamer. This is equivalent to a description of the molecular arrangement in water by a continuous random network of tetrahedrally coordinated H-bonds [31, 32].

The main results of recent experiments can be compared with those of molecular dynamics simulations. It is noted that the liquid structure is qualitatively reproduced from several independent water potential models when the model incorporates the high directionality of intermolecular interactions [18].

2.4 Dynamics of liquid water

In this section, we report on scattering experiments such as light and neutron scattering which are currently used to characterize the dynamics of water at the microscopic level.

Inelastic neutron scattering appears to be a unique tool to study the individual motions and the collective motions of liquid water, due to the fact that the respective scattering cross-sections of H_2O and D_2O are very different.

In the case of H_2O, the incoherent scattering cross-section is very high when compared to the coherent one, the opposite of D_2O where the coherent scattering cross-section is only a little more than twice higher (see Table 2). The space time pair correlation function $G(r, t)$ [33] for H_2O and D_2O can be expressed in terms of the self and distinct terms, labelled s and d respectively.

For H_2O, $G(r, t)$ can be expressed as:

$$G^{H_2O}(r, t) = 13.0G_H^s + 0.34G_O^s + 0.56G_{HH}^d - 0.87G_{OH}^d + 0.34G_{OO}^d$$

$$(14)$$

In the above expression the self terms dominate, which allows us to study:

- The diffusive motions (translation and rotation) by quasielastic incoherent neutron scattering (QENS) by H_2O.
- The vibrational density of states including the intermolecular and intramolecular parts by inelastic incoherent neutron scattering (IINS) from H_2O (for energy transfers between 0 and 600 meV).

For D_2O, $G(r, t)$ can be expressed as:

$$G^{D_2O}(r, t) = 1.22G_D^s + 0.34G_O^s + 1.78G_{DD}^d + 1.55G_{OD}^d + 0.34G_{OO}^d \quad (15)$$

where the coherent contributions dominate. Coherent inelastic neutron scattering by D_2O allows us to study collective motions in water. In the following, we will report on the first experimental evidence of the existence of short-wavelength (between 10 and 20 Å) collective excitations in liquid water.

2.4.1 Individual motions of water: Numerous studies on the diffusive motions of water molecules in liquid water have been reported previously [34–37]. Analyses of these early experimental data were often made without considering the rotational motions, which constitute an important contribution [38–45].

The classical way to analyse the quasielastic neutron scattering spectra assumes decoupling between vibrations, rotations, and translational motions. This is not correct at room temperature when rotational and translational motions are coupled. However, this assumption can be justified at low temperatures when the time scales of these two motions are different. Within this approximation, the scattered intensity $I(Q, \omega)$ depends on the momentum transfer Q and energy transfer ω in the following way:

$$I(Q, \omega) = \exp(-Q^2 \langle u^2 \rangle / 3) T(Q, \omega) \otimes R(Q, \omega) \qquad (16)$$

$$T(Q, \omega) = \frac{1}{\pi} \frac{\Gamma_s(Q)}{(\omega^2 + (\Gamma_s(Q))^2)} \qquad (17)$$

$$R(Q, \omega) = \frac{1}{\pi} \sum_{l=1}^{\infty} (2l+1) j_l^2(Qa) \frac{l(l+1)D_r}{\omega^2 + (l(l+1)D_r)^2} \qquad (18)$$

$$\Gamma_s(Q) = \frac{DQ^2}{1 + DQ^2 \tau_0} \qquad (19)$$

$$\tau_1 = \frac{1}{6D_r} \qquad (20)$$

$$L = \sqrt{6D\tau_0} \qquad (21)$$

where the first term of $I(Q, \omega)$ is the Debye–Waller factor, $R(Q, \omega)$ accounts for the hindered rotations, and $T(Q, \omega)$ for the translation movements.

The symbol $\otimes$ means frequency convolution and $j_l(x)$ are the spherical Bessel functions. The quantity $\langle u^2 \rangle$ is the mean square amplitude of the vibrations (mainly the librational movements – see legend to Fig. 13), Γ_s is the width of the translational line, D is the self diffusion (low Q limit of Γ_s), τ_0 is the residence time (inverse of the high Q limit of Γ_s), τ_1 is the hindered rotations characteristic time, a the oxygen-hydrogen intramolecular distance (a = 0.98 Å), and L is the mean square jump length, assuming a random jump diffusion mechanism.

An example of the spectra obtained is given in Fig. 10, where the solid lines are the best fits deduced from the proposed model (relations

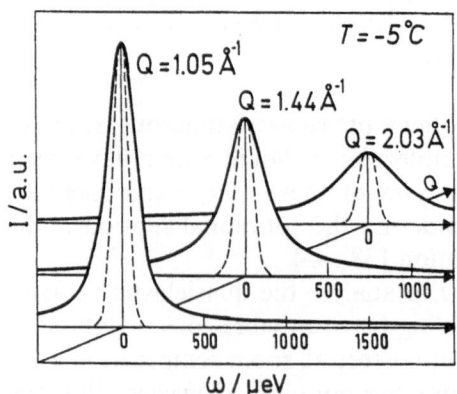

Fig. 10. Typical quasielastic spectra, obtained at $-5°C$ for three values of Q, Q = 1.05 Å^{-1}, 1.44 Å^{-1}, 2.03 Å^{-1}. The energy of the incident neutrons was 3.14 meV. It is noted that the peak heights decrease rapidly as Q gets larger. The dashed lines are the shape of the resolution function of the time-of-flight measurement (IN6, ILL). a.u.: arbitrary units. (From Ref. [42]).

16–21). The typical error is 1% using the three free parameters: $\langle u^2 \rangle$, $\Gamma_s(Q)$, and D_r.

Figure 11 shows the variation of $\Gamma_s(Q)$ versus Q^2 at different temperatures. The solid lines represent the best fit using the random jump diffusion model (relation 19), from which it is possible to deduce the behaviour of the mean jump length L and the residence time τ_0 as a function of the temperature [38–45]. Figure 12 shows that the behaviour of τ_0 is non-Arrhenius, while that of τ_1 is approximately Arrhenius with an activation energy $E_A = 1.85$ kcal/mole. In particular τ_0 is about 1.25 ps at 20°C and increases rapidly to more than 10 ps at $-20°C$. The temperature dependence of the mean jump length L leads to an average value close to 1.6 Å which is the distance between the two protons in the water molecule. Finally the Debye-Waller parameter determined is, within the experimental error, temperature-independent. The value of the vibrational amplitude is $\langle u^2 \rangle^{1/2} = 0.484$ Å.

From these features, it is possible to justify a mechanism for the proton motion. As pointed out in section 2.1, water molecules are instantaneously connected with most of their neighbours through hydrogen bonds. However, each hydrogen bond is very directional and the libration amplitude is very large. Molecular dynamics simulations of Impey et al. [46] indicate that, when the angle of libration is above 28

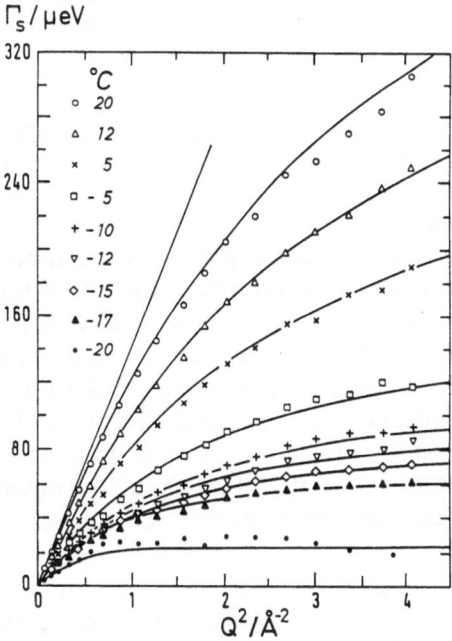

Fig. 11. $\Gamma_s(Q)$ vs Q^2 plots of a series of 9 temperatures ranging from 20°C down to $-20°C$. It is noteworthy that $\Gamma_s(Q)$ tends to a constant value at large Q for temperatures below $-15°C$. For T = 20°C, the self-diffusion line is shown. (From Ref. [42]).

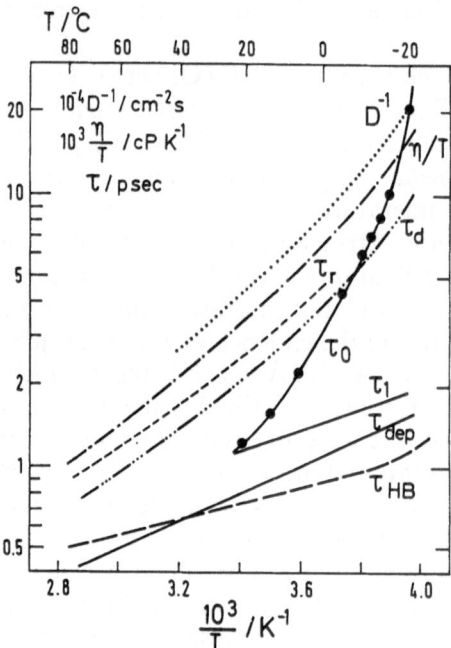

Fig. 12. Arrhenius plot of τ_0 and τ_1. Different transport properties (self diffusion constant D, shear viscosity η, dielectric relaxation time τ_d and orientational relaxation time τ_r) are plotted versus temperature. τ_{dep} is the hydrogen-bond lifetime from depolarized light scattering. τ_{HB} is the hydrogen-bond lifetime from dielectric relaxation time and the percolation model. (From Ref. [41]).

degrees, the bond is broken. This estimation of the angle is in very good agreement with the value $\langle u^2 \rangle^{1/2} = 0.484$ Å which is deduced from the Debye-Waller factor. Moreover, the characteristic time τ_1 has an Arrhenius temperature dependence. Since τ_1 is associated to hindered rotations, this time is interpreted as the typical hydrogen-bond lifetime. The breaking of the hydrogen bonds is due to the librational motions which have a large amplitude and occurs when the vibration angle becomes superior to the critical value of 28 degrees as mentioned above. When at least three bonds are simultaneously broken, diffusion is possible. It then appears that the residence time τ_0 is a direct measure of the average duration which is required for a water molecule to break loose from its hydrogen bondage by rotational excitation; it should be noted that τ_0 is a time constant that no other technique except neutron scattering can access.

2.4.2 Intermolecular and intramolecular vibrations of water: As stated before, neutrons are uniquely suited for probing the single-particle motions of protons in hydrogen-containing substances. Although traditionally Raman scattering has been used extensively to study hydrogen-bond dynamics in water [2, 48], the quantitative theoretical prediction of the resulting Raman spectra has been far from straightforward; moreover it is not possible to extract a true density of states because the ω-dependent coupling function is not known.

Incoherent inelastic neutron scattering allows us to study the intermolecular and intramolecular vibrations of water because the incoherent inelastic neutron spectra are directly related to the Q-dependent density of vibrational states of atoms $G_s(Q, \omega)$ in the molecular system [47] and equally to the true density of states $f_H(\omega)$.

The basic principle of the method is to extract $G_s(Q, \omega)$ through the relation

$$G_s(Q, \omega) = \frac{\omega^2}{Q^2} S_s(Q, \omega), \tag{22}$$

$S_s(Q, \omega)$ being directly related to the intensity of neutron spectra.

The $Q \to 0$ limit of $G_s(Q, \omega)$ allows us to extract the true density of states of the hydrogen atom $f_H(\omega)$, by the relation:

$$\lim_{Q \to 0} G_s(Q, \omega) = f_H(\omega) \cdot \frac{k_B T}{M_{H_2O}} \tag{23}$$

In practice the $Q \to 0$ limit is satisfied when the product $Qa < 1$ [47]. For water $a = 0.98$ Å and measurements of $f_H(\omega)$ are possibly only for $E = \hbar\omega$ less than about 50 meV.

The vibrational spectrum of water stretches from 0 to ≈ 500 meV. It is impossible to cover this wide energy range with a single spectrometer. The results described in the following correspond to two sets of experiments: one set of experiments used the IN6 spectrometer of the Institut Laue Langevin at Grenoble and covered the intermolecular vibrations. The second set of experiments used the Intense Pulsed Neutron Source at Argonne (USA) and allowed energy transfers equal to 600 meV to be reached, thus covering the intramolecular vibrations.

a) Intermolecular vibrations.

Figure 13 shows a proton density of states $f_H(\omega)$ extracted from an inelastic neutron scattering spectrum for water at $-15°C$ [49].

Three intermolecular bands are present: a sharp band at about 6 meV, called A, a very weak band at around 25–35 meV, called B and an intense band in the range 50–130 meV, called L (cf. Table 3).

The bands A and B correspond to the bending and stretching modes while the L band is librational in character. All these bands have their counterparts in the Raman spectrum [2, 48]. The temperature variation between 20°C and $-20°C$ is similar to that obtained by Raman scattering. The peak at 8 meV shows almost no variation with temperature and the librational band moves to higher energies with decreasing temperature.

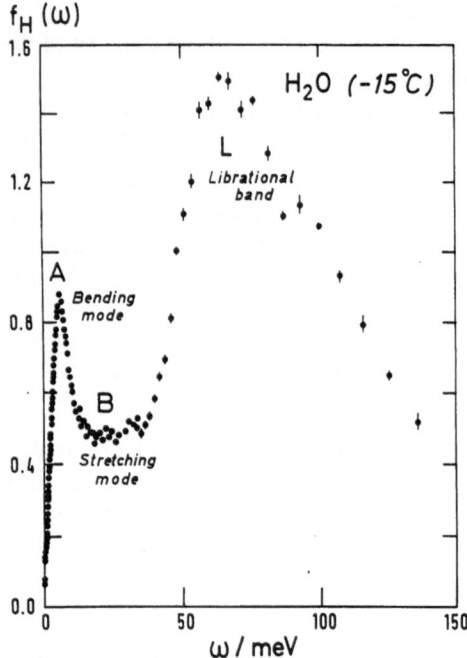

Fig. 13. Intermolecular part of the density of states of hydrogen atoms in water at $-15°C$, deduced from the INS data (IN6, ILL). The three bands A, B, L are mentioned. The librational band L is due to the rotation of the water molecule around the centre of mass (nearly coincident with the oxygen atom) which results in large amplitude motion of H-atoms. A water molecule has three different moments of inertia, and librations around each of the axes have different frequencies. This leads to the broad and intense band L in the range 50–130 meV. (From Ref. [39]).

Table 3. Vibrational energies in water (1 meV = 8.0668 cm^{-1})

T(°C)	Intermolecular[a] O-O-O bending (meV)		Intermolecular[a] stretching (meV)		E_L(meV)[b]	E_B(meV)[c]	E_S(meV)[c]
	IINS	Raman scattering	IINS	Raman scattering			
Ice −20	6	8	25.5	24.3	82	207	407
H$_2$O −15	6	8	22	25	74	207	418
H$_2$O 20	6	8	22	23	—	—	—
H$_2$O 40	—	—	—	—	—	207	441
H$_2$O 52	—	—	—	—	65[d]	228[d]	429, 448[d]
H$_2$O 80	—	—	—	—	—	207	443
Gas 25	—	—	—	—	—	198, 205[d]	454, 465
(Raman)	—	—	—	—	—	—	475, 490[d]
Q (Å^{-1})	—	—	—	—	2.6	3.5	6.6

[a]Data for 3.14 eV incident neutron energy.
[b]Data for 500 meV incident neutron energy.
[c]Data for 800 meV incident neutron energy.
[d]Derived from CMD simulations using the SPC model [55].

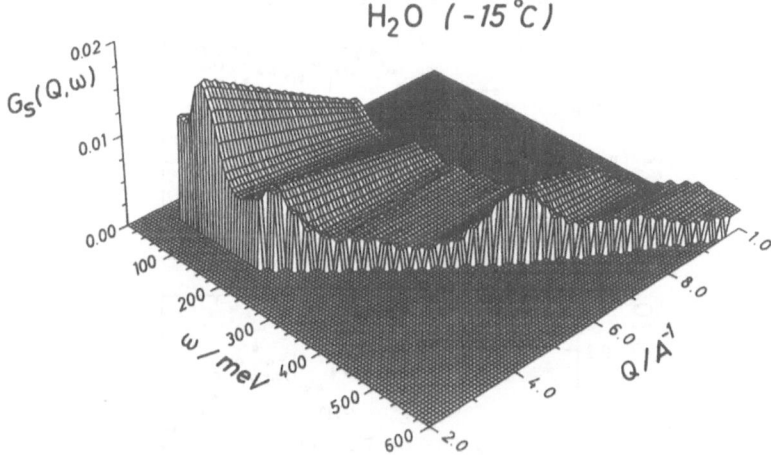

H_2O $(-15°C)$

Fig. 14. Q-dependent density-of-states $G_S(Q, \omega)$ for supercooled water at $-15°C$ plotted in (Q, ω) space. The bands L, BEN and STR as described in the text are visible. (From Ref. [39]).

b) Intramolecular vibrations.

Figure 14 presents a Q-dependent proton density-of-states [38, 39, 50–52] in a two-dimensional plot, as a function of Q and ω, obtained for water at $-15°C$.

The first peak at about 80 meV is the L band, the second peak at 220 meV is the OH bending mode (BEN) and the broad peak at around 420 meV is the OH stretch vibration (STR) (cf. Table 3).

Figure 15 shows two constant Q plots of $G_s(Q, \omega)$ for H_2O at $T = 258$ K. In Fig. 15a, the L-mode peak at about 100 meV and the BEN-mode peak at 230 meV are visible. There is a satellite peak visible at 300 meV. In Fig. 15b, the STR-mode peak at 430 meV and a satellite peak at 530 meV are identified. These satellite peaks have been interpreted as a combination band [53] in Raman spectroscopy literature. Recent infrared measurements [54] have identified these modes, with the exception of the satellite peak at 530 meV because of the limited range of frequency.

A computer molecular dynamics simulation (CMD) performed by Toukan and Rahman [55] predicted these two lines at respectively 427 meV and 510 meV corresponding to stretching vibrations without and with simultaneously breaking of the hydrogen bond [2, 48a]. The energy difference between the stretching vibrations is close to the librational energy, which demonstrates that there is some coupling between the breaking of H-bonds and librational motions; this argument is in favour of the description of molecular motions in liquid water previously proposed.

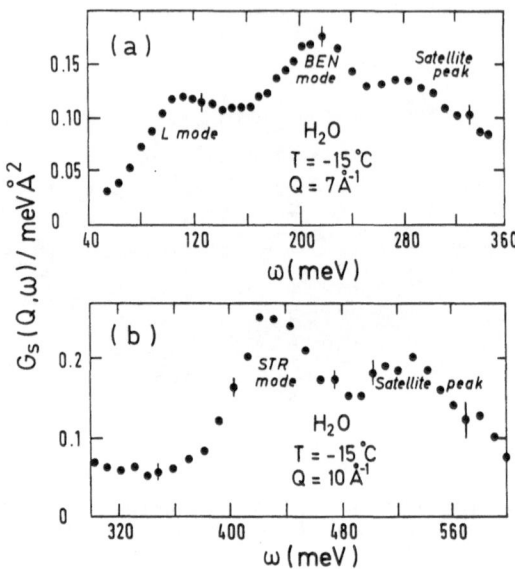

Fig. 15. Constant Q plots of $G_S(Q, \omega)$ for water at $T = -15°C$ in two energy ranges. a) Energy range covering L and BEN modes. Notice the clearly visible satellite peak at 300 meV. b) Energy range covering the STR-mode and its satellite peak at 530 meV. (From Ref. [38], reprinted with permission of Kluwer Academic Publishers).

2.4.3 Collective motions of water: In this section, the first experimental evidence of the existence of short-wavelength collective excitations [56, 57] in liquid D_2O as observed by coherent inelastic neutron scattering is reported.

Evidence of short-wavelength collective excitations in dense atomic liquids is given by the existence of a shoulder on each side of the main central peak of the inelastic spectrum $S(Q, \omega)$ for Q values ranging between 0 and $1.2 Å^{-1}$. The damping of these modes depends on the steepness of the repulsive part of the potential and on the depth of the well of the attractive part of the potential. These modes have been observed in liquid Ne, Rb, and Pb [58–60]. In a molecular dynamics simulation of water by Rahman and Stillinger [61] and Impey et al. [46] two peaks were observed suggesting the existence of two distinct propagating modes. A similitude can be found with Brillouin lines observed by light scattering at very small Q values and lower energies.

Measurements were performed for several different wave-number transfers ranging from $0.35 Å^{-1}$ to $0.6 Å^{-1}$ on the IN8 spectrometer of the Institute Laue Langevin. Typical spectra are shown in Fig. 16 with the "side" peaks corresponding to sound propagations. The best fit of the curves, within a χ^2 criterion, was obtained by using a damped harmonic oscillator. The relation between the ω_s harmonic frequency of the oscillator and Q is shown in Fig. 17. As seen from Fig. 17 the side

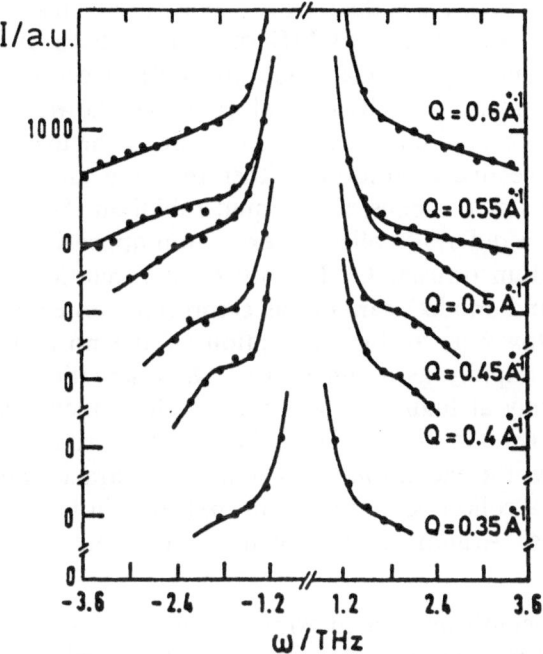

Fig. 16. Scattered intensity I, plotted against the energy transfer ω for all the measured wavevector transfers Q. The solid lines represent the best fits with damped harmonic oscillator-type functions. (From Ref. [56]).

peaks at different Q values can be interpreted as a manifestation of collective excitations propagating at a speed c given by:

$$c = \frac{\omega_s}{Q} \tag{24}$$

The data can be represented by a straight line passing through the origin with a slope $c = 3310 \pm 250 \, \text{m s}^{-1}$. It is worth noting that this

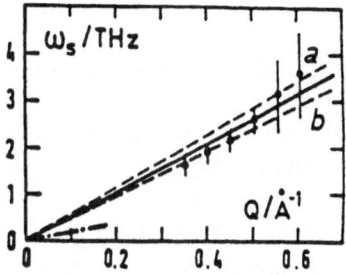

Fig. 17. Representation of the dispersion curve which gives a velocity equal to $3310 \, \text{m s}^{-1}$. Dots: experimental points. Solid line: best linear fit according to Eqn. 24. Dashed lines: predictions of CMD from Ref. [61] (a) and from Ref. [46] (b). Dot-dashed line: dispersion curve of the ordinary sound wave. The corresponding velocity ($1390 \, \text{m s}^{-1}$) is measured up to 5 GHz by Brillouin light scattering (From Ref. [65]).

velocity is more than twice the sound velocity (1400 m s^{-1}) and in good agreement with predictions of CMD simulations both by Rahman and Stillinger [61] and Impey et al. [46], and more recently confirmed by Wojcik and Clementi [62], Ricci et al. [63], and Kowall et al. [64].

Essentially two possibilities appear to explain the existence of short-wavelength collective excitations: either they are an extension of the hydrodynamic sound modes, as in liquid rubidium [59] or lead [60], or they represent a different collective excitation appearing at high values of the momentum transfer Q. These two interpretations of the "high-frequency sound mode" cannot be completely tested in the present experiment. However, as the absorption of the normal sound mode increases very rapidly as a function of Q, observation of the normal sound is unlikely at high Q. Thus, this "high-frequency sound mode" cannot be interpreted as the high-Q limit of the ordinary sound wave. Experimentally, the excitation has a solid-like character and it has been interpreted as a collective excitation propagating in the local hydrogen-bond structure. Arguments in favour of the interpretation are the following:

- the wavelength associated with the observed Q range is around 20 Å, comparable with the extension of the patches of hydrogen-bonded molecules [31];
- the damping Γ of the modes is proportional to Q^2 and much smaller than the equivalent damping of the ordinary sound extrapolated to this Q range [65];
- the observed velocity of the collective excitations is similar to the velocity of sound in ice.

2.5 Conclusions

From the temperature-dependence studies done by X-rays and neutron scattering and computer dynamics simulation, it is possible to draw some conclusions.

The local structure of liquid water is close to hexagonal ice; when the temperature is decreased, there is an opening of the O-O-O angle which tends towards the ideal value ($109°$) of the tetrahedral angle. As a consequence, the local density decreases with decreasing temperature giving rise to the density maximum.

As demonstrated in section 2.3, the ITD gives a direct measurement of the structural change. It has been shown that $\Delta d_L(r, \Delta T)/\Delta T$ has an almost universal shape emphasizing the systematic changes occurring over a range of 10 Å or more. The only difference appears in the region of 2 Å and can be explained by saturation effects for the O-D intermolecular distance at 1.8 Å corresponding to the hydrogen bond.

The structural measurements, at low temperature, confirm the increasing spatial correlations in deeply supercooled water as the temperature is decreased, and the tendency to evolve towards the structure of low-density amorphous ice.

The molecular motions in liquid water are described in terms of two dynamical processes; the residence time τ_0 and the characteristic lifetime τ_1 of a hydrogen bond have been given. It appears that the breaking of hydrogen bonds is due to large amplitude librational movements. An argument in favour of this description is deduced from inelastic incoherent neutron scattering in the OH-stretching region and from predictions of computer dynamics simulations. In fact the energy difference between the stretching vibrations is close to the librational energy.

The existence of a "high-frequency sound mode" in liquid water is demonstrated and interpreted as a collective excitation propagating in the local hydrogen-bond structure.

Recent structural data should provide an important basis for a comparison with predictions based on different forms of the interaction potential between water molecules.

Moreover, it now appears that it is possible to reproduce in a molecular dynamics simulation a wide range of measurable properties of water from thermodynamics to structure and dynamics. The "flexible" SPC model potential for the interaction between water molecules seems to be the "best effective potential" [66].

3. Aqueous solutions

3.1 Introduction

In section 2 interest was focused on the properties of water. However, pure water is an idealization. Even after very high purification water contains impurities. Therefore it is more a solution than a pure liquid. In reality, we are concerned with aqueous solutions containing one or several types of particles. When the particles are charged, the solution consists of at least three different kinds of particles: the solvent, cations, and anions. For instance, sea water is itself a very complicated aqueous solution. It contains many different salts, among which NaCl and $MgSO_4$ are dominant. As sea water is the environment of many life processes the study of its properties is directly related to biology. But also within biological systems the solution of small ions like Na^+, K^+, and Cl^-, or Mg^{2+} and Ca^{2+}, plays an important role. For instance, these ions participate in the generation of action potentials by transport processes through membranes.

In biochemical and biophysical processes very complex molecules like proteins or polyacides are involved. Their stability and biological activity are influenced by the aqueous environment. Smaller molecules showing some characteristics of these macromolecules are often investigated in order to understand these complex systems. For example, aqueous solutions of urea serve as a probe for the interactions between water and the $-CONH_2$ group so as to mimic hydration of peptide groups [67]. Another example is the study of highly symmetric and highly charged polyelectrolytes [68]. In this case, the influence of the Coulombian forces are under consideration. These forces dominate all other interactions between the highly charged ions and the counter ions or the water molecules.

In the last few years reliable information has been obtained about the structure and the dynamics of macromolecules in solution. This has become possible because of a combination of refined experimental and theoretical methods [69].

The interaction of water with complex biological structures will be the topic of section 4. In the present section basic phenomena occurring in aqueous solutions will be discussed. We have selected experimental and theoretical methods which are of fundamental importance and which can also be applied to the study of complicated solutions.

In section 3.2 thermodynamic properties of solutions are discussed. Thermodynamic quantities globally reflect the structure around ions. Simple models of association and hydration are sometimes good enough to explain experimental data. Even qualitative pictures like that of hydrophobic effects are useful for the interpretation of some thermodynamic data.

When more quantitative models of the structure around solute particles are required, statistical mechanics is often used. The basic quantities of statistical mechanics are the distribution functions which can be related to thermodynamic properties. In some cases the calculated correlation functions can even be directly compared to correlation functions obtained by scattering techniques. Interactions between particles in solution are also detectable by force measurements (section 3.3).

The distribution of ions in solution influences not only the thermodynamic properties of the solution but also the dynamic behaviour of all the particles including the solvent molecules. Two kinds of motions can be distinguished: the overall movement of the particles through the solution relative to other particles and the internal motions of ions and water such as vibrations or rotations. Some examples are discussed in section 3.4.

Finally, let us point out one important feature which is common to "simple" and complex solutions: the properties of the system are determined by a subtle competition of various effects: electrostatic forces, polarizabilities, geometrical restrictions like the covolumes or hydrophobic effects, van der Waals interactions, the static and dynamic electroneutrality conditions, and so on. One of the aims of this section is to show the consequences of this competition. A collection of papers dealing with solution effects can be found, for example, in Refs. [70–75].

3.2 Thermodynamics

3.2.1 Introduction: The basic quantities for describing the thermodynamic properties of solutions are the chemical potentials $\mu_i(p, T)$ of every component i. (An introduction to the thermodynamics of solutions is given in [76], a more complete description in [77].) As indicated they are functions of pressure p and temperature T. The chemical potential μ_w of water (i = w) in solution is defined as follows:

$$\mu_w = \mu_w^* + RT \ln a_w = \mu_w^* + RT \ln x_w f_w$$

$$= \mu_w^* + RT \ln x_w + \mu^{excess} = \mu_w^{ideal} + \mu_w^{excess} \qquad (25)$$

R is the gas constant, a_w the activity of water, x_w its molar fraction, and f_w the corresponding activity coefficient. For an ideal solution the activity coefficient is equal to unity and hence $\mu_w^{excess} = 0$. The quantity μ_w^* is the chemical potential of pure water

$$\mu_w^* = \lim_{x_w \to 1} \mu_w; \quad \lim_{x_w \to 1} f_w = 1. \qquad (26)$$

The chemical potential of a solute i = k is

$$\mu_k = \mu_k^\infty + RT \ln a_k = \mu_k^\infty + RT \ln m_k \gamma_k. \qquad (27)$$

For convenience the solute concentration is given in molality m (mol/kg of solvent) with its corresponding activity coefficient γ_k. The reference state is the infinitely dilute solution

$$\mu_k^\infty = \lim_{x_w \to 1} (\mu_k - RT \ln m_k \gamma_k), \quad \lim_{x_w \to 1} \gamma_k = 1. \qquad (28)$$

When the component k is a salt consisting of a cation $K_{\nu+}^{z+}$ and an anion $A_{\nu-}^{z-}$ (z_+ and z_- are the charges and ν_+ and ν_- the valencies of the cation and anion, respectively) its chemical potential can be written

$$\mu_k = \mu_k^\infty + \nu RT \ln m_k \gamma_\pm \qquad (29)$$

where $\nu = \nu_+ + \nu_-$ and $\gamma_\pm$ is the mean molal activity coefficient of the salt,

$$\gamma_\pm = \gamma_+^{\nu+} \gamma_-^{\nu-}. \qquad (30)$$

γ_+ and γ_- are the activity coefficients of the cation and the anion, respectively, and

$$\mu_k^\infty = \nu_+ \mu_+^\infty + \nu_- \mu_-^\infty \qquad (31)$$

where μ_+ and μ_- are the individual chemical potentials of the cation and the anion at infinite dilution, respectively.

Chemical potentials are important because they are related to the free-energy change ΔG of reactions. For example, let us look at the following reaction

$$v'_A A + v'_B B \rightarrow v'_C C + v'_D D \tag{32}$$

As a convention, the stoichiometric reaction coefficients v'_A and v'_B of the reactants A and B are equal to the stoichiometric coefficients v_A and v_B, respectively, whereas the reaction coefficients of the products have a negative sign: $v'_C = -v_C$ and $v'_D = -v_D$.

Let the chemical potentials of the compounds A to D be defined in a general way

$$\mu_i = \mu_i^0 + v_i RT \ln a_i. \tag{33}$$

For the reaction given in Eqn. 32, the free energy at constant pressure and temperature, $\Delta G_{p,T}$, is

$$\Delta G_{p,T} = \sum_{i=A}^{D} v'_i \mu_i^0 + RT \sum_{i=A}^{D} v'_i \ln a_i \tag{34}$$

With the definition of the *standard free-energy change* ΔG^0,

$$\Delta G_{p,T}^0 = \sum v'_i \mu_i^0 \tag{35}$$

expression (34) can be rewritten as

$$\Delta G_{p,T} = \Delta G_{p,T}^0 + RT \sum v'_i \ln a_i. \tag{36}$$

Thus, the $\Delta G_{p,T}$ of a reaction depends not only on the nature of the reactants and on their concentrations, but also on specific interactions which cause deviations from ideality. As far as ionic solutions are concerned, the solvation (in water called hydration) and the association of ions are phenomena which are particularly responsible for these deviations. Ions which are neither associated with other ions nor strongly solvated by the solvent molecules can be highly reactive. An example is given by the so-called phase transfer reactions [78]: strongly hydrated halide ions are "transported" from the aqueous phase into an organic solvent which does not strongly solvate these ions. In this solvent the halide ions are very reactive and are particularly appropriate for nucleophilic substitutions.

Once a reaction has come to equilibrium, $\Delta G_{p,T}$ vanishes and

$$\Delta G_{p,T}^0 = -RT \ln \prod_i [a_i]_{eq}^{v_i} = RT \ln K_{eq} \tag{37}$$

where $[a_i]_{eq}$ is the activity of component i *at equilibrium* and K_{eq} is the equilibrium constant for the reaction at constant temperature and pressure.

Instead of discussing the activity in terms of an activity coefficient it is sometimes more convenient to introduce an osmotic coefficient ϕ

$$\phi = -\frac{1}{M_w v_k m_k} \cdot \ln a_w \tag{38}$$

where M_w is the molecular weight of water.

The quantities ϕ and $\gamma_\pm$ are related by the Gibbs-Duhem equation

$$\ln \gamma_\pm = \phi - 1 + 2 \int_0^{\sqrt{m_k}} \frac{\phi - 1}{\sqrt{m_k}} d\sqrt{m_k}. \tag{39}$$

Both coefficients are appropriate to discuss ion-ion and ion-solvent interactions. The osmotic coefficient is more directly related to the activity of the solvent, $\gamma_\pm$ is more directly related to the salt activity. The choice of the thermodynamic coefficient depends on the experimental or theoretical method. Nevertheless, Eqn. 39 allows us to convert one coefficient to the other. In Fig. 18 the osmotic coefficients of some aqueous solutions are given as a function of concentration.

Roughly speaking, the osmotic coefficient can be understood as a measure of the amount of ion-solvent interactions. The following features influence the osmotic coefficient when measured as a function of concentration:

• When the concentration of ion pairs increases, the concentration of free ions in the solution decreases. Therefore the overall ion-solvent interaction becomes smaller and the osmotic coefficient has smaller values.
• When the hydration of the ions is strong, the osmotic coefficient has high values.

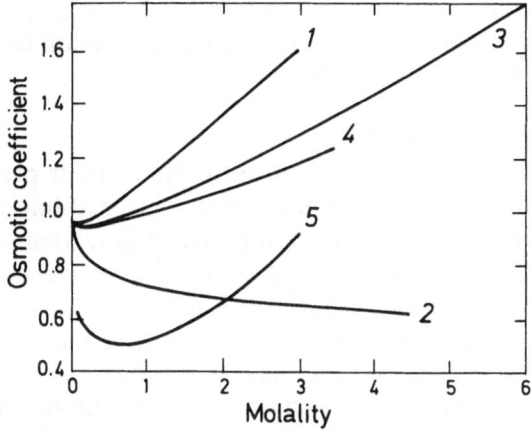

Fig. 18. Osmotic coefficients of various aqueous electrolyte solutions as a function of molality (moles of electrolyte per kilogram of solvent) at 25°C. (1) Tetramethylammonium fluoride, (2) tetrabutylammonium bromide, (3) lithium chloride, (4) sodium iodide, (5) magnesium sulfate. The parameters permitting the calculation of the given values are taken from [89].

- When the concentration of the ions increase, the osmotic coefficient also increases. This effect is due to an increasing volume occupied by the ions. For the same reason, solutions containing larger ions have higher osmotic coefficients than solutions containing small ions at the same concentration, provided that association and hydration, as discussed above, remain comparable.

Of course, these are only very rough and intuitive rules. Before discussing some simple but more quantitative models, we will show how to measure the activity of solution components.

3.2.2 Some experimental methods:

(a) Osmotic pressure [79]
The most direct way to determine an osmotic coefficient is to measure the osmotic pressure of a solution. The osmotic pressure Π is the pressure necessary to maintain an equilibrium across a membrane when the membrane is only permeable to the solvent. For aqueous solutions of macro-ions the membrane may also be permeable to small ions. The quantities ϕ and Π are related as follows

$$\phi = \frac{\bar{V}_w \Pi}{RTM_w \nu m_k} \tag{40}$$

where $\bar{V}_w$ is the partial molar volume of water. In practice, however, it is difficult to find membranes which are only permeable to water. Nevertheless, this technique was successfully applied for solutions of large biological molecules and polymers. For example Wittmann and Gros [80] measured the osmotic pressure of solutions of human haemoglobin under physiological conditions. They used these data to explain volume changes of red blood cells as a function of haemoglobin concentration and ionic strength.

(b) Vapour pressure studies
Instead of measuring the osmotic pressure, the vapour pressure of pure solvent (p*) and that of a solution (p) with a given solute concentration can be determined. The ratio between p and p* is related to the osmotic coefficient [81]. This technique gives an absolute value for the osmotic coefficients because no external standard is required.

Once precise data for one solution are known, they can be used as standards for other relative measurements. Among them, isopiestic distillation [82] is one of the techniques most commonly used. With this method the activity coefficients of several hundred single-salt solutions and aqueous mixtures of salts have been determined with high precision.

(c) Other techniques [83]

A great number of other techniques can be used to measure the activity of solution components. Whereas the methods quoted in (a) and (b) measure directly the activity of water, these methods yield information about the activity of the solute. Examples are the measurement of electromotive force (e.m.f.), and cryoscopic techniques or measurements of transport properties like electrical conductivity and diffusion.

3.2.3 Theories of the Debye-Hückel type: In 1923 Debye and Hückel (DH) [84] established a theory to evaluate activity coefficients. The theory is based on the combination of the Poisson equation (41)

$$\frac{1}{r^2}\frac{\partial}{\partial r}\left(r^2\frac{\partial\Psi(r)}{\partial r}\right) = -\frac{\rho_c(r)}{\varepsilon\varepsilon_0}, \tag{41}$$

with a Boltzmann distribution for the charge density $\rho_c(r)$ around a central ion (Eqn. 42):

$$\rho_c(r) = \sum_i^n \rho_i^0 z_i e_0 \exp\left(\frac{-z_i e\Psi(r)}{k_B T}\right) \tag{42}$$

Here $\Psi(r)$ is the interaction potential, ε and ε_0 are the permittivities of pure water and vacuum, respectively, e is the electron charge, n is the number of different types of ions, and ρ_i^0 is the average number density of ion i in the solution.

The combination of both equations, known as the Poisson-Boltzmann equation, can be solved with appropriate boundary conditions. Debye and Hückel further simplified their theory by expanding the exponential function appearing in (42) as a series and dropping quadratic and higher terms. The result is an analytical expression for the unknown interaction potential $\Psi(r)$. Only electrostatic interactions are considered for the derivation of $\Psi(r)$. Once $\Psi(r)$ is known, two important quantities can be calculated: the ionic distribution around a central ion (cf. Eqn. 42) and the activity coefficient of the ions,

$$\ln\gamma_\pm^{DH} = z_+ z_- A \sum m_i z_i^2, \tag{43}$$

where A is a function of the permittivity ε of pure water and of the temperature T. The activity coefficient $\gamma_\pm^{DH}$ takes into account the difference in free energy between a neutral particle and a charged particle in the presence of all the other ions around the charged particle under consideration. The distribution of the ions given by the corresponding distribution function $\rho_c(r)$ can be interpreted as an "ionic cloud" around the central ion. Due to several crude approximations in the derivation of $\Psi(r)$, the result is only valid for very dilute solutions. Experimental results confirm its validity in this concentration range.

Thus, more sophisticated theories must be in agreement with the DH theory for infinite dilution.

In section 3.2.1 we mentioned phenomena like association and hydration. Many attempts have been made to introduce these chemical concepts in the framework of the DH theory. For example, ionic association can be taken into account in using Bjerrum's model [85] or a related chemical model [86]. Comprehensive studies have been carried out [87] showing that one parameter set from chemical model calculations can correctly reproduce a variety of properties of the investigated electrolyte solutions. This parameter set may be used for the prediction of other properties at low and moderate concentrations. In the chemical model calculation, the finite size of the ions (the "excluded volume") is explicitly introduced.

Recently a new approach has been able to predict activity coefficients of simple electrolyte solutions for concentrations up to saturation without any adjustable parameter [88]. In this approach, also based on Poisson-Boltzmann equations, the excess chemical potential of an ion i is written

$$kT \ln \gamma_i = B_{ii} + B_{is} - B_{is}^0 \tag{44}$$

where B_{ii} is the interaction energy of the ionic atmosphere with the central ion i, B_{is}^0 is the solvation energy at infinite dilution, and B_{is} is the solvation energy for a given concentration. Ion-solvent interactions and hence "hydration" are explicitly taken into account as a function of concentration. In the framework of this theory the description of the molecular structure of the solvent requires a knowledge of the variation of the dielectric constant as a function of particle distance and salt concentration. Although based on a simple model, this theory illustrates the competition between ionic association and the hydration of isolated ions.

A different concept was introduced by Pitzer and his co-workers [89]. Semi-empirical equations based on simple chemical models were proposed. These equations are able to describe osmotic and activity coefficients over both a wide concentration and temperature range with few adjustable parameters. The advantage of this approach is that the activity and also the solubility of salts in complicated mixtures of many electrolytes in water can be predicted with high precision. Data for single salt solutions and few mixing parameters are required. Pitzer's equations have also been extended to other thermodynamic properties such as enthalpies, heat capacities, molar volumes, and compressibilities.

Approaches based on Poisson-Boltzmann equations are not restricted to solutions of spherical ions. They also play an important role in the description of charged linear polymers so as to mimic polyacids (DNA) or proteins [90, 91]. The strong electric field in the vicinity of the chain causes the small counter ions to "condense" at the surface of the macro-ion. In order to describe this case, the Poisson-Boltzmann equation has to be solved for a cylindrical symmetry. The Poisson-Boltzmann equation as well as a related model known as Manning's theory [92] can yield information about the environment, for example of the DNA chain: they predict that even at very low concentrations the polyacid chain is surrounded by molar concentrations of counter ions. Victor [93] speaks of a "biological trick" to trap ions along the DNA chain, where they are "biologically useful".

3.2.4 The energy of hydration and the hydrophobic effect: Let us consider the case where an aqueous solution of a salt is in contact with the pure salt lattice. Ions will go into solution up to saturation. When the dissolution has reached equilibrium, the chemical potential of the salt $\mu^0_{salt\ lattice}$ in the lattice is equal to the chemical potential of its ions in solution

$$\mu^0_{salt\ lattice} = \mu^{\infty(m)} + \nu RT \ln m_{ion,sol}\gamma_{\pm\ ion,sol}. \tag{45}$$

Provided that the molality and the activity coefficient of the salt in the solution are known, *the standard free energy of solution ΔG^0_{sol} can be obtained.*

$$\Delta G^0_{sol} = \mu^{\infty(m)} - \mu^0_{salt\ lattice} \tag{46}$$

The standard free energy of solution ΔG^0_{sol} must not be confused with the *standard free energy of solvation $\Delta G^0_{solvation}$.* The latter describes the energy change when isolated ions *in the gas phase* (g) are dissolved in water at infinite dilution. As an example we consider the reaction

$$H_2 + Cl_2 \rightarrow 2(H^+Cl^-)_{aq}$$

The standard free energy of the reaction $\Delta G^0_{reaction}$ can be obtained by an e.m.f. measurement. $\Delta G^0_{H_2}$ and $\Delta G^0_{Cl_2}$ are the standard free energies of dissociation of gaseous (g) H_2 and Cl_2 into radicals in the gas phase H_g and Cl_g, and I_{H^+} and e_{Cl} are the ionization energy of H and the electron affinity of Cl, respectively. H_g^+ and Cl_g^- are the ions in the gas phase. With these known values the free energy of solvation, $\Delta G^0_{solvation}$, of this particular process can be derived.

When the measurements are done at different temperatures, the entropy of solvation $\Delta S^0_{solvation}$ can be determined and also the enthalpy of solvation $\Delta H^0_{solvation}$ according to the following relations

$$\Delta S^0_{solvation} = \frac{\partial \Delta G^0_{solvation}}{\partial T} \qquad (47)$$

$$\Delta G^0_{solvation} = \Delta H^0_{solvation} - T\Delta S^0_{solvation}. \qquad (48)$$

For a detailed discussion of the energy and entropy of solvation, the reader is referred to an excellent monograph [77].

We will restrict our discussion to an effect which is of particular importance for biological systems. In order to explain this effect we consider the solution process of methane in water. The enthalpy of solution is slightly negative due to favourable interactions between the methane molecule and neighbouring water molecules. The entropy of solvation, however, is very negative and finally responsible for the fact that methane is poorly soluble in water (cf. Eqn. 48).

When considering an ion with hydrophobic parts comparable to methane, for example $N(CH_3)_4^+$, the salt is soluble, but only because the long-range charge-solvent effects dominate over the unfavourable solvent effect.

What happens when the concentration of hydrophobic particles in water is increased? In this case, these molecules have a tendency to associate in order to avoid contact with water molecules. This "passive" association is called the *hydrophobic effect* [94]. Several models are established to explain this mainly entropy-driven effect. One picture is summarized by Finney [69] as follows:

"Although there will be a weak interaction between a water molecule and an apolar (e.g. methyl) group, the strength of this interaction will generally be perhaps an order of magnitude less than a typical water-water hydrogen bond (about $4-5$ kcal mol^{-1}). Thus, for an ensemble of one apolar group and several water molecules, the overall energy is likely to be lower if the waters can arrange to hydrogen bond with each other rather than 'wasting' a hydrogen-bond site on an apolar group. This however, will have unfavourable entropic consequences in that the configurational space volume available to the water molecules will be reduced. The water molecules close to the solute are thus in some way restricted.

For an ensemble of two apolar solute molecules in a bath of water, this entropic penalty can be reduced if the two solute molecules come into contact. In doing so, some of the previously restricted waters will be released to the bulk solvent with a consequent entropic gain."

Although this picture is extremely simplified it has often been used to explain the structural stability of biomolecules like globular proteins.

Figure 19 [95] displays the scheme which is commonly used to explain the hydrophobic effect. Perhaps the most obvious evidence of the hydrophobic effect is the formation of micelles [96].

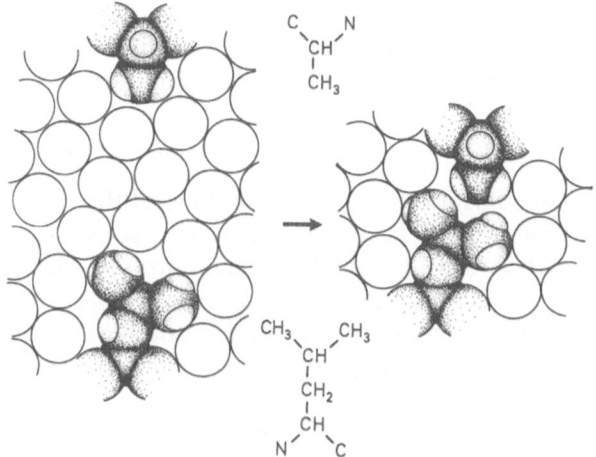

Fig. 19. Schematic representation of the hydrophobic effect. Two isolated side chains (alanine and leucine) come together in order to reduce the number of nearest neighbours. (The water molecules are only schematically outlined as circles.) The picture is adapted from [95].

However, it is difficult to estimate the contribution of hydrophobic effects in quantitative terms. In some cases a fine balance exists between "hydrophobic association" and the hydration of hydrophobic particles; this can lead to a structure where the ions are separated by a water molecule rather than building contact ion pairs [97, 98].

3.3 The local order around ions – theory, experiment, simulation

3.3.1 Statistical-mechanics theories:

Introduction. In the previous section the thermodynamic properties of aqueous solutions were briefly discussed. The relation with structural properties remained qualitative. In this section we will consider more elaborate theories based on statistical mechanics [99–101]. Two quantities are of fundamental interest:

(i) The particle-particle correlation function $g_{ij}(r)$
The radial pair correlation $g_{ij}(r)$ describes the probability of finding a particle j at a distance r from a particle i (cf. section 1.3). The probability is normalized by the average probability given by the number densities ρ_i and ρ_j of the particles i and j (i.e. the molar concentration multiplied by Avogadro's constant).

(ii) The interaction potential $u_{ij}(r)$
The interaction potential $u_{ij}(r)$ is the energy acting between two isolated particles i and j at a certain distance r.

A fundamental task of statistical mechanics is to find relations between $u_{ij}(r)$ and $g_{ij}(r)$ and also between these quantities and macroscopic quantities. Furthermore, convenient models for the direct interaction potentials $u_{ij}(r)$ must be established.

In the case of solutions, two levels of description can be distinguished:

- At the so-called McMillan-Mayer (MM) level the solvent is treated as a continuum with a certain dielectric constant ε. Only the interactions between the ions themselves are considered. In consequence the interaction potentials $u_{ij}(r)$ are replaced by solvent-averaged ion pair potentials at infinite dilution $\bar{u}_{ij}(r)$. Calculations at the MM level are normally restricted to the concentration range from 0 to 1 M.
- At the Born-Oppenheimer (BO) level the interactions ion-solvent and solvent-solvent are explicitly taken into account.

McMillan-Mayer level. The general relation between the ion-pair correlation function $g_{ij}(r)$ and the solvent-averaged ion pair potentials $\bar{u}_{ij}(r)$ is [99]

$$g_{ij}(r) = \exp(-\beta\bar{u}_{ij}(r) + Y_{ij}(r)). \qquad (49)$$

Within some approximations, this equation can be solved numerically together with the so-called Ornstein-Zernike equation, provided $\bar{u}_{ij}(r)$ is known.

The term $Y_{ij}(r)$ can be considered as an additional concentration-dependent potential taking into account the *indirect* interactions between two particles i and j via other solute particles. Although $Y_{ij}(r)$ is a very complicated function, it depends only on the temperature and sums of integrals over $\bar{u}_{ij}(r)$.

The quantity $\bar{u}_{ij}(r)$ is derived either from intuitive models or from BO calculations [102–104]. A current intuitive model is the following: the potential consists of the Coulombian potential, a core potential $\bar{u}_{ij}^{COR}$ so as to mimic the repulsion due to the finite size of the ions, and an additional term taking into account specific effects such as hydrophobic interactions, e.g. $\bar{u}_{ij}^{GUR}$ [105, 106]

$$\bar{u}_{ij}(r) = \frac{z_i z_j e^2}{4\pi\varepsilon\varepsilon_0} \cdot \frac{1}{r} + \bar{u}_{ij}^{COR} + \bar{u}_{ij}^{GUR} \qquad (50)$$

The so-called Gurney term $\bar{u}_{ij}^{GUR}$ describes the energy change when two hydrated ions come so close together that their hydration spheres overlap and thus part of the hydrating water molecules leaves the hydration shell back to the bulk. Hence, hydrophobic effects can be included within a rigorous statistical theory, although ion-solvent interactions are not explicitly considered [107]. Figure 20 illustrates the so-called Gurney process. An example of a calculated $g_{ij}(r)$ in an

Fig. 20. Schematic representation of the Gurney process. Two hydrated ions i and j come so close togather that their hydration shell overlap and thus part of the hydrating water molecules leave the hydration shell back to the bulk. Note the analogy to the hydrophobic effect displayed in Fig. 19.

aqueous tetraalkylammonium salt solution is given in Fig. 21. The peak in the $g_{++}(r)$ function can be interpreted in terms of a cation-cation association due to hydrophobic interactions. However, the drawback of intuitive models is that potentials are adjusted so as to reproduce thermodynamic data. The relation between $g_{ij}(r)$ and $\bar{u}_{ij}(r)$ on the one hand and the osmotic coefficient on the other is rather indirect and hence does not allow strict verification of the underlying potential model. As long as no reliable structural data are accessible, the result remains somewhat speculative.

The situation is slightly more encouraging when the potentials are derived from BO calculations. In this case they are the result of rigorous statistical mechanical calculations.

Recently some small angle neutron scattering experiments (cf. section 3.3.3) have been carried out on tetrabutylammonium bromide in water [108]. The experimental results differ from the picture given in Fig. 21. The peak of calculated $g_{++}(r)$ is overestimated. Nevertheless there is experimental evidence for the hydrophobic effect in this system: the cations overlap, which can be explained by a mutual penetration of the alkyl chains of neighbouring cations.

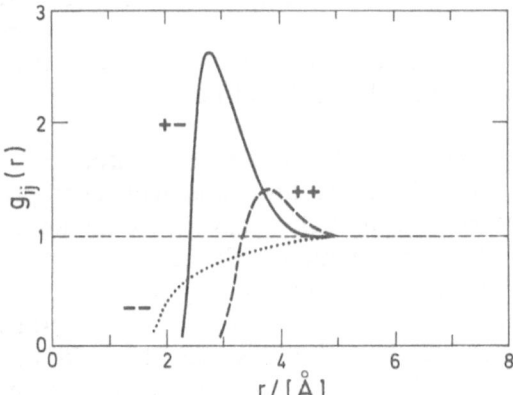

Fig. 21. Estimated ion-pair correlation functions $g_{ij}(r)$ of a 0.4 M tetraethylammonium bromide solution in water at 25°C. The results were obtained by the integral equation theory HNC in combination with a Gurney potential [106]. The only experimental reference was the osmotic coefficient. Neutron scattering results revealed a different cation-cation correlation function which is similar to the anion-anion (– –) correlation function [108].

The intrinsic problem for solutions of low-charged ions of small size is that Coulombian and non-Coulombian interactions are of the same order of magnitude. Therefore the osmotic coefficient is very sensitive to the balance between various attractive and repulsive forces. In this regard, it can be easier to handle strongly-charged polyelectrolytes, since there the Coulombian interactions are dominant. An example is the calculation of ionic distributions around DNA chains [109] in an aqueous mixture of NaCl and $MgCl_2$. The results relative to the competitive association of Na^+ and Mg^{2+} at the surface of the "rod-like polyion" DNA are in agreement with experimental structural information.

Born-Oppenheimer level. It would be beyond the scope of this review to give a detailed introduction to theories dealing explicitly with ion-solvent and solvent-solvent interactions. The reader is referred to the literature [110–119].

Roughly speaking, the techniques discussed above can be extended to the calculation of ion-solvent and solvent-solvent correlation functions. The interactions include dipole, quadrupole, and higher terms as well as angular-dependent correlations in order to approach the geometry of the molecules as realistically as possible. This leads to considerable mathematical and computational problems even for aqueous solutions of very simple ions. Although this technique is perhaps the most rigorous and therefore the most satisfying approach to ionic hydration, it has not yet reached a level where it allows precise structural predictions [117].

3.3.2 Computer simulations: Nowadays, progress in computer techniques such as the high speed of calculation and the increased available disk space allows models for electrolyte solutions to be treated by means of computer simulation methods. By analogy with the theoretical approaches, two levels may be distinguished.

Born-Oppenheimer level. At BO level, molecular dynamics (MD) simulations, yielding the space and time-dependent properties for an ensemble of interacting particles, have become an almost classical method. An excellent introduction to the basic principles of MD simulations on water and aqueous solutions is given by Bopp [120].

At BO level the molecular nature both of water and solute molecules is taken into account. This fact restricts the field of application of simulation methods, since only few particles can be considered with reasonable computing costs. That means that concentrated solutions are preferentially simulated because in this case the number of water molecules per ion is small.

Two further problems arise: the Coulombian forces of the ions are long-range. Several procedures like the Ewald summation have to be used in order to handle this problem correctly. These methods are very time-consuming and represent a considerable complication when compared to simulations of pure water. Another problem is the variety of interaction potentials proposed for pure water. Depending on the model, the structure obtained for the hydration of ions can be different. Experimental results are needed for comparison. As will be discussed in the next section, scattering experiments are most suitable for a direct comparison. In Fig. 22 weighted Cl^--D_2O radial correlation functions are given both from experiments and MD simulation.

In recent years MD has also been applied to simulate the hydration of biological macromolecules. Beginning with the work of Hagler and Moult [121] and van Gunsteren et al. [122] the link was established between simulations of macromolecules and aqueous solutions. Wong and McCammon [123] studied aqueous solutions of enzyme bovine trypsin, of the inhibitor benzamidine and a solution of the complex formed by these molecules. The number of water molecules was sufficient to mimic fairly dilute solutions. The authors simulated not only the influence of the water molecules on the structure and on the free energy of binding of the complelx, but also the perturbation of the solvent itself caused by the macromolecule. Wong and McCammon

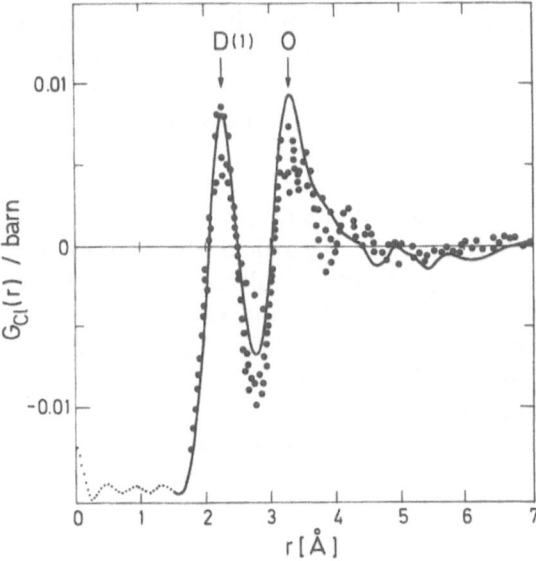

Fig. 22. An example of the agreement of a molecular dynamics simulation and neutron scattering experiments. The solid line represents the weighted total radial correlation function $\Delta G_{Cl^-}^{D_2O}$ (cf. Eqn. 51) obtained by a MD simulation of a 1.1 molal $MgCl_2$ solution, the points are results from different neutron scattering studies of various chloride solutions in deuterated water. The picture is adapted from [120].

believe ". . . that simulation studies have great potential as tools to help design new proteins and new ligands that have preselected activities".

For more information about MD simulations on biological systems in solutions the reader is referred to the literature [71, 124, 125].

In contrast to MD simulations, Monte-Carlo simulations (MC) [126] do not explicitly use the time variable; they are restricted to space-dependent properties. For the simulation of biological systems MC simulations are often used. Goodfellow [127] et al. studied the hydration shell of the crystallized vitamin B_{12} coenzyme. The atoms of the coenzyme were kept fixed and 56 water molecules were moved around the macromolecule. From the subsequent configurations of the water molecules after equilibrium of the whole system, the average properties can be calculated; for example, projections of the solvent networks are given in Fig. 23. The advantage of this particular system is that its structure is experimentally well-known. Therefore the parameters of the potential model of water can be adjusted in order to account for the experimental result. In this way, appropriate water potentials can be selected so as to predict hydration structures around other biological macromolecules.

McMillan-Mayer level. In principle, MD and MC simulations can be used at the MM level if the solvent molecules are not explicitly considered and the ion-ion potentials are solvent-averaged. By complete analogy with the theories at the MM level (section 3.3.1), the influence of the water molecules is partially introduced in the inter-ionic potentials.

At the MM level MD simulations provide information only on static structural properties, because the underlying Newton equation does not take into account the influence of the solvent on ion dynamics. In a so-called Brownian Dynamics (BD) simulation [128–130], the Newton equation is replaced by a Langevin equation in order to get reliable information about time-dependent properties of the solute particles. Since the BD approximation is particularly favourable when the solute particles are much bigger than the solvent, this simulation technique has also been used to study the diffusion of biological macromolecules in water [131]. Furthermore the Langevin equation allows simulations at longer time scales (> 1 ns) neglecting the motions of the water molecules which are much faster.

For a given solution, MD simulations at the MM level and BD simulations yield exactly the same structural results.

3.3.3 Neutron scattering experiments: Whereas in the preceding sections we discussed theoretical and computational methods, this section will deal with direct experimental evidence of ionic hydration structures. X-ray scattering studies on hydrated protein crystals are an example. However, this technique suffers from a relatively small scattering intensity and the fact that X-rays destroy biological systems within a short

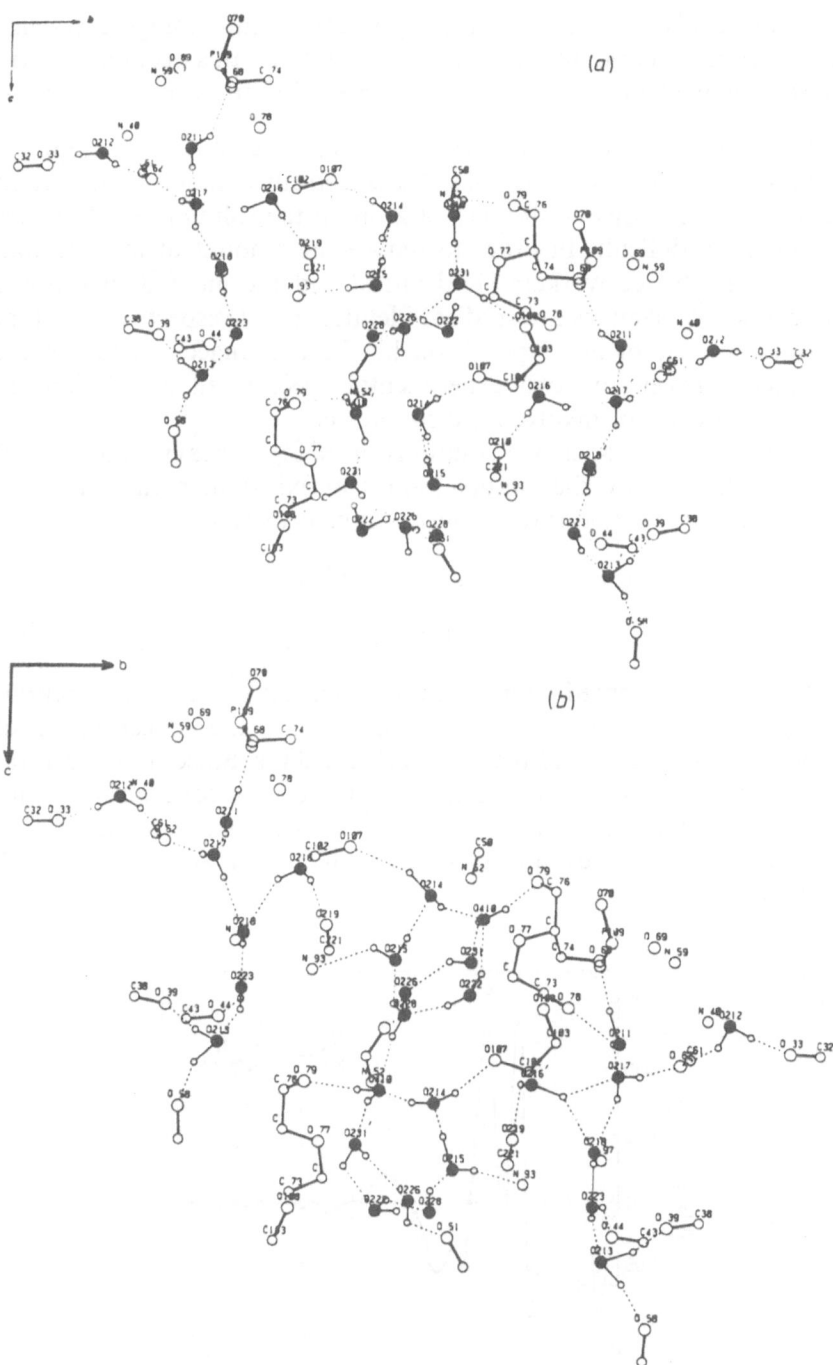

Fig. 23. Comparison of the solvent networks in vitamin B$_{12}$ coenzyme crystals. a) experimental data, b) results from a simulation. The pictures are reprinted from [127] with permission.

time. For details of X-ray scattering by solutions and biological systems the reader is referred to the literature [132]. In this section we will discuss only neutron scattering experiments [133] on aqueous solutions.

(a) Structural results on the microscopic scale (≤ 10 Å) [134, 135]
In the case of aqueous solutions, the spectrum contains information about the correlations of all types of atom in the solution which makes interpretation difficult. In order to reduce the amount of information. Enderby and his co-workers [134] introduced the method of isotopic differences: two samples are studied, identical in all respects except that the isotopic state of one type of ion has been changed. By taking the difference between the two experimental spectra, all correlations in which this ion is not involved are suppressed.

Example: We consider two solutions of $NiCl_2$ in heavy water (D_2O) which differ only by the isotopic states of Ni^{2+}; then the difference function can be written (after Fourier transformation) as follows

$$\Delta G_{Ni^{2+}}^{D_2O}(r) = A[g_{Ni\text{-}O}(r) - 1] + B[g_{Ni\text{-}D}(r) - 1]$$

$$+ C[g_{Ni\text{-}Cl}(r) - 1] + D[g_{Ni\text{-}Ni}(r) - 1] \qquad (51)$$

$\Delta G_{Ni^{2+}}^{D_2O}(r)$ is a weighted sum of the pair partial correlation functions $g_{a-b}(r)$. The factors A, B, C, and D depend on the concentration of the two atoms a and b and their scattering lengths. Since the concentration of water in the solution is dominant even for concentrated solutions, the terms involving the oxygen and the deuterium dominate over the other terms. Two examples are given in Fig. 22 and Fig. 24 [136], respectively.

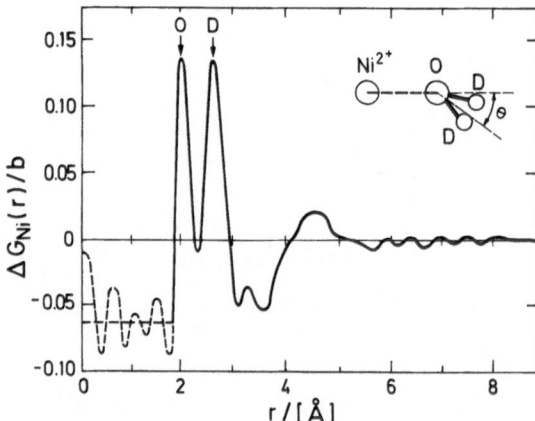

Fig. 24. The weighted correlation function $\Delta G_{Ni^{2+}}^{D_2O}$ (cf. Eqn. 51) for a 4.35 molal solution of $NiCl_2$ in D_2O and the geometry of the water molecule in the nickel hydration shell as derived from the correlation function. 1b (barn) = 10^{-28} m^2. The picture is adapted from [136].

Two important results can be obtained:

(i) The distances Ni^{2+}-O and Ni^{2+}-D which lead to the geometry of the hydration sphere (cf. Fig. 24).

(ii) The hydration number which is obtained by evaluating the integral over the correlation peaked Ni-D and Ni-O. This is the most direct way of defining hydration numbers. In the literature there are a great variety of hydration numbers for a given system. The method of isotopic difference helps to overcome some ambiguity.

The second order isotope difference method allows one to extract ion-ion correlation functions [136]. The results of the two methods, the hydration and the ionic distribution, give a detailed picture of the competition between solvation and ionic association. In this way, for example, the appearance of inner-sphere or outer-sphere complexes can be inferred.

(b) Small-angle neutron scattering (SANS) [137, 138]

Small-angle neutron scattering experiments cover only small Q ranges typically between 10^{-3} and $0.5 \, \text{Å}^{-1}$ (r-range $> 10 \, \text{Å}$). Therefore this technique is applied to the study of the structure of large molecules like polyelectrolytes or biological macromolecules. After normalization to a standard, the neutron intensity I(Q) scattered by a spherically symmetric macromolecule is given by

$$I(Q) = V \cdot F(Q) \cdot S(Q) \qquad (52)$$

The factor V depends on volumes and the scattering lengths of the macromolecule and of the solvent. When I(Q) is measured for solutions with different D_2O/H_2O compositions – the so-called "contrast variation method" – the following information can be derived: the partial molar volume of the macromolecule in solution and the number of protons exchanged between the macromolecule and the solvent. This exchange can only occur when two conditions are fulfilled: first, the protons must be attached to acid or basic groups in the molecule; second, the water molecules must come into contact with these groups. This is not possible when these groups are hidden on the inside of the molecule. Therefore proton-deuterium exchange gives an estimate of the percentage of water molecules which are in contact with groups like R-O-H or R_3-$\overset{+}{N}$-H [140].

The form factor F(Q) can deliver information about the shape and the size of the macromolecule. Usually one tries to approximate the overall shape of the macromolecule by a sphere or a cylinder but this approximation is often very crude. The radii obtained are sometimes in disagreement with the volume derived from the factor V. In Ref. 139 this difference has been attributed to the hydration shell of the macromolecule.

S(Q) is the structure factor. It reflects the interactions between the macromolecules. In contrast to F(Q), S(Q) depends on the concentration. It can be approximated by theories at the McMillan-Mayer level (cf. 3.3.1). For example, Wu and Chen [139] presented a simple electrostatic theory to describe S(Q) of a protein. As a result, quantities like the charge of the macromolecule can be derived.

3.3.4 Forces between surfaces in aqueous solutions: As discussed in the preceding section, neutron scattering techniques directly give an insight into solution structure. Force measurements between surfaces in water and in aqueous solutions are another example of methods which yield information about time-averaged particle interactions at the microscopic level [141]. Models both at the McMillan-Mayer and the Born-Oppenheimer level are used to interpret these forces. The results can help us to understand the hydration structure of ions as well as the subtle interactions between water, small ions, and macromolecules.

The experimental device. During the last twenty years the experimental techniques for measuring forces between surfaces have been considerably improved. Figure 25 shows an example of a surface force apparatus (SFA) which has been developed by Israelachvili and his co-workers [141–143]. The force is measured as a function of the distance between two smooth surfaces from the deflection of a spring which has a variable stiffness. The force sensitivity is about 10 nN and the distance between the surfaces can be varied and measured up to less than 1 Å. Different materials can be used for the surface. Mica which is smooth even at the molecular level is most often chosen for force measurements in aqueous electrolyte solutions [142].

The measured forces. Several types of force exist between surfaces:
(i) Van der Waals forces
Van der Waals forces operate between the solute and the solvent molecules as well as between two surfaces. At a very small surfaces separation they overbalance the other forces and lead to strong adhesion at contact. Van der Waals forces are always attractive.

(ii) Electrostatic forces
When the surfaces are equally charged, repulsive double-layer forces occur. The strength of these forces depends on the surface charge density. In an electrolyte solution these forces are screened. Their range and the steepness of their decay depend on the electrolyte concentration. The non-linear Poisson-Boltzmann equation is most often used to describe double-layer forces in aqueous electrolyte solutions. For a long time these two kinds of forces were the only ones that were considered and experimentally detected. The theoretical description of both forces

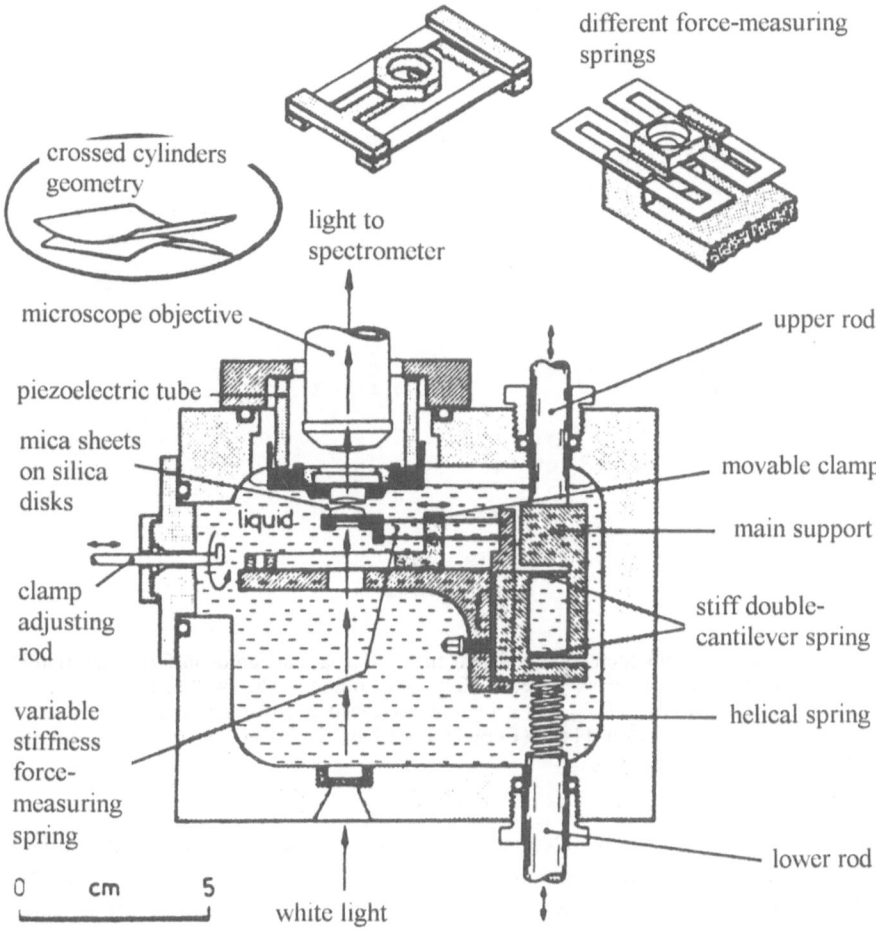

Fig. 25. Schematic representation of a surface force apparatus. The picture is reprinted from [141]. (Copyright 1988 by the American Association for the Advancement of Science.)

was combined in the "Derjaguin, Landau, Verwey, Overbeek" (DLVO) theory [144–146]. A schematic picture of the corresponding potential energy E of DLVO interactions is shown in Fig. 26.

(iii) Oscillatory or "hydration" forces

Usually, any deviation of the measured force from that predicted by the DLVO theory is attributed to so-called hydration forces. One of the most striking features of these deviations is their oscillatory behaviour measured at surface distances smaller than five or ten solvent diameters, cf. Fig. 27. At these distances the molecular aspect of water and of the solute particles appears. Furthermore the water molecules can be strongly oriented close to the surfaces. Therefore a proper description of

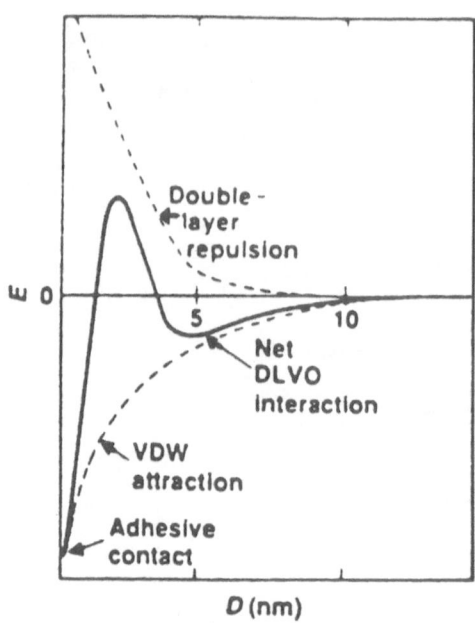

Fig. 26. Classical DLVO interaction potential energy E as a function of surface separation D between two flat surfaces in an aqueous electrolyte solution. The two interactions are an attractive van der Waals (VDW) and a repulsive screened electrostatic double-layer energy. The picture is reprinted from [141]. (Copyright 1988 by the American Association for the Advancement of Science.)

forces at such small distances requires models and simulations at the BO level, whereas at higher distances theories at the MM level are sufficient. The DLVO theory is essentially a MM level theory which in most cases correctly describes the long-range behaviour of the total force. The situation may be compared with the different techniques of neutron scattering measurements which have been discussed in the previous section. On the microscopic scale (3.3.3a) the results must be interpreted in terms of molecular ion-solvent correlations including the geometry of the water molecule, whereas results from small-angle neutron scattering can be interpreted using solvent-averaged models.

It should be noticed that the expression "hydration forces" is somewhat ambiguous. It is used here only to describe the oscillatory behaviour of forces at small surface distances, indicating a short-range structure.

A general introduction to the problem of hydration forces can be found in [146].

(iv) Hydrophilic and hydrophobic forces
Sometimes the oscillatory forces at small distances exhibit a smoothly varying component which can be either repulsive or attractive (Fig. 28).

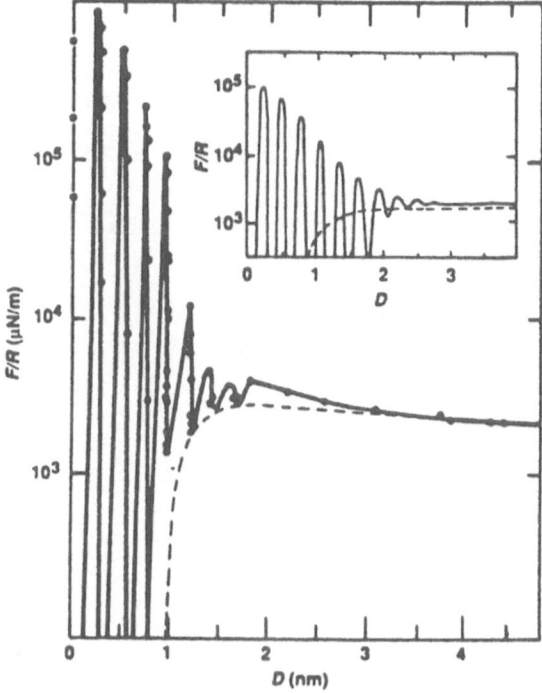

Fig. 27. Short range oscillations of a surface-surface interaction potential. The system is an aqueous solution of KCl (10^{-3} M) between two mica surfaces. The periodicity is about the diameter of a water molecule. The inset shows the result of a computer simulation. The picture is reprinted from [141]. (Copyright 1988 by the American Association for the Advancement of Science.)

This additional force is not yet fully understood. Usually it is ascribed to cooperative structural phenomena in the solution. Its long-range behaviour involving the arrangement of many water molecules makes a detailed MD simulation expensive. Therefore simple models have been proposed in order to give, at least, a qualitative explanation of the phenomena. One of these models is closely related to hydrophilic and hydrophobic effects (cf. section 3.2.4). A schematic description of these effects is given on the right of Fig. 28: close to hydrophilic surfaces the water molecules are almost perpendicular to the surfaces (Fig. 28A). This leads to an additional repulsive force when antiparallel water molecules of both layers come into contact at the midplane.

On the contrary, water molecules are almost parallel to hydrophobic surfaces (Fig. 28B), Then an additional attractive force is induced. Of course, this hydrophobic effect can occur not only between mica surfaces but also between macromolecules. An example is DNA double helices which in solution can be condensed into ordered arrays of parallel molecules by adding certain biological or small inorganic

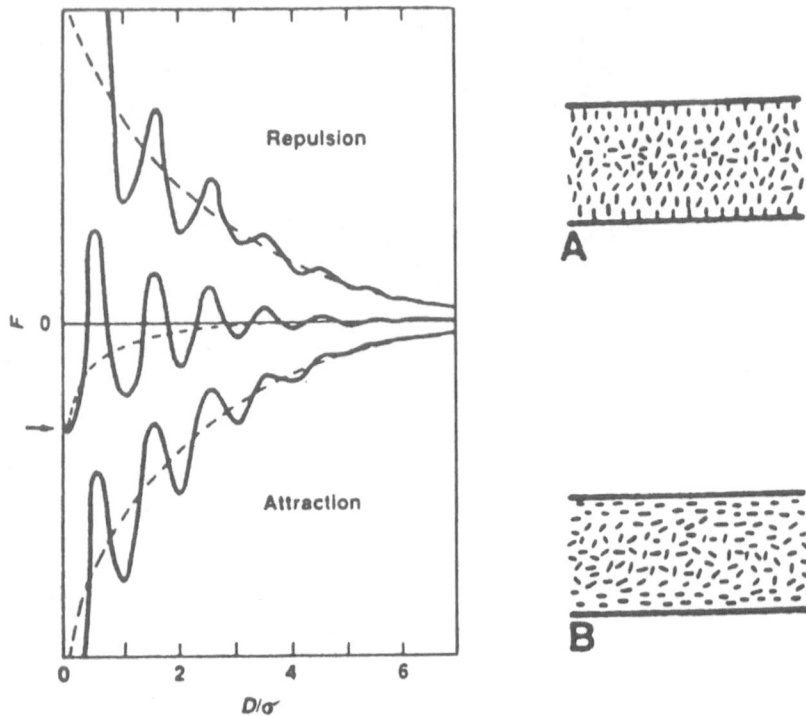

Fig. 28. Additional attractive or repulsive components corresponding to hydrophilic (A) and hydrophobic (B) interactions, respectively. The picture is reprinted from [141]. (Copyright 1988 by the American Association for the Advancement of Science.)

cations. These cations are supposed to be "adsorbed" at the macro-ion surface. Then, these patches of the macromolecule surface become more hydrophobic leading to an additional attractive interaction between the macromolecules [147].

(v) Ion-correlation forces

In some cases, for example in an aqueous solution of $CaCl_2$, several water layer oscillations are missing (Fig. 29) whereas in other solutions (for example KCl in water) all the oscillations can be observed (Fig. 27). Kjellander, Marčelja and their co-workers recently postulated that the lack of oscillation at certain surface distances can be ascribed to the distribution of the ions in solution [148]. They used the hypernetted chain (HNC) approximation in order to calculate ion-ion and surface-ion correlation functions at the MM level [149]. By contrast with Eqn. (49) and the correlation functions given in Fig. 21, anisotropic HNC calculations must be carried out owing to the presence of the surfaces. For the ions a charged hard-sphere model has been used. From the calculated solvent-averaged correlation functions an additional attractive force is

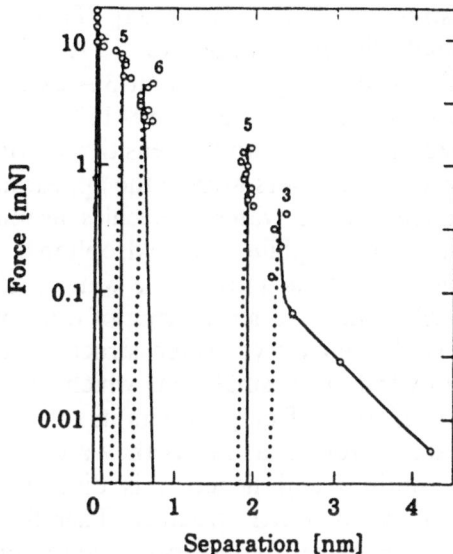

Fig. 29. Forces measured between mica surfaces in 0.15 M $CaCl_2$. The oscillatory structure exhibits a gap between 0.7 and 1.8 nm which is ascribed to ion-correlation forces. The picture is adapted from [148b] with permission.

inferred. This is in agreement with the attractive gap between 0.7 and 1.8 nm. (For a detailed discussion see Ref. 148.) However, it should be stated that other effects can crucially influence this result, such as the pH value of the solution, a preferential substitution of the hydrated calcium ions by hydronium ions between the surfaces or a specific adsorption of calcium ions at the surface. Nevertheless, the approximate calculation of the attractive forces due to ion-correlations of divalent ions is a significant advance. For instance, it can plausibly explain the limited swelling of calcium clays.

(vi) .In the preceding paragraphs the surface was assumed to be smooth and rigid. However, biological systems have a rather "rough" and fluctuating surface such as membranes made of lipid bilayers. Because of both the surface roughness and the thermal fluctuation of the hydrophilic headgroups no oscillating force can be detected at short distances. The force between the membranes is smeared out and only a monotonic repulsive hydrophilic force can be observed [141, 150–152].

It is difficult to separate the contribution of the roughness of the surface from that due to its fluidity. Furthermore both contributions can be dominated by ion-correlation forces or attractive forces due to some hydrophobic areas at the molecule surface.

(vii) Repulsive hydration forces
In the last twenty years Parsegian and his co-workers have investigated the forces between various types of biopolymers by means of the

so-called osmotic stress method [150, 153a]. They found a strong repulsive force at small distances (about 10–15 Å). This exponentially increasing force with a decay length of about 3 Å exists between DNA molecules [153b], between lipid bilayers [150, 153a], and even between polysaccharides [153c]. It is remarkably insensitive to ionic strength and surface charge density. The authors ascribe the appearance of this force to a collective arrangement of water molecules around biopolymers similar to the picture of hydrophilic and hydrophobic interactions (cf. Fig. 28A,B) [153d,e]. The interactions of the water molecules with the biopolymer at its surface induce a structuring of other water molecules in the vicinity. It is this collective water structure and its possible modification by the hydration of small ions which seem to be of great importance for the properties of biopolymers in solution.

However, the measured forces, as far as they are not oscillating, are essentially global quantities and hence it is difficult to relate them directly to detailed models of water structure. Therefore, spectroscopic and scattering measurements will be indispensable to obtain insight into this structure. Furthermore, we are still far away from understanding the relations between the water structure around biopolymers and the macroscopic properties, such as solution viscosity, the transport of solvent and ions through membranes, and the biological functions of biopolymers. Nevertheless, osmotic stress measurements have shown that neither a proper description of the interaction between few water molecules and the biopolymer nor the model of biopolymers in water considered merely as a continuous solvent are sufficient for a proper understanding of the structure, of the correlations, and hence of the function of biopolymers in solution.

3.4 The dynamics of water and solute particles

The intermolecular forces in aqueous solutions influence not only the structure and thermodynamics but, of course, also the movements of the solute and of the water molecules. In this section we will describe some experimental techniques which are appropriate for the study of either external (mobility) or internal (rotation, vibration) motions. Finally, we will give a short survey of the theoretical methods dealing with solution dynamics.

3.4.1 Radioactive tracer techniques: The so-called capillary method [154, 155] is the technique which is most often used for measuring diffusion coefficients. A capillary tube with known geometry is filled with a solution containing a radioactive tracer of the ion under consideration (tracer concentration C_0). The capillary is immersed in a container with an inactive solution. After a time t a certain concentration C_f

of the tracer can be detected in the container. From the ratio C_r/C_0 and the diffusion time t a diffusion coefficient D* can be derived.

In cases where the initial solutions in the container and in the capillary tube are identical in all respects except for the labelling of the tracer ions, the measured diffusion coefficient is the ionic self-diffusion coefficient, which is representative of Brownian motion. Normally, cations and anions have different self-diffusion coefficients.

On the other hand, when a concentration gradient between the capillary and the container exists, a so-called mutual diffusion occurs: only one global diffusion coefficient can be detected. This coefficient describes the correlated motion of the cation and the anion whatever the difference may be between the individual self-diffusion coefficients. The reason is the dynamic electroneutrality condition which prevents opposite charges from separation due to different velocities of the ions. As an example, a polyelectrolyte with a very slow polyion and fast counter ions will exhibit the same mutual diffusion coefficients if one determines the diffusion coefficient of either the polyion or of the counterions. For heparin solutions, the mutual diffusion coefficients of heparin and its counter ions Na^+ were determined independently by isotopic labelling and were found to be identical [155].

3.4.2 NMR measurements: Nuclear magnetic resonance (NMR) [156–160] is particularly suitable for studying the dynamics of aqueous solutions. It provides information both about internal and external motions.

Two basic quantities can be measured: the spin-lattice relaxation time T_1 and the self-diffusion coefficient D of the particle under consideration. The latter is usually measured in an NMR spin-echo experiment in the presence of a pulsed field gradient [161]. The basic restriction for the determination of D is the spin-spin relaxation time T_2 which must be sufficiently long. For a detailed introduction to the NMR technique the reader is referred to the literature.

With NMR measurements the self-diffusion coefficients of a variety of salts in water have been studied. Since NMR spin-echo is a slow-time-scale method, the measured diffusion coefficients are the same as those obtained by radioactive tracer techniques. The advantage of NMR is its sensitivity to atoms like H, D, and C which makes it possible to study organic molecules. No radioactive isotopes are needed. The precision of the self-diffusion coefficients deduced from NMR spin echo measurements has nearly reached the precision of tracer measurements.

As far as biological systems are concerned, NMR measurements are of particular use, for example for the study of the mobility of water inside a cell or of the permeability of water through a membrane [162, 163].

Whereas the relation between the measured decay of the spin echo amplitude and the diffusion coefficient is straightforward, models are needed to deduce the correlation times of the internal motions of the particles from the measured relaxation rates. These models may not be unique so that the result can be somewhat questionable. This is for example the case for interpretations of the dynamics of water close to macromolecules [69]. Nevertheless, qualitative interpretations are at least possible. As a simplified picture, the relaxation rates T_i^{-1} of water molecules in a solution of macromolecules can be approached by [159]

$$\frac{1}{T_i} = \frac{x}{T_{ib}} + \frac{1-x}{T_{if}} \quad \text{(where i = 1, 2)} \tag{53}$$

where x is the mole fraction of bound (b) water and $1-x$ is the mole fraction of free (f) water. Even if the quantity of bound water x is small, T_i can significantly deviate from T_{if} because T_{ib} can be reduced by several orders of magnitude. Furthermore, T_{ib} and T_{if} have a different frequency dependence. For example, the increase of T_1 of tumor tissues (compared to a normal tissue) can be explained by an additional amount of water which reduces the relative amount of bound water represented by x.

For additional information on NMR measurements of biological systems and a critical survey see Ref. 164.

3.4.3 Quasielastic neutron scattering (QENS): The basic principles of this technique as well as examples of its application to pure water are given in section 2.4. A general introduction can be found in [165]. Here we will briefly show some applications to aqueous solutions. Two cases may be distinguished.

(a) Water dynamics in solution. Some quasielastic neutron scattering measurements have been done in aqueous solutions [166, 167]. As in the case of pure water the motions are approximated by Lorentzians representing rotations and translations. The dependence on the concentration of the added salt yields information about the hydration of the ions. A particular example is a study of a concentrated $NiCl_2$ solution where protons bound to the ion are in slow exchange with the remaining water. Due to this slow exchange the translational motion of the water molecules in the first solvation shell is strongly coupled with that of the ion. This is reflected in a similar self-diffusion coefficient of hydration water and the ion, the latter being measured, for example, by tracer techniques.

(b) The dynamics of the ions in solution. Since this technique is mainly sensitive to hydrogen, hydrogenated organic ions can be studied in deuterated water. When either the cations or the anions are deuterated,

the motion of the other hydrogenated ions can be followed individually, for example in a solution of polysterene with tetramethylammonium ions as counter ions in deuterated water [168].

Finally we will point out two peculiarities of quasielastic neutron scattering when compared to NMR. First, the spectra are measured at different scattering angles. Therefore a Q-dependent "effective" diffusion coefficient $D(Q)$ can be inferred. Second, quasielastic neutron scattering is a technique revealing motions on the pico-second time-scale. Therefore, NMR and QENS may not give the same values for the diffusion coefficients. It may happen that some interactions are "turned on" after some nanoseconds so that the effect on the dynamics cannot be detected by QENS while it is seen by the NMR technique [169]. In this case, a combination of both techniques can help to separate these dynamical effects in solution because of their different time-scales.

3.4.4 Raman spectroscopy: As far as the structure of water in aqueous solutions is concerned, Raman spectroscopy is of particular interest. A Raman signal arises when vibrations due to the polarizability of molecules or molecular groups can be excited [170]. In the low-frequency region ($\Delta\hat{v} < 350\ cm^{-1}$) two broad bands at about 170 and $60\ cm^{-1}$ can be found. The band at about $170\ cm^{-1}$ can serve as a measurement of hydrogen-bonding since it is directly related to the formation of O-H $\cdots$ O-R bonds [171]. For example, an increase of temperature lowers the integrated intensity of this Raman band, because the concentration of intermolecular O-H bonds decreases.

In aqueous solutions the $170\ cm^{-1}$ band can also be used as a probe for the so-called *structure-making* and *structure-breaking* effects. When an ion is structure-breaking, the amount of hydrogen bonds O-H $\cdots$ O will be decreased and hence also the integrated intensity of the $170\ cm^{-1}$ band (normalized by the actual water concentration in the solution). An example is given in Fig. 30(a). The case of an aqueous solution of sucrose is slightly more difficult. Sucrose is a structure-maker. The increase of the $170\ cm^{-1}$ band is interpreted either as an increase in the number of hydrogen bonds between water molecules or as the formation of hydrogen bonds between water and solute particles (cf. Fig. 30(b)).

Further and more subtle information can be obtained from the OH-stretching region of the Raman spectrum [171]. A comparison between pure water and the sucrose solution shows that the increase of the O-H $\cdots$ O bonds is due to sucrose-water interactions rather than to the formation of new hydrogen bonds between water molecules.

3.4.5 Other techniques: In the previous sections we selected only four experimental techniques which are frequently used to determine the dynamical properties of solutions. Of course, a great number of other

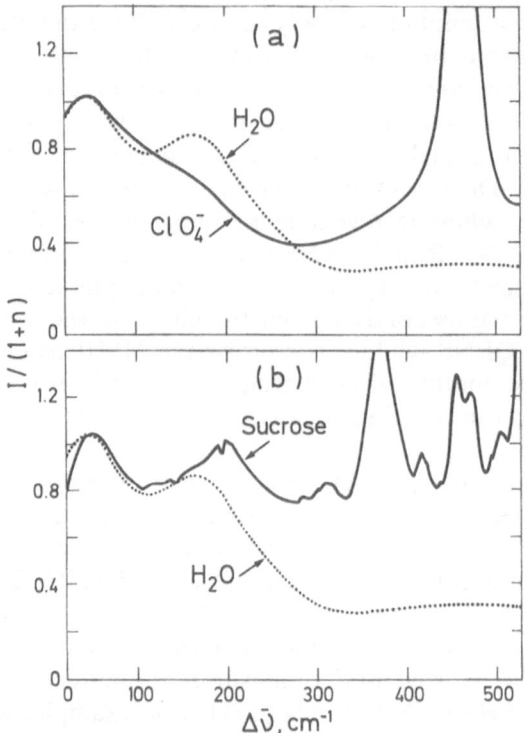

Fig. 30(a). Raman spectra from pure water and an aqeous 5.3 M NaClO₄ solution. The decrease in the intensity near 170 cm⁻¹ for the NaClO₄ solution is related to the breakdown of the water structure caused by the ClO₄⁻ anion. (b) Raman spectra from water and a 2.1 M solution of sucrose in water. The increase in the intensity near 170 cm⁻¹ for the sucrose solution comes from additional hydrogen bonds between sucrose and water molecules. The picture is adapted from [171].

techniques also contribute to our present knowledge of solution dynamics. For example, conductivity [172] and dielectric relaxation studies [173], quasielastic light scattering [174, 175], infrared [176], Mössbauer spectroscopy [177], and viscosity measurements [178]. For these techniques the reader is referred to the literature.

3.4.6 Theories: It would be beyond the scope of this section to give a detailed introduction to the current theories concerning the dynamics of water and solute molecules in solution. We restrict our survey mainly to an introduction to the current literature on this topic. As far as "simple" electrolyte solutions are concerned, the basic approaches to dynamics are summarized by Wolynes in an excellent review article [179]. In the last ten years the theories based on the McMillan-Mayer level of description have been extended to dynamical properties such as diffusion or electric conductivity [180–182]. The interpretation of the diffusive motion of polyelectrolytes is often done by means of the so-called normal-mode analysis [155], yielding analytical expressions for the observation of the system. As far as simulation techniques (molecular dynamics, Brownian dynamics) are concerned, the methods of calculation of dynamical properties are given in the references quoted in section 3.3.2.

As regards solutions of biomolecules, the situation is much more complex and little is known about theories on particle dynamics. An elegant approach is constituted by the work of Szabo et al. [184] who established diffusion-reaction equations for oxygen bonding to haemoglobin and myoglobin. A survey of the results obtained from computer simulations is given by Brooks and Karplus [183] (see also [125]).

4. The solvation of biomolecules

4.1 Introduction

The almost pure water considered in section 2 can only be obtained in the laboratory. However, it is difficult to keep it in this state since it is immediately susceptible to contamination by many different impurities. Section 3 dealt with the salt solutions of known composition.

In nature "water" is always an aqueous solution that contains many different kinds of solutes and . . . living organisms. Even in extreme conditions, such as low or high temperature or pH, high salinity or pressure, adapted living organisms can be found there [185]. On the other hand, dryness is not always lethal but always cancels metabolism and enzyme activities. Therefore water appears to be absolutely necessary to life. This should probably be ascribed to the exceptional physico-chemical properties of this rather unusual liquid whose molecules are hydrogen-bonded.

As a result, water is also the most abundant component of living matter. A cell, which is the basic structural and functional unit of living organisms, is made up of water, mineral ions and organic molecules such as nucleic acids, polypeptides, carbohydrates and lipids . . . However 70% or more of the total cell weight is water.

A cell is a tremendously complex system. In a prokaryotic cell the protoplasm contains a single chromosome and the cytoplasm where many ribosomes, various RNAs and proteins and other smaller molecules can be found. Eukaryotic cells are much more intricate. They have many chromosomes and a nucleus surrounded by the cytoplasm which is subdivided into many compartments by numerous membranes. Many different sorts of organelles are bound to this endomembrane system. Between this system and the plasma membrane lies the matrix

which contains ribosomes, microtubules, microfilaments, vesicles, vac-
uoles, soluble proteins and RNAs ... Owing to cell differentiation other
substructural parts may also be present. A schematic drawing of a
typical plant cell is shown in Fig. 31.

This brief description of cell morphology shows that there is a very
extensive boundary surface area separating intracellular water from the
other cell components. Studies of aqueous solutions of "simple" molecular components of the cell are very important to understand how these
hydrated molecules function. The purpose of this section is mainly to
describe briefly and explain how water and salts interact with proetins
and nucleic acids to form the bases of living matter.

4.2 Proteins

Proteins are biopolymers generally made up of 20 different kinds of
α-amino acids covalently linked by peptide bonds. In such a polypeptide
the amino acids are arranged in a particular sequence which is controlled by gene activity and depends on the protein function. This
primary structure leads to a particular three-dimensional conformation
for the native polypeptide chain. The secondary structure corresponds
to the formation of helical and sheetlike superstructures. For globular
proteins the polypeptide chain further folds into a very compact and
unique structure endowing it with a specific biological function. An
example is shown in Fig. 32. It should be emphasized that the primary

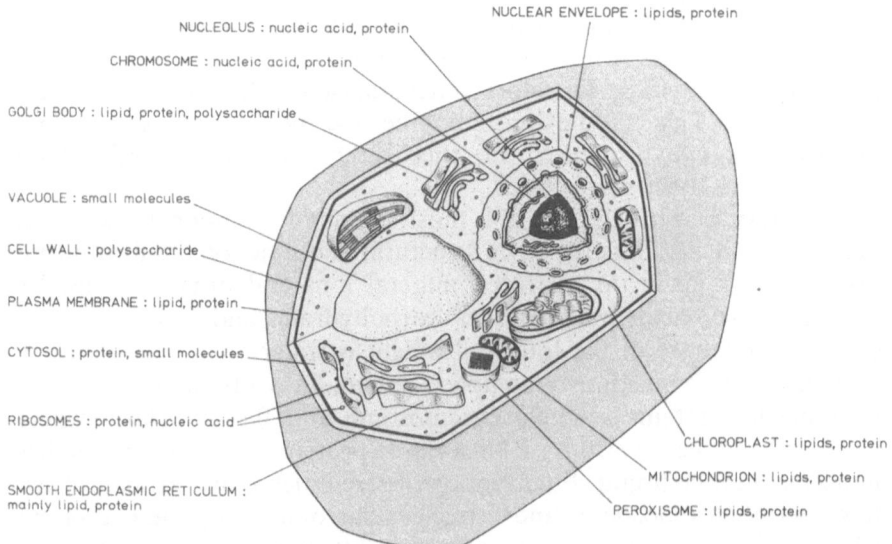

Fig. 31. Distribution of biomolecules in a typical plant cell.

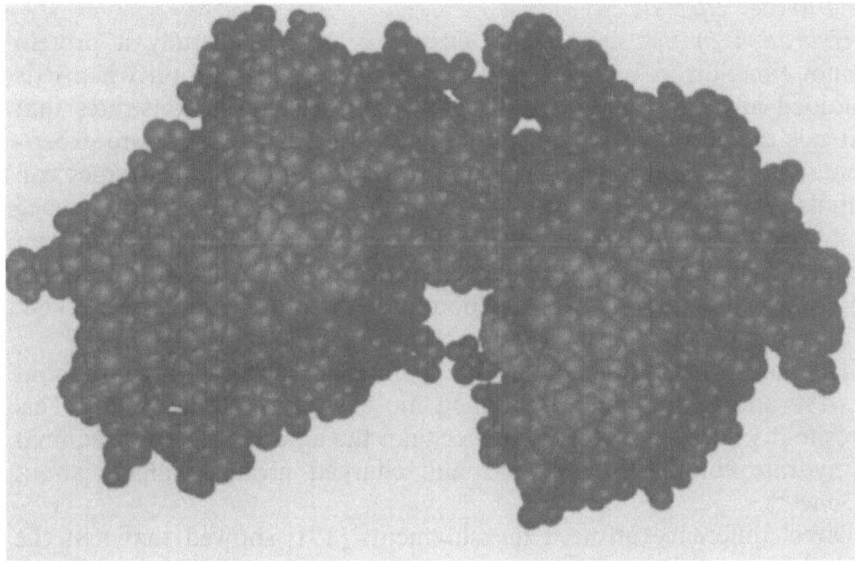

(a)

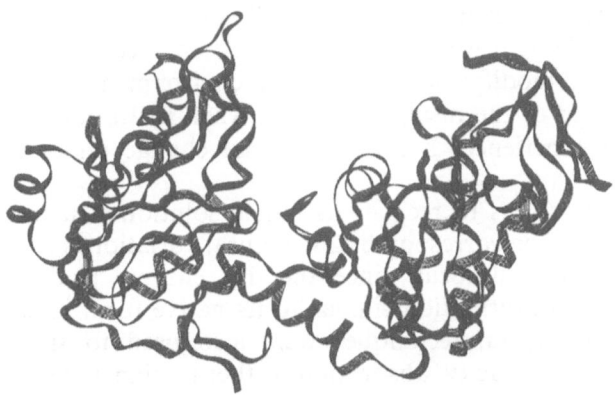

(b)

Fig. 32. The three dimensional structure of phosphoglycerate kinase. (a) A space-filling model. (b) A more schematic representation clearly showing helices and β-sheets. (By courtesy of M Desmadrie.)

structure carries all the information necessary for further folding. As a matter of fact, proteins that are denatured in such a way that they do not aggregate can refold into their native form when restored to physiological conditions [186–189]. This means that the overall free-energy change ΔG on folding is negative. Enzyme-substrate binding and protein association or aggregation also are exergonic. All these processes strongly depend on environmental conditions.

4.2.1 Water effects:

Hydration in the solid state: enzyme activity. Generally a protein cannot function in two different circumstances: in solution when it is unfolded and when it is dry. The last case clearly demonstrates that water is directly involved in physiological processes. Therefore experimental studies dealing with the progressive hydration of enzymes can help us to understand the role of water in restoring the activity. Hen egg white lysozyme was studied in this manner by means of many different techniques. Experiments were carried out with enzyme films or powders. To control the amount of water bound to them the samples were kept in an atmosphere of known relative humidity.

The onset of activity occurs when hydration reaches a value of about 20 wt% and further increases when more water is added [190]. This threshold value is significantly lower than the amount of water required to hydrate completely all polar and charged groups, namely about $\simeq 30$ wt%.

Direct difference infrared measurements [191] showed that first the acid groups ionize and return to their normal pK. This process is achieved at a hydration of about 10 wt%. At the onset of enzyme activity hydration is almost complete for the polar side chain and the amid NH groups but not for peptide CO ones.

Structural changes were followed by Raman spectroscopy [192]. The spectra showed modifications in the local environments of both a buried tryptophan and of the disulphide bonds when hydration increases. The α-helix also experiences a conformational change under these conditions.

The enzyme seems to recover its native solution structure before its activity begins. The corresponding structural variations can be regarded as rather small and local, since the dry enzyme probably keeps a compact state conformation similar to its native form in solution [69].

Other experimental techniques were also used to study lysozyme hydration. Heat capacity, absorption isotherm, diamagnetic susceptibility, and U.V. absorption measurements show that enzyme properties change on hydration [193].

The dynamics of the biomolecule have also been investigated using an electron spin resonance probe [190] and deuterium-hydrogen exchange investigated by nuclear magnetic resonance [192]. Dielectric and neutron inelastic scattering [191] measurements were also performed. All these experiments have shown that enzyme flexibility increases with increasing hydration. This change in flexibility allows many buried amide groups to become accessible to the solvent water molecules and occurs before the enzyme recovers its activity.

A similar experimental study of the homologous protein α-lactalbumin has given analogous results but with slightly different hydration levels [194].

Hydration in solution. Many different experimental methods have been employed to estimate the degree of protein hydration in solutions. However, the results strongly depend on the techniques used since they are concerned with different effects resulting from protein-solvent interactions. The degree of hydration is a relative measure of the amount of water in the protein solution which differs from bulk water because it is affected by the solute. Generally the degree of hydration is a thermodynamic quantity which has to be inferred from appropriate measurements.

The results of water binding at isopiestic equilibrium [195] give a typical protein hydration of about 30 ± 5 wt%. For lysozyme the value is about 25 wt%, close to the one corresponding to the onset of activity of the dry enzyme. These values vary with the solution pH and comparisons between different measurements are sometimes difficult.

The addition of a third component, either a salt or a cosolvent, to the solution can help to estimate protein hydration [196, 197]. In these conditions densitometry [198] or other experimental techniques [199] allow the preferential hydration coefficient of the biomolecule to be obtained. These methods can only give unambiguous values of both hydration and solvation if they remain constant when the concentration of the cosolvent or the added salt varies. Furthermore the additives must neither alter the conformation of the protein nor specifically bind to it.

Many different salts and cosolvents have been used in such experiments. The results obtained with stabilizing cosolvents or salts show that they are preferentially excluded from the protein hydration water, at least for sufficiently low concentrations [200] (cf. section 4.2.2). Hydration values can be estimated in this way [198].

Small-angle X-ray or neutron scattering experiments have also been performed [201, 202] because the forward scattering intensity gives results similar to those deduced from density measurements. In addition, these techniques also give values of the radius of gyration of the hydrated molecule. In this way it has been shown that glycerol does not enter the hydration water of ribonuclease, whose hydration was found to be 23 ± 5 wt% [202].

Small-angle neutron scattering (SANS) was also used to measure the amount of water and salt associated with a halophilic protein. In this case the third component of the solution was either NaCl, KCl, $MgCl_2$, or potassium phosphates. In addition, sedimentation measurements were also performed: these give information similar to that obtained from the neutron forward scattered intensity but contributions from the associated salt and water are weighted according to their molecular masses instead of their scattering lengths [203, 138]. This allows both the amount of water and of salt associated with the protein to be inferred from the data without any hypothesis regarding their possible dependence on salt concentration [138].

It has been found that the hydration of halophilic malate dehydrogenase is about 85 wt% in NaCl, KCl, and $MgCl_2$, and half of this for a pH 7.0 mixture of KH_2PO_4 and K_2HPO_4 [204]. This rather high hydration is typical of halophilic proteins.

Although comparisons of hydration values are difficult owing to both the different techniques and the solvent conditions used it seems that the hydration of non-halophilic proteins in solution approximately corresponds to the values measured for the onset of activity of a progressively hydrated dry enzyme. These hydration values are much lower than the ones required to cover the protein surface completely with a single monolayer of water molecules. They correspond rather to the almost full hydration of polar and charged groups of the polypeptide chain.

Protein folding. Protein folding in aqueous solvents has been studied by calorimetry. Orders of magnitude for the enthalpy and entropy changes are $\Delta H \simeq -200 \text{ kJ/mol}$ and $\Delta S \simeq -500 \text{ J/(K mol)}$, respectively. Consequently, the corresponding variation of the free energy $\Delta G \simeq -50 \text{ kJ/mol}$ is relatively small and corresponds to the energy of about three hydrogen bonds. It should be noted that the previous values depend on the protein investigated and on the solvent conditions [205], but in any case ΔG is small while both ΔH and ΔS are large.

Contributions to ΔH mainly arise from electrostatic and van der Waals interactions and internal hydrogen-bonding. The decrease in entropy results from the polypeptide chain folding, but it is partially balanced by an entropy increase due to water interacting with the protein molecule. These features are to be ascribed to hydrophobic effects [206] resulting from the particular properties of apolar solutes.

Electrostatic interactions occur between two side-chain groups carrying opposite charges. If these groups are close enough they can form salt bridges within the protein. In addition globular proteins are very compact. As a result the relatively weak van der Waals interactions between molecular groups can also make a significant contribution to ΔH.

A polypeptide chain has many sites capable of forming hydrogen bonds. In the unfolded state most of these bonds probably occur between the polar residues of the protein molecule and its hydration water. In the native state only about 40–50% of these polar sites remain bound to water while almost all the other ones are involved in internal hydrogen bonds. The corresponding enthalpy change comes from the remaining polar sites which cannot form hydrogen bonds, and the differences in strength between the different kinds of hydrogen bonds occurring in the system. The first contribution clearly works against folding, while the sign and the magnitude of the second are difficult to estimate since they result from the differences between large and nearly identical quantities [207].

Apolar groups tend to induce ordering of the water molecules surrounding them, reducing the randomness of the solvent [208]. When a polypeptide chain folds, some of its hydrophobic residues become buried within it. Then the number of solvent molecules ordered by the unfolded protein decreases, leading to a large entropy increase for the system. This entropy increase on folding will stabilize the folded state of the protein since the corresponding enthalpy variation is very small.

It should be emphasized that a folded polypeptide chain cannot be regarded as having almost all its apolar groups shielded from solvent. Crystallographic structures of small proteins have shown that about 50% of the surface area accessible to the solvent is occupied by apolar residues [209, 210]. In addition the mean fractional area loss on protein folding has been estimated. While lower for polar residues than for apolar ones, it ranges from 62% to 91% [211] as shown in Fig. 33.

There is some evidence from both theory [97] and Monte Carlo simulation [98, 212] that two apolar particles can be either in close contact or separated by a distance allowing a water molecule to sit between them. The high-resolution neutron and X-ray crystal structures of coenzyme B_{12} show that methyl groups can be separated by a water molecule [132], as already discussed in sections 3.2.4 and 3.3.2.

All these results suggest that in native proteins both polar and apolar amino-acid residues can be accessible to water molecules. Furthermore conformational fluctuations of the polypeptide chain will increase this accessibility. As a matter of fact 80–90% of the hydrogen atoms involved in both hydrogen and amide bonds exchange with deuterium atoms when proteins are dissolved in heavy water.

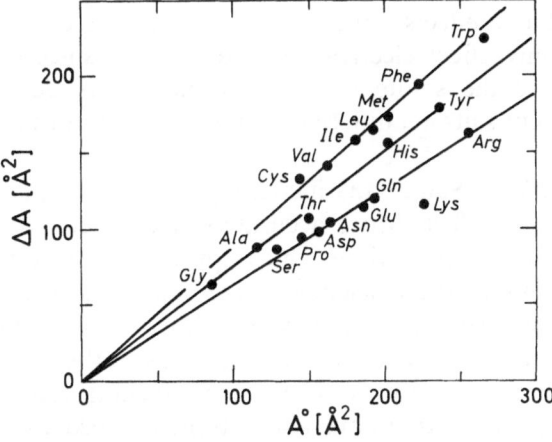

Fig. 33. Average area buried upon folding, ΔA, versus the standard area, A°, for all twenty residues. For each residue, the mean fractional area loss upon folding is given by the slope of the straight line passing by the corresponding point and the origin. (Data from Ref. [211].)

Protein interactions. Water effects on either ligand or protein associa-tion with proteins can be described in a way similar to those on protein-folding stability [69].

Typical orders of magnitude of ΔS, ΔH, and ΔG for protein-ligand association are about two to three times less than for protein folding [213]. These values are strongly dependent on both the protein and the ligand natures and on solvent conditions. The free-energy changes are either lower or slightly higher than the enthalpy and entropy contribu-tions [213]. Generally the conformational entropy change of the system works against protein-ligand association.

For protein-protein aggregation calorimetric measurements show that the free energy change ΔG is also small, approximately ranging from -15 to -60 kJ/mol. However, the corresponding entropy and enthalpy changes can be either positive or negative and their absolute values very different [213]. Again, these results strongly depend on both the system studied and the experimental conditions. This suggests that there is no universal explanation of protein association.

4.2.2 Salt and cosolvent effects: Proteins are polyampholytes: they carry on their surface patches of positive and negative charges. At the isoelectric pH proteins have no net charge but for other pH values they are charged positively or negatively and therefore repel each other. Then they are more soluble than at their isoelectric point where they may tend to aggregate because of attractions between their oppositely charged surface regions.

Such long-range electrostatic interactions can be strongly changed by adding salt to the solution. Then the charged patches become sur-rounded by a counter-ion atmosphere which reduces the range of such interactions. These effects have been discussed in section 3. However, salts do not only affect electrostatic interactions, since some of them tend to increase the solute solubility while others do the contrary. Similarly different salts act differently on protein conformation stability.

Protein solubility. Some salts increase protein solubility while others have the opposite effect. More accurately this effect is to be ascribed separately to anions and cations which approximately add their respec-tive contributions to the efficiency of the salt. Ions which increase or decrease the solute solubility are termed salting-in or salting-out agents, respectively. In order of effectiveness in salting-out the anions follow the Hofmeister series inferred from the experimental study of globulin solubility [214]. For both anions and cations this order of effectiveness is given in Table 4.

Solubilities of simple model compounds have also been studied such as hydrocarbons [214, 215] or complexes of apolar residues and peptide

Table 4. Relative effectiveness of various ions in stabilizing or destabilizing the native form of collagen, ribonuclease, and DNA (from [219])

	Stabilizing or salting-out	Destabilizing or salting-in
Collagen-Gelatin	$SO_4^{2-} < CH_3COO^- < Cl^- < Br^- < NO_3^- < ClO_4^- < I^- < CNS^-$ $(CH_3)_4N^+ < NH_4^+ < Rb^+, K^+, Na^+, Cs^+ < Li^+ < Mg^{2+} < Ca^{2+} < Ba^{2+}$ $(CH_3)_4N^+ < (C_2H_5)_4N^+ \lll (C_3H_7)_4N^+, (C_4H_9)_4N^+$	
Ribonuclease	$SO_4^{2-} < CH_3COO^- < Cl^- < Br^- < ClO_4^- < CNS^-$ $(CH_3)_4N^+, NH_4^+, K^+, Na^+ < Li^+ < Ca^{2+}$ $(CH_3)_4N^+ < (C_2H_5)_4N^+ < (C_3H_7)_4N^+, (C_4H_9)_4N^+$	
DNA	$Cl^-, Br^- < CH_3COO^- < I^- < ClO^- < CNS^-$ $(CH_3)_4N^+ < K^+ < Na^+ < Li^+$	

groups [216, 217]. It has been found that the experimental results can be represented by the empirical Setchenow [214] equation

$$\log \frac{S_0}{S} = K_S C_S, \qquad (54)$$

where S_0 is the compound solubility in pure water and S the actual solubility in the presence of the added salt. C_S is the salt molarity and K_S the salting-out coefficient which is positive for salting-out agents.

The relative effectiveness of various ions as salting-out agents follows the series of Table 4, but there are significant differences in the strength of the effect so that salts which have nearly no effect on protein solubility appear to be effective salting-out agents for these model compounds.

From such experiments it has been inferred that almost all salts salt-out apolar groups but generally salt-in amide groups [216]. As a result the potency of such an agent to salt-in or -out a protein will depend on the ratio of apolar to polar groups accessible to it.

Protein stability. The effects of different salts on protein stability has also been studied extensively [218, 219]. To this end the average denaturation temperatures of numerous globular proteins have been measured as a function of the salt concentration for many different salts. The main result of these studies is that salting-out ions tend to stabilize the polypeptide chain conformation while salting-in ones are potent destabilizers. Examples are given in Fig. 34. The specific effects of the different ions primarily result from their ability to change the solubility of apolar amino acids [219] by altering the balance between solvophobic and solvophilic interactions.

As a matter of fact, stabilization of protein structures can also be induced by cosolvents. A variety of proteins with a broad range of

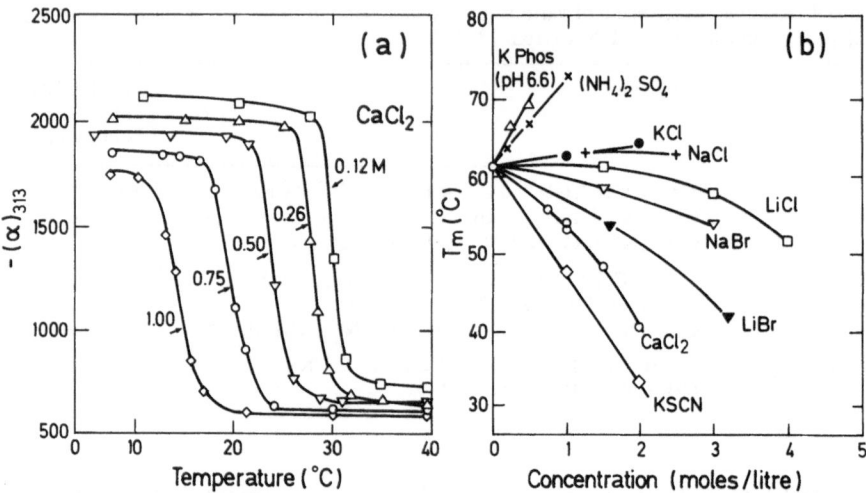

Fig. 34. (a) Melting curves for collagen in solutions containing various concentrations of CaCl$_2$. (b) Average melting temperatures T$_m$ for ribonuclease as a function of concentration of different added salts. These data were inferred from optical rotatory measurements. (Adapted from Ref. [218] and [219].)

specific surface, polarity, and structural hydrophobicity have been studied in different stabilizing or destabilizing solvents [220, 221]. The preferential interaction parameters have been inferred from densitometry or differential refractometry carried out on solutions which were kept in two different thermodynamic equilibrium states [198].

A general result emerged from these studies: whatever a salt or a cosolvent, all solvent stabilizing additives are preferentially excluded from the vicinity of the protein [200, 222] which therefore is preferentially hydrated. This means that adding protein or a stabilizing agent to the solution increases their respective chemical potentials. Such a situation is thermodynamically unfavourable and induces protein stabilization or salting-out in order to minimize the total contact surface area between the polypeptide and the mixed solvent. Preferential hydration of proteins in stabilizing solvents can be simply ascribed to unfavourable interactions of the stabilizing additives with the protein surface, or to changes of the solvent properties induced by the salt or the cosolvent [222]. For instance protein stabilization in aqueous sucrose solvents is related to their surface tension [200]. Similar effects occur with monovalent salts [223].

Similarly the converse is valid for structure destabilizing solvents. All the denaturants added to the solutions have been found to be preferentially bound to the unfolded polypeptide chain. For instance guanidinium hydrochloride [224] and urea [225] interact with peptide groups and hydrophobic side chains [226]. Denaturing alcohols [227] and detergents [228, 229] bind to apolar groups.

It has been shown that the important guanidinium cation behaves as others do [230]. The actual effect of a salt on a protein results from a balance between specific ion binding and hydration which reflects non-specific interaction of the protein surface with the salt.

4.3 Nucleic acids

Deoxyribonucleic acid (DNA) and ribonucleic acid (RNA) are the two kinds of nucleic acid. These are heteropolymers made up of nucleotides which contain a cyclic sugar, a phosphate residue, and a base. The backbone of the nucleic acid molecule is formed by the sugar moieties connected by phosphodiester links. Along the backbone genetic information is carried by the sequence of bases. In DNA the major bases are the purines adenine (A) and guanine (G) and the pyridines cytosine (C) and thymine (T), while in RNA uracil (U) replaces thymine.

Except for some RNA's which are globular the secondary structure of nucleic acids is helical. Such a helix can be defined by the "rise" h and the "twist" t, which are the axial distance and the rotation angle about the axis between two neighbour nucleotides, respectively. In a double-helix arrangement the two strands are connected by hydrogen bonds between complementary bases: A always pairs with T or U, and G always with C. Base pairs are usually centred off the helix axis. They are inclined about it with a "tilt" angle Θ_T and a "roll" angle Θ_R. Furthermore, if the two bases of a pair are not coplanar they have a "propeller twist" angle Θ_P about the roll axis. Typical values of these double-helix parameters will be given later in Table 5.

Generally DNA is double-stranded and RNA single-stranded. However single-stranded DNAs and double-stranded RNAs can be found in some viruses. In addition some nucleic acid samples can be triple- [231–234] or quadruple-stranded [235].

As for proteins, water and salts are responsible for structure stabilization of nucleic acids. Hydrophobic and hydrophilic effects are also

Table 5. Helix parameters from single crystal X-ray diffraction adapted from [242]

Parameter	A-DNA	B-DNA	Z-DNA
h (Å)	$2.92 \pm .39$	$3.36 \pm .42$	G-C: $3.52 \pm .22$ C-G: $4.13 \pm .18$
t (°)	33.1 ± 5.9	35.9 ± 4.3	G-C: -51.3 ± 1.6 C-G: -8.5 ± 1.1
Θ_R (°)	5.9 ± 4.7	-1.0 ± 5.5	3.4 ± 2.1
Θ_T (°)	13.0 ± 1.9	-2.0 ± 4.6	8.8 ± 0.7
Θ_P (°)	15.4 ± 6.2	11.7 ± 4.8	4.4 ± 2.8

involved. In addition the electrostatic repulsion between the phosphate groups can be screened by counter ions.

4.3.1 Water effects:

Structure and hydration in the solid state. Because usual preparations of native double-helical polymeric nucleic acids are not homogeneous it is not possible to obtain single crystals suitable for structure determination. As a consequence early X-ray studies of DNA were carried out on fibres [236, 237] drawn from gels produced by precipitation of the nucleic acid by the addition of alcohol. These fibres were further stretched and dried to orient them as much as possible. During the recording of the diffraction pattern the fibre was kept in an atmosphere of controlled humidity. Such samples can only give limited information because, in addition to inhomogeneous sequences of the molecules, the axial orientation of the fibre crystallites is not perfect. Especially their rotational orientation is nearly random. As a result an adjustable model has to be chosen to describe the diffraction pattern. However the choice of such a model is not unequivocal. This means that a wrong model cannot be ruled out. An interesting example concerns the double-helical DNA model of Watson and Crick [238–243]. Nevertheless three main different models have been inferred from X-ray diffraction studies of DNA fibres [240]. These are schematically shown in Fig. 35.

- A-DNA (Fig. 35-a): its bases are strongly tilted about the helix axis and lie well off it. The diameter of the molecule is about 2.3 nm. The major groove is deep and narrow while the minor one is wide and shallow.
- B-DNA (Fig. 35-b): its diameter is about 1.9 nm. The grooves are shallow but with different widths, the major groove being wider than the other. The bases are nearly perpendicular to the helix axis.
- Z-DNA (Fig. 35-c): unlike the two other forms, this is a left-hand helix with a backbone following a zigzag path. The phosphate residues are alternately close and far apart so that the repeating unit is a dinucleotide. This DNA forms seems to be present only in molecules containing alternate purine and pyrimidine bases.

The occurrence of these different families depends on preparation conditions. In fibres A-DNA is observed at low relative humidity while B-DNA is favoured at higher humidity. It should be stressed that within the classical DNA forms previously described wide variations of conformation are possible [241–244], depending on the base composition, the ionic conditions, and the degree of hydration. This is reflected by the standard deviations for the helix parameters listed in Table 5. On the other hand, double-helical RNA, either natural or synthetic, remains in the A-form. Hydration of DNA fibres depends on their environmental relative humidity. Above 80% humidity, the DNA double helix is

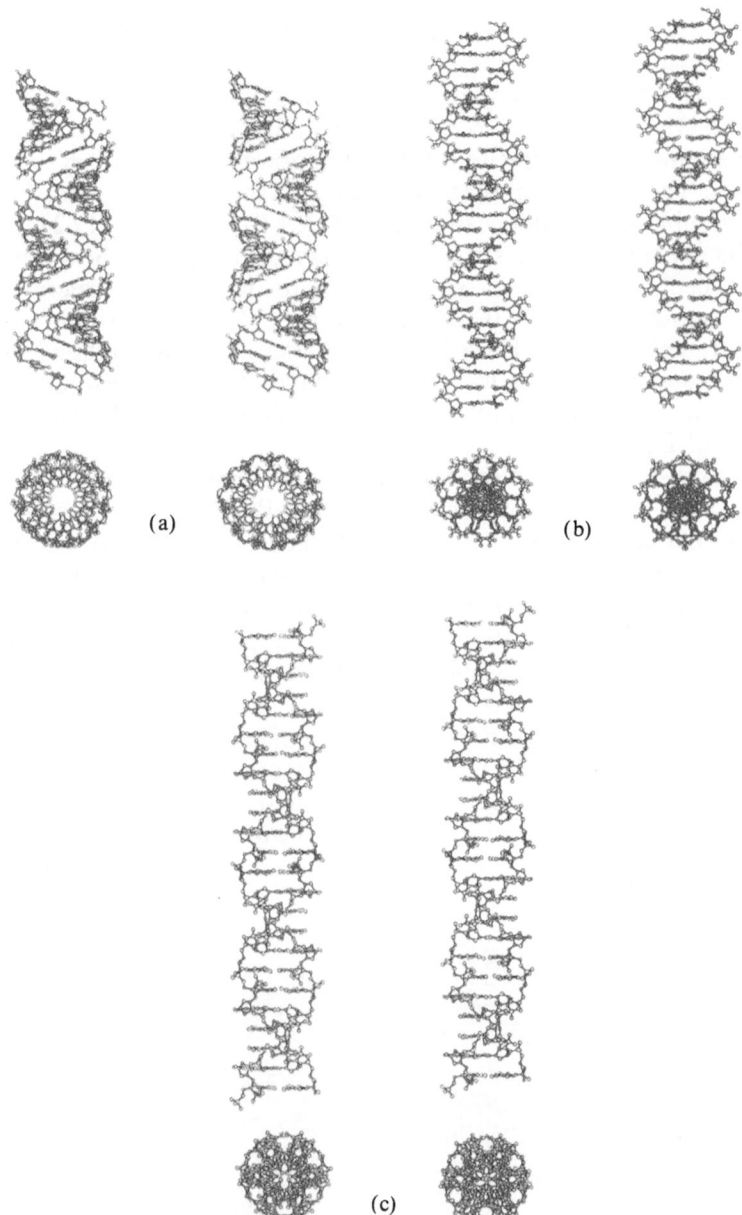

Fig. 35. The helical structures of DNA (stereoscopic views). (a) A form, (b) B form, and (c) Z form. (From Ref. [240], copyright 1981 Adenine Press.)

completely hydrated with about 20 water molecules per nucleotide. Then DNA is in the B-form. At a relative humidity lower than about 65% the bases are no longer hydrated and each nucleotide is then bound to 11 to 12 water molecules. At lower humidity only the more polar phosphate oxygens remain hydrated with about 3.6 water molecules per nucleotide. This decrease in overall hydration induces the B → A transition.

A major advance was made when it became possible to crystallize synthetic DNA fragments. Molecular crystallography of such fragments has given much more precise information about the secondary structure of DNA. The most comprehensive study of this kind concerns crystals of the double-strand dodecamer CGCGAATTCGCG [241, 242, 245–250]. Its crystallographic structure, displayed in Fig. 36, shows that the helix is bent and that "no single structure or class of structures can define this molecule" [243] for which very large local variations in the helix parameters occur. These local properties are mainly related to base sequence and sometimes to molecular environment. Helix parameters from X-ray crystallography are given in Table 5 for the three main families of DNA [242].

Single crystal X-ray diffraction has also been able to prove how DNA is hydrated. A high-resolution study of the B-form dodecamer showed that approximately every accessible hydrogen-bond donor or acceptor is hydrated [241, 242, 248–250]. In particular the minor groove can accommodate a spine of water molecules in its A-T rich portions. This hydration feature may stabilize the B-form under high relative humidity [248]. Such a hydration spine has not yet been observed for A-form molecules. In the only A-form crystal so far studied the minor groove is not accessible because of intermolecular packing [251, 252]. Nevertheless, in contrast with B-DNA, the sugar and base moieties are not hydrated while water molecules show a well-ordered structure between phosphate groups. These differences are shown in Fig. 37.

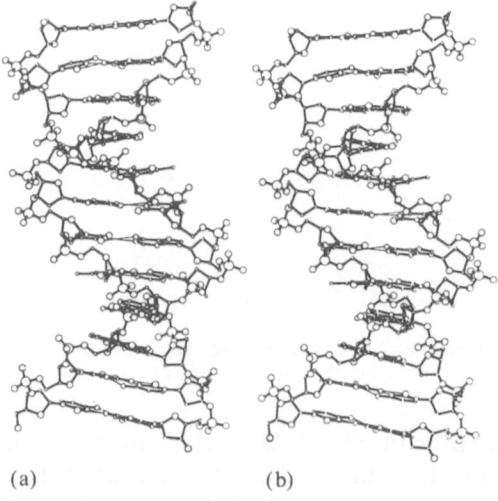

(a)   (b)

Fig. 36. Three-dimensional structure of the double-strand dodecamer CGCGAATTCGCG. (Reprinted from Ref. [250] with permission. Copyright 1981, Academic Press, Inc.).

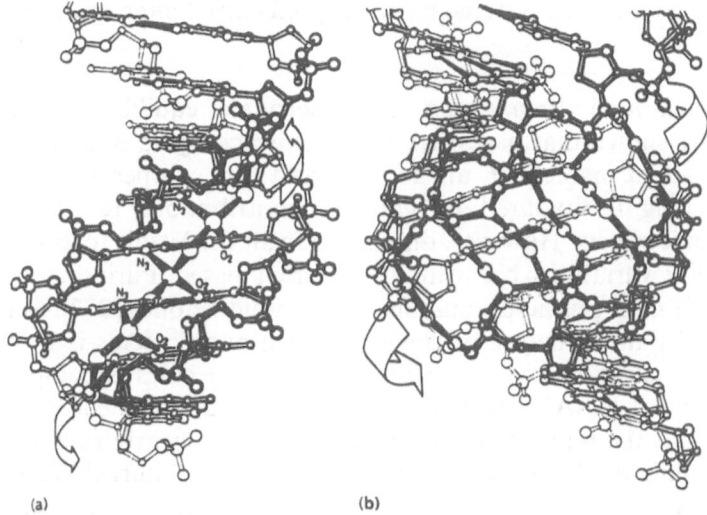

(a)   (b)

Fig. 37. Water molecules: (a) In the minor groove of an AATT region for B-DNA. (b) across the major groove in A-DNA. (Reprinted from Ref. [244], Copyright 1984, Springer Verlag, New York.)

Structure and hydration in solution. As early as 1971 X-ray scattering experiments showed that DNA in dilute solutions might have a conformation different from the B-form [253]. Since then other techniques, such as electrophoretic mobility of superhelical DNA [254], enzymatic digestion with DNase 1 [255], and energy-minimization calculations [256], have corroborated that B-DNA in dilute solution is slightly underwound by about one-half base pair per turn. This difference of conformation can be ascribed to a phase transition occurring at moderately high DNA concentrations [257, 258]. Under these conditions long-range electrostatic interactions between the charged macromolecules induce the formation of ordered nucleic acid clusters. When the solution concentration is increased liquid crystalline phases appear. The moderately concentrated phase is cholesteric [259, 260]. It becomes columnar hexagonal at higher concentrations [261]. Such organizations of DNA double helices are observed both in vivo and in vitro [259–267].

Experiments involving techniques similar to the ones used for proteins lead to the conclusion that nucleic acids are strongly hydrated [268–274]. The hydration is not homogeneous around the macromolecule and can be regarded as consisting of two hydration domains. The first domain corresponds to the bound water molecules which are detected by crystallography. The second domain is the usual hydration shell which surrounds the nucleid acid molecules and which differs from the bulk solvent. A value of about 0.22 g of water per g of DNA has been inferred from measurements of the preferential interaction parameters [275].

The double-helix denaturation. Osmometry and equilibrium sedimentation have shown that bases self-associate, stacking up in an isodesmic way [276–278], namely in an additive and not co-operative manner. Base stacking is favoured by entropic contributions resulting from hydrophobic effects. However, dipolar and van der Waals forces explain the stability variations between the different bases. Purine stacks are more stable than purine-pyrimidine and pyrimidine ones [277]. Stacking explains why the structure of single-stranded nucleic acids is helical.

When the surface area accessible to water molecules is compared in single-stranded DNA with that of A- or B-type double helices [279], it is found that the phosphate oxygen atoms remain nearly fully exposed to water while the bases become approximately 80% buried. Therefore, double helices are more polar than single strands. This effect is analogous to that of proteins which also have their polar groups mainly exposed to solvent while apolar ones are more buried in their interior [206, 211].

For double helices, the helix-coil transition is a cooperative phenomenon. The transition temperature depends on the length of the chains [280, 281], on their base composition, and also on the ionic strength of the solvent [282]. This last point is discussed in the next section. For A-T- or A-U-rich domains the transition temperature, or "melting temperature", is lower than for G-C-rich regions. The correlation between the base composition of the helices and the melting temperature is so good that it is possible to predict it with good accuracy for any B-DNA form of known base composition [283]. The converse is also possible [282]. Stability of RNA double helices is different because they pertain to the A-family. For a given base composition longer double helices have both a higher melting temperature and a higher cooperativity. The influence of the solvent ionic strength on thermal denaturation is discussed in the next paragraph.

4.3.2 Salt and cosolvent effects: A-DNA and A-RNA have more rigid molecular conformations than B-DNA. Consequently they are rather insensitive to environmental conditions, in contrast to the latter. In fibres B-DNA salts can adopt different helical structures with different counter ions. In solution the structure is sensitive to temperature and to both the type and the concentration of cations but not to anions [284, 285].

For nucleotides the potential binding sites for metal ions are oxygen and nitrogen atoms. Cations which bind to phosphate groups tend to stabilize the double-helical structure of nucleic acids while those binding to bases promote an opposite effect [242]. An example is given in Fig. 38

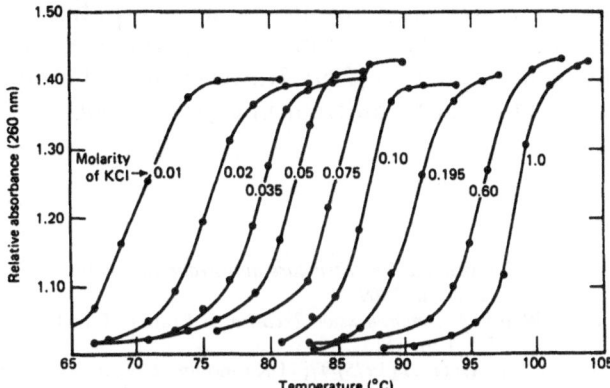

Fig. 38. Variation of the denaturation temperature of DNA as a function of the ionic strength. (Adapted from Ref. [282].)

for potassium. In general alkali metal ions increase the melting temperature of the double helix [282], unlike transition metal ions. These metal ions can cocrystallize with nucleic acids and their complexes have been studied by crystallography [286].

Salt or cosolvent effects can also be less specific. They are directly related to water activity which is lowered by increasing concentrations of these additives. Alcohols and salts that are used for fibre preparations dehydrate the solute macromolecules. At physiological pH each residue carries a negative charge on the phosphate group which is partly neutralized by small counter ions. The addition of salts to the solution screens electrostatic interaction between the charged groups, although some counter ions and co-ions can be excluded from the vicinal hydration shell by a Donnan mechanism. As a result the melting temperature of double-helical nucleic acids increases with the ionic strength of the solvent [282] (Fig. 38).

In aqueous solution DNA is generally in the B-form. However, transitions to other forms have been observed. The first example of such a conformational change was detected in concentrated ethanol solutions [287]. This transition was further shown to be of the B→A type [288–290]. The A form in solution is very similar to the one observed in fibres at low relative humidity.

The B → Z transition also occurs in solutions at high salt or ethanol concentrations [291]. Even at physiological ionic strength (200 mM NaCl) CG domains of a negatively supercoiled double helix can adopt the Z form [292–295], in order to relieve torsional stress. Stabilization of this form can be induced in different ways. Thus 5-methylation of the C-residues in poly (dG · dC) · poly(dG · dC) leads to a huge decrease in the concentration of Mg^{2+} necessary to induce the B→Z transition [296, 297]. According to X-ray diffraction experiments the methyl groups are much less accessible to water in the Z form than in the B one [298]. Polyamines, such as spermine and spermidine, also strongly stabilize the Z form in solution [299]. However the stabilization mechanism is unclear and could be different in crystals [250].

References

1. D Eisenberg and W Kauzmann, *The Structure and Properties of Water*, Oxford University Press, Ely House, London, 1969.
2. GE Walrafen in *Water: A Comprehensive Treatise*, F Franks (Ed), Plenum Press, New York-London, 1972, Vol. 1, p. 151.
3. GS Kell, J. Chem. Eng. Data 20 (1975) 97; H Kanno and CA Angell, J. Chem. Phys. 73 (1980) 1940.
4. HE Stanley, J Teixeira, A Geiger and RL Blumberg, Physica 106A (1981) 260.
5. EW Lang and HD Lüdemann, Ang. Chem. Int. Ed. Engl. 21 (1982) 315.
6. JB Hasted and M Shahidi, Nature 262 (1976) 777.
7. JM Hodge and CA Angell, J. Chem. Phys. 68 (1978) 1363.
8. GT Gillen, DC Douglas and MR Hoch, J. Chem. Phys. 57 (1972) 5117; YA Osipov, BV Zheleznyi and NF Bondarenko, Zh. Fiz. Khim. 51 (1977) 1264.
9. HR Pruppacher, J. Chem. Phys. 56 (1972) 101.
10. HE Stanley and J Teixeira, J. Chem. Phys. 73 (1980) 3404.
11. GJ Montrose, JA Bucaro, JJ Marshall-Coakley and TA Litovitz, J. Chem. Phys. 60 (1974) 5025.
12. W Danninger and G Zundel, J. Chem. Phys. 74 (1981) 2769; O Conde and J Teixeira, J. Phys. France 44 (1983) 525; O Conde and J Teixeira, Mol. Phys. 53 (1984) 951.
13. D Bertolini, M Cassettari and G Salvetti, J. Chem. Phys. 76 (1982) 3285; OA Nabokov and YA Lubimov, Mol. Phys. 65 (1988) 1473.
14. AH Narten and HA Levy in *Water: A Comprehensive Treatise*, F Franks (Ed), Plenum Press, New York-London, 1972, Vol. 1, p. 311.
15. DI Page in *Water: A Comprehensive Treatise*, F Franks (Ed), Plenum Press, New York-London, 1972, Vol. 1.
16. PA Egelstaff, Adv. Chem. Phys. 53 (1983) 1.
17. SH Chen and J Teixeira, Adv. Chem. Phys. 64 (1985) 1.
18. JC Dore in *Water Science Reviews*, 1, F Franks (Ed), Cambridge University Press, 1985, p. 3.
19. M-C Bellissent-Funel in *the Proceedings of NATO ASI on Hydrogen Bonded Liquids*, JC Dore and J Teixeira (Eds), Series C, Vol. 329, 1990.
20. AH Narten and HA Levy, J. Chem. Phys. 55 (1971) 2263.
21. L Bosio, SH Chen and J Teixeira, Phys. Rev. A27 (1983) 1468.
22. L Bosio, J Teixeira and HE Stanley, Phys. Rev. Lett. 46 (1981) 597.
23. L Bosio, J Teixeira and MC Bellissent-Funel, Phys. Rev. A39 (1989) 6612.
24. L Bosio, J Teixeira, JC Dore, D Steytler and P Chieux, Mol. Phys. 50 (1983) 733.
25. MAM Sufi, PhD. Thesis, University of Kent, 1986.
26. M-C Bellissent-Funel, L Bosio, JC Dore, J Teixeira and P Chieux, Euro-phys. Lett. 2 (1986) 241.
27. M-C Bellissent-Funel, J Teixeira, L Bosio and JC Dore, J. Phys.: Condens. Matter 1 (1989) 7123.
28. MR Chowdhury, JC Dore and JT Wenzel, J. Non Cryst. Solids 53 (1982) 247.
29. MC Bellissent-Funel, J Teixeira and L Bosio, J. Chem. Phys. 87 (1987) 2231.
30. A Geiger, FH Stillinger and A Rahman, J. Chem. Phys. 70 (1979) 4185.
31. A Geiger and HE Stanley, Phys. Rev. Lett. 46 (1982) 1749.
32. P Boutron and R Alben, J. Chem. Phys. 62 (1975) 4898; MG Sceats, M Stavola and SA Rice, J. Chem. Phys. 70 (1979) 3927.
33. W Marshall and SW Lovesey, *Theory of Thermal Neutron Scattering*, Clarendon Press, Oxford, 1971.
34. BN Brockhouse, Nuovo Cim. Suppl. 9 (1958) 45; Phys. Rev. Lett. 2 (1959) 287.

35. M Sakamoto, BN Brockhouse, RG Johnson and NK Pope, J. Phys. Soc. Jpn. 17, Suppl. B-II (1962) 370.
36. JH Simpson and HY Carr, Phys. Rev. 111 (1958) 1201.
37. P von Blanckenhagen, Ber. Bunsenges. Phys. Chem. 76 (1972) 891.
38. SH Chen in *the Proceedings of NATSO ASI on Hydrogen Bonded Liquids*, JC Dore and J Teixeira (Eds), Series C, Vol. 239, 1990.
39. M-C Bellissent-Funel and J. Teixeira, J. of Molecular Structure 250 (1991) 213.
40. SH Chen, J Teixeira and R Nicklow, Phys. Rev. A 26 (1982) 3477.
41. J Teixeira, M-C Bellissent-Funel, SH Chen and AJ Dianoux, J. Phys. France, Coll. C7, Suppl. 9, 45 (1984) 65.
42. J Teixeira, M-C Bellissent-Funel, SH Chen and AJ Dianoux, Phys. Rev. A 31 (1985) 1913.
43. SH Chen and J Teixeira, Structure and Dynamics of Low Water as Studied by Scattering Techniques, in *Advances in Chemical Physics*, I. Prigogine and SA Rice (Eds), J Wiley and Sons 1986, Vol. LXIV, pp. 1–45.
44. J Teixeira, M-C Bellissent-Funel and SH Chen, J. Mol. Liquids 48 (1991) 123.
45. J Teixeira, M-C Bellissent-Funel and SH Chen, J. Phys.: Cond. Matter 2 (1990) SA 105.
46. RW Impey, PA Madden and IR McDonald, Mol. Phys. 46 (1982) 513.
47. SH Chen and S Yip, Physics Today 29 (1976) 32.
48. a) GE Walrafen, J. Chem. Phys. 40 (1964) 3249; b) G d'Arrigo, G Maisano, F Mallamace, P Migliardo and F Wanderlingh, J. Chem. Phys. 75 (1981) 4264; c) R Bansil, J Wiafe-Akenten and JL Taafe, J. Chem. Phys. 76 (1982) 2221; d) Y Yek, JH Bilgram and W Kanzig, J. Chem. Phys. 77 (1982) 2317; e) S Krishnamurthy, R Bansil and J Wiafe-Akenten, J. Chem. Phys. 79 (1983) 5863.
49. J Teixeira, M-C Bellissent-Funel and SH Chen, unpublished data.
50. SH Chen, K Toukan, C-K Loong, DL Price and J Teixeira, Phys. Rev. Lett. 53 (1984) 1360.
51. MA Ricci, SH Chen, DL Price, C-K Loong, K Toukan and J Teixeira, Physica 136B (1986) 190.
52. K Toukan, MA Ricci, SH Chen, C-K Loong, DL Price and J Teixeira, Phys. Rev. A 37 (1988) 2580.
53. GE Walrafen and LA Blatz, J. Chem. Phys. 59 (1973) 2646.
54. Y Marechal, J. Chem. Phys. 95 (1991) 5565.
55. K Toukan and A Rahman, Phys. Rev. B 31 (1985) 2643.
56. J Teixeira, M-C Bellissent-Funel, SH Chen and B Dorner, Phys. Rev. Lett. 54 (1985) 2681.
57. J Teixeira, M-C Bellissent-Funel, SH Chen and B Dorner in *Water and Aqueous Solutions*, GW Neilson and JE Enderby (Eds), Adam Hilger, Bristol, 1986, p. 99.
58. HG Bell, H Moeller-Wenghoffer, A Kollmar, R Stockmeyer, T Springer and H Stiller, Phys. Rev. A 11 (1975) 316.
59. JRD Copley and JW Rowe, Phys. Rev. Lett. 32 (1974) 49; Phys. Rev. A 9 (1974) 1656.
60. O. Söderström, JRD Copley, JB Suck and B Dorner, J. Phys. F 10 (1980) L151.
61. A Rahman and FH Stillinger, Phys. Rev. A 10 (1974) 368.
62. M Wojcik and E Clementi, J. Chem. Phys. 85 (1986) 6085.
63. MA Ricci, D Rocca, G Ruocco and R Vallauri, Phys. Rev. A 40 (1989) 7226.
64. Th Kowall, P Mausbach and A Geiger, Ber. Bunsenges. Phys. Chem. 94 (1990) 279.
65. J Teixeira and J Leblond, J. Phys. (Paris), Lett. 39 (1978) L83.
66. A Geiger in *the Proceedings of NATO ASI on Correlations and Connectivity*, HE Stanley and N Ostrowsky (Eds), Series E, 1990, Vol. 188.
67. J Turner, JL Finney, JP Bouquière, GW Neilson, S Cummings and J Bouillot in [75], p. 277.
68. L Belloni, J. Chem. Phys. 85 (1986) 519.
69. JL Finney, in [75], p. 229.
70. F Franks and SF Mathias (Eds), *Biophysics of Water*, John Wiley & Sons, Chichester, 1982.
71. S. Lifson and M Levitt (Eds), *Structure and Dynamics of Macromolecules*, Israel, J. Chem. 27, No. 2 (1986).
72. Workshop on Water, Structure and Dynamics of Water and Aqueous Solutions: Anomalies and their Possible Implications in Biology, J. Physique-Colloque, C7 (1984).

73. M-C Bellissent-Funel and GW Neilson (Eds), *The Physics and Chemistry of Aqueous Ionic Solutions*, NATO ASI Series. Series C, D. Reidel, Dordrecht, Netherl., 1987, Vol. 205.
74. SP Colowick and NO Kaplan (Eds), *Biomembranes, Part O, Methods in Enzymology*, Academic Press, Orlando, 1986, Vol. 127.
75. GW Neilson and JE Enderby (Eds), *Water and Aqueous Solutions*, Colston Papers No. 37, Adam Hilger, Bristol, 1986.
76. RH Stokes in [83], Vol. 1, p. 1.
77. BE Conway, *Ionic Hydration in Chemistry and Biophysics*, Elsevier, Amsterdam, 1981.
78. EV Dehmlow and SS Dehmlow, *Phase Transfer Catalysis*, Verlag Chemie, Weinheim, 1980.
79. JE Desnoyers, in [83], Vol. 1, p. 139.
80. B Wittmann and G Gros, in [70], p. 121.
81. J Barthel and R Neueder, GIT Fachz. Lab. 28 (1984) 1002.
82. RF Platford, in [83], Vol. 1, p. 65.
83. RM Pytkowicz (Eds), *Activity Coefficients in Electrolyte Solutions*, CRC Press, Boca Raton, Florida, 1979.
84. P Debye and E Hückel, Physik. Z. 24 (1923) 185.
85. N Bjerrum, Kgl. danske Videnskab. Selskab. math.-fysiske Medd. VIII. 9 (1926) 1.
86. J Barthel, Pure Appl. Chem. 51 (1979) 2093.
87. J Barthel, R Wachter, G Schmeer and H Hilbinger, J. Solution Chem. 15 (1986) 531.
88. J M'Halla, J. Chim. Phys., Phys.-Chim. Biol. 86 (1989) 507.
89. KS Pitzer in [83], p. 157.
90. M Fixman, J. Chem. Phys. 70 (1979) 4995.
91. RW Wilson, DC Rau and VA Bloomfield, Biophys. J. 30 (1980) 317.
92. GS Manning, J. Chem. Phys. 51 (1969) 924.
93. JM Victor in [73], p. 291.
94. W Kauzmann, Adv. Prot. Chem. 14 (1959) 1.
95. G Nemethy and HA Scheraga, J. Phys. Chem. 66 (1972) 1773.
96. KL Mittal (Ed), *Micellization, Solubilization and Microemulsions*, Plenum Press, New York, 1977.
97. LR Pratt and D Chandler, J. Chem. Phys. 67 (1977) 3683.
98. C Pangali, M Rao and BJ Berne, J. Chem. Phys. 71 (1979) 2975.
99. HL Friedman, *A Course in Statistical Mechanics*, Prentice Hall, Englewood Cliffs, New York (1985).
100. JP Hansen and IR McDonald, *Theory of Simple Liquids*, Academic Press, London, 2nd ed., 1986.
101. DA McQuarrie, *Statistical Mechanics*, Harper and Row, London, 1976.
102. PG Kusalik and GN Patey, J. Chem. Phys. 88 (1988) 7715.
103. BM Pettitt and PJ Rossky, J. Chem. Phys. 84 (1986) 5836.
104. F Hirata, PJ Rossky and BM Pettitt, J. Chem. Phys. 78 (1983) 4133.
105. PS Ramanathan and HL Friedman, J. Chem. Phys. 54 (1971) 1086.
106. PS Ramanathan, CV Krishnan and HL Friedman, J. Solution Chem. 1 (1972) 237.
107. PJ Rossky and HL Friedman, J. Phys. Chem. 84 (1980) 587.
108. P Calmettes, W Kunz and P Turq, Physica B 180 & 181 (1992), 868.
109. RJ Bacquet and PJ Rossky, J. Phys. Chem. 92 (1988) 3604.
110. L Blum, J. Chem. Phys. 61 (1974) 2129.
111. D Levesque, JJ Weis and GN Patey, J. Chem. Phys. 72 (1980) 1887.
112. PH Fries, JS Perkyns and GN Patey, Mol. Phys. 57 (1986) 529.
113. PG Kusalik and GN Patey, J. Chem. Phys. 89 (1988) 5843.
114. PG Kusalik and GN Patey, J. Chem. Phys. 89 (1988) 7478.
115. BM Pettitt and PJ Rossky, J. Chem. Phys. 77 (1982) 509.
116. PJ Rossky, Pure Appl. Chem. 57 (1985) 1043.
117. WL Jorgensen, JK Buckner, SE Huston and PJ Rossky, J. Am. Chem. Soc. 109 (1987) 1891.
118. BM Pettitt and M Karplus, Chem. Phys. Lett. 121 (1985) 194.
119. BM Pettitt and M Karplus, Chem. Phys. Lett. 136 (1987) 383.
120. P Bopp in [73], p. 217.
121. AT Hagler and J Moult, Nature 272 (1978) 222.

122. WF van Gunsteren, HJC Berendsen, J Hermans, WGJ Hol and JPM Postma, Proc. Natl. Acad. Sci. 80 (1983) 4315.
123. CF Wong and JA McCammon in [71], p. 211.
124. JA McCammon and SC Harvey, *Dynamics of Proteins and Nucleic Acids*, Cambridge University Press, Cambridge, 1988.
125. CL Brooks, III, M Karplus and BM Pettitt, *Advances in Chemical Physics*, I Prigogine and SA Rice (Eds), Wiley and Sons, New York, 1988, Vol. 71.
126. JC Rasaiah, DN Card and JP Valleau, J. Chem. Phys. 56 (1972) 248.
127. JM Goodfellow, F Vovelle, JE Quinn, HFJ Savage and JL Finney in [75].
128. P Turq, F Lantelme and HL Friedman, J. Chem. Phys. 66 (1977) 3039.
129. P Turq, F Lantelme and D Levesque, Mol. Phys. 37 (1979) 223.
130. DL Ermak and JA McCammon, J. Chem. Phys. 69 (1978) 1352.
131. JA McCammon, BR Gelin, M Karplus and PG Wolynes, Nature 262 (1976) 325.
132. H Savage and A Wlodawer in [74], p. 162 and references quoted there.
133. GL Squires, *Introduction to the Theory of Thermal Neutron Scattering*, Cambridge University Press, 1978.
134. AK Soper, GW Neilson, JE Enderby and RA Howe, J. Phys. C: Solid State Phys. 10 (1977) 1793.
135. JE Enderby and GW Neilson, Rep. Prog. Phys. 44 (1981) 593.
136. JE Enderby in [73], p. 129.
137. G Zaccai and B Jacrot, Annu. Rev. Biophys. Bioeng. 12 (1983) 139.
138. P. Calmettes, H Eisenberg and G Zaccai, Biophys. Chem. 26 (1987) 279.
139. CF Wu and SH Chen, J. Chem. Phys. 87 (1987) 6199.
140. SH Chen, EY Sheu, J Kalus and H Hoffmann, J. Appl. Cryst. 21 (1988) 751.
141. JN Israelachvili and PM McGuiggan, Science 241 (1988) 795.
142. JN Israelachvili and GE Adams, J. Chem. Soc. Faraday Trans. 1, 74 (1978) 975.
143. JN Israelachvili and J Marra, Methods Enzymol. 127 (1986) 353.
144. BV Derjaguin and L Landau, Acta Physicochim. URSS 14 (1941) 633.
145. EJW Verwey and JThG Overbeek, *Theory of the Stability of Lyophobic Colloids*, Elsevier, Amsterdam, 1948.
146. JN Israelachvili, *Intermolecular and Surface Forces*, Academic Press, London, 1991, pp. 260–288.
147. VA Parsegian, RP Rand and DC Rau, Chem. Script. 25 (1985) 28.
148. a) R Kjellander, S Marčelja, RM Pashley and JP Quirk, J. Phys. Chem. 92 (1988) 6489; b) J. Chem. Phys. 92 (1990) 4399.
149. R Kjellander and S Marčelja, J. Coll. Interf. Sci. 126 (1988) 194.
150. DM LeNeveu, RP Rand and VA Parsegian, Nature 259 (1976) 601.
151. J Marra and JN Israelachvili, Biochemistry 24 (1985) 4608.
152. J Marra, J Colloid Interface Sci, 107 (1985) 446; ibid 109 (1986) 11.
153. a) RP Rand and VA Parsegian, Biochim. Biophys. Acta 988 (1989) 351. b) DC Rau, B Lee and VA Parsegian, Proc. Natl. Acad. Sci. USA 81 (1984) 2621. c) DC Rau and VA Parsegian, Science 249 (1990) 1278. d) VA Parsegian and DC Rau, J. Cell. Biol. 99 (1, Pt. 2) (1984) 196s. e) VA Parsegian, RP Rand, DC Rau, Lessons from the direct measurement of forces between biomolecules in *Physics of Complex and Supermolecular Fluids*, SA Safran and NA Clark (Eds), Wiley, New York, 1987, p. 115.
154. JS Anderson and KJ Saddington, J. Chem. Soc. 2 (1949) 381.
155. JP Simonin, P Tivant, P Turq and E Soualhia, J. Phys. Chem. 94 (1990) 2175.
156. M Holz, in *Progress in NMR Spectroscopy*, Vol. 18, pp. 327–403, 1986.
157. E Lang, W Fink and HD Lüdemann in [72], p. 173.
158. HG Hertz, J. Chimie Physique 82 (1985) 557.
159. BM Fung in [74], p. 151.
160. EW Lang and H-D Lüdemann, High Pressure NMR Studies on Water and Aqueous Solutions in *NMR Basic Principles and Progress*, Springer, Berlin, Heidelberg (1990), Vol. 24, p. 129.
161. EO Stejskal and JE Tanner, J. Chem. Phys. 42 (1965) 288.
162. JE Tanner, Biophys. J. 28 (1979) 107.
163. T Conlon and R Outhred, Biochim. Biophys. Acta 511 (1978) 408.
164. RG Bryant and B Halle in [70], p. 389.
165. T Springer, *Quasielastic Neutron Scattering for the Investigation of Diffusive Motions in Solids and Liquids*, Springer Tracts in Modern Physics, Springer, Berlin (1972), Vol. 64.

166. NA Hewish, JE Enderby and WS Howells, Phys. Rev. Lett. 48 (1982) 756.
167. M-C Bellissent-Funel, R. Kahn, AJ Dianoux, MP Fontana, G Maisano, P Migliardo and F Wanderlingh, Mol. Phys. 52 (1984) 1479; PS Salmon, M-C Bellissent-Funel and GJ Herdman, J. Phys.: Condens. Matter 2 (1990) 4297.
168. F Nallet, G Jannink, JB Hayter, R Oberthür and C Picot, J. Physique 44 (1983) 87.
169. W Kunz, P Turq, M-C Bellissent-Funel and P Calmettes, J. Chem. Phys. 95 (1991) 6902.
170. GE Walrafen in *Water*, F Franks (Ed), Plenum, New York, 1972, Vol. 1, pp. 151–214.
171. GE Walrafen and MR Fisher in [74], p. 91.
172. RM Fuoss and F Accascina, *Electrolytic Conductance*, Interscience Publishers, New York, 1959.
173. R Pethig, *Dielectric and Electronic Properties of Biological Materials*, Wiley and Sons, New York, 1979.
174. B Chu, *Laser Light Scattering*, Academic Press, New York, 1974.
175. M Drifford and JP Dalbiez, J. Physique Lett. 46 (1985) L-311.
176. J Paquette and C Jolicoeur, J. Solution Chem. 6 (1977) 403.
177. F Parak in [74], p. 196.
178. RB Gregory and A Rosenberg in [70], p. 238.
179. P Wolynes, Ann. Rev. Phys. Chem. 31 (1980) 345.
180. EZ Zhong and HL Friedman, J. Phys. Chem. 92 (1988) 1685.
181. TS Thacher, JL Lin and CY Mou, J. Chem. Phys. 81 (1984) 2053.
182. W Ebeling and H Krienke in *The Chemical Physics of Solutions*, RR Dogonadze, E Kalman, AA Kornyshev and J Ulstrup (Eds), Elsevier, Amsterdam, 1988, Part C, Chapter 2, pp. 113–160.
183. CL Brooks III and M Karplus in [74], p. 369.
184. A Szabo, D Shoup, SH Northrup and JA McCammon, J. Chem. Phys. 77 (1982) 4484.
185. R Jaenicke in [70], pp. 352–356.
186. ML Ansen and AE Minsky, J. Physiol. 60 (1925) 50.
187. CB Anfinsen, E Haber, M Sela and FH White, Proc. Natl. Acad. Sci. USA 47 (1961) 1309.
188. ME Golberg, Trends Biochem. Sci. 10 (1985) 388.
189. R Jaenicke, Prog. Biophys. Molec. Biol. 49 (1987) 117.
190. JA Rupley, P-H Yang and G. Tollin in *Water in Polymers*, SP Rowland (Ed), American Chemical Society, 1980, ACS Symposium 127, pp. 111–132.
191. PL Poole, J. Phys. Colloq. 45 (1984) C7-249.
192. PL Poole and JL Finney, Int. Jour. Biol. Macromol. 5 (1983) 308.
193. G Careri, E Gratton, P-H Yang and JA Rupley, Nature 284 (1980) 572.
194. PL Poole and JL Finney in [70], pp. 36–38.
195. HB Bull and K Breese, Arch. Biochem. Biophys. 128 (1968) 484.
196. JG Kirkwood and RJ Golberg, J. Chem. Phys. 18 (1950) 56.
197. WH Stockmayer, J Chem. Phys. 18 (1950) 58.
198. K Gekko and SN Timasheff, Biochem. 20 (1981) 4667.
199. G Cohen and H Heisenberg, Biopolymers 6 (1968) 1077.
200. SN Timasheff, JC Lee, EP Pittz and N Tweedy, J. Colloid. Interfac. Sci. 55 (1976) 658.
201. V Luzzati and A Tardieu, Annu. Rev. Biophys. Bioeng. 9 (1984) 1.
202. MS Lehmann and G. Zaccai, Biochemistry 23 (1984) 1939.
203. G Zaccai, E Wachtel and H Eisenberg, J. Mol. Biol. 97 (1986) 97.
204. G Zaccai, F Cendrin, Y Haik, N Borochov and H Eisenberg, J. Mol. Biol. 208 (1989) 491.
205. CN Pace, Trends Biochem. Sci. 15 (1990) 14.
206. W. Kauzmann, Adv. Prot. Chem. 14 (1959) 2313.
207. JL Finney in [70], pp. 55–58.
208. HS Frank and MW Evans, J. Chem. Phys. 13 (1945) 507.
209. B Lee and FM Richards, J. Mol. Biol. 55 (1971) 379.
210. FM Richards, Annu. Rev. Biophys. Bioeng. 6 (1977) 151.
211. GD Rose, AR Geselowitz, GJ Lesser, RH Lee and MH Zehfus, Science 229 (1985) 834.
212. C Pangali, M Rao and BJ Berne, J. Chem. Phys. 71 (1979) 2982.

213. PD Ross and S Subramanian, Biochemistry 30 (1981) 3096.
214. FA Long and WT McDerit, Chem. Rev. 51 (1952) 119.
215. RL Balwin, Proc. Nat. Acad. Sci. USA 83 (1986) 8069.
216. DR Robinson and WP Jencks, J. Am. Chem. Soc. 87 (1965) 2470.
217. EE Schrier and EB Schreier, J. Phys. Chem. 71 (1967) 1851.
218. PH von Hippel and KY Wong, J. Biol. Chem. 240 (1965) 3909.
219. PH von Hippel and T Schleich, Accounts of Chem. Research 2 (1969) 257.
220. SN Timasheff, T Arakawa, H Inoue, K Gekko, MJ Gorbunoff, JC Lee, GC Na, EP Pittz and V Prakash in [70], pp. 48–50.
221. T Arakawa and SN Timasheff, Biochemistry 23 (1984) 5912.
222. EP Pittz and SN Timasheff, Biochemistry 17 (1978) 615.
223. T Arakawa and SN Timasheff, Biochemistry 21 (1982) 6545.
224. JC Lee and SN Timasheff, Biochemistry 13 (1974) 257.
225. V Prakash, C Loucheux, S Scheufele, MJ Gorbunoff and SN Timasheff, Arch. Biochem. Biophys. 210 (1981) 455.
226. Y Nozaki and C Tanford, J. Biol. Chem. 245 (1970) 1648.
227. M Inoue and SN Timasheff, Biopolymers 11 (1972) 737.
228. S Makino, Adv. Biophys. 12 (1979) 131.
229. SH Chen and J Teixeira, Phys. Rev. Lett. 57 (1986) 2583.
230. T Arakawa and SN Timasheff, Biochemistry 23 (1984) 5924.
231. RD Blake, J Massoulié and JR Fresco, J. Mol. Biol. 30 (1967) 291.
232. J Massoulié, Eur. J. Biochem. 3 (1968) 439.
233. S Arnott, PJ Bond, E Selsing and PJC Smith, Nucl. Acids Res. 3 (1976) 2459.
234. S Arnott, R Chandrasekaran and CM Marttila, Biochem. J. 141 (1974) 537.
235. S Arnott and E Selsing, J. Mol. Biol. 88 (1974) 509.
236. W Fuller, F Hutchinson, M Spencer and MHF Wilkins, J. Mol. Biol. 27 (1967) 507.
237. R Olby, *The Path of the Double Helix*, Univ. of Washington Press, Seattle, 1974.
238. FMC Crick and JD Watson, Proc. Roy. Soc. (London) A223 (1954) 80.
239. S Arnott and DWL Hukins, Biochem. J. 130 (1972) 453.
240. S Arnott and R Chandrasekaran in *Proceedings of the 2nd SUNYA Conversation in the Discipline Biomolecular Stereodynamics*, R. Sarna (Ed), Adenine Press, 1981, Vol. 1, pp. 99–122.
241. RE Dickerson, J. Mol. Biol. 166 (1983) 419.
242. RE Dickerson, Sci. Am. (December 1983) 94.
243. KE van Holde, *Chromatin*, Springer Verlag, New York, 1989.
244. W Saenger *Principles of Nucleic Acid Structure*, Springer Verlag, New York, 1988, second corrected printing.
245. RM Wing, HR Drew, T Tanako, C Broka, S Tanaka, K Itakura and RE Dickerson, Nature (London) 287 (1980) 755.
246. HR Drew, RM Wing, T Tanako, C Broka, S Tanaka, K Itakura and RE Dickerson, Proc. Natl. Acad. Sci. USA 78 (1981) 2179.
247. RE Dickerson and HR Drew, Proc. Natl. Acad. Sci. USA 78 (1981) 7318.
248. ML Kopka, AV Fratini, HR Drew and RE Dickerson, J. Mol. Biol. 163 (1983) 129.
249. RE Dickerson, ML Kopka and P Pjura, Proc. Natl. Acad. Sci. USA 80 (1983) 7099.
250. HR Drew and RE Dickerson, J. Mol. Biol. 151 (1981) 535.
251. BN Conner, T Tokano, S Tanaka, K Itakura and RE Dickerson, Nature 295 (1982) 294.
252. BN Conner, C Yoon, J Dickerson and RE Dickerson, J. Mol. Biol. 174 (1984) 663.
253. S Bram, J. Mol. Biol. 58 (1971) 277.
254. JC Wang, Proc. Natl. Acad. Sci. USA 76 (1979) 200.
255. D Rhodes and A Klug, Nature (London) 286 (1980) 573.
256. M Levitt, Proc. Natl. Acad. Sci. USA 75 (1978) 640.
257. M Mandelkern, N Dattagupta and DM Crothers, Proc. Natl. Acad. Sci. USA 78 (1981) 4294.
258. A Patkowski, E Gulari and B Chu, J. Chem. Phys. 73 (1980) 4178.
259. F Livolant, Eur. J. Cell. Biol. 33 (1984) 300.
260. F Livolant, J. Physique 47 (1986) 1605.
261. F Livolant, AM Levelut, J Doucet and JP Benoit, Nature (London) 339 (1989) 724.
262. Y Bouligand, MO Soyer and S Puiseux-Dao, Chromosoma 24 (1968) 251.
263. RL Rill, F Livolant, HC Aldrich and MW Davidson, Chromosoma 98 (1989) 280.

264. M Feughelman, R Langridge, WE Seeds, AR Stokes, HR Wilson, MHF Wilkins, RK Balclay and LD Hamilton, Nature (London) 175 (1955) 834.
265. V Luzzati and A Nicolaieff, J. Mol. Biol. 1 (1959) 127.
266. V Luzzati and A Nicolaieff, J. Mol. Biol. 7 (1963) 142.
267. J Lepault, J Dubochet, W Bashong and E Kellenberger, EMBO J. 6 (1987) 1507.
268. JE Hearst and J. Vinograd, Proc. Natl. Acad. Sci. USA 47 (1961) 825.
269. M Falk, KA Hartman Jr and RC Lord, J. Am. Chem. Soc. 84 (1962) 3843; 85 (1963) 387 and 85 (1963) 391.
270. JE Hearst, Biopolymers 3 (1965) 57.
271. MJ-B Tunis and JE Hearst, Biopolymers 6 (1968) 1325 and 1345.
272. G Cohen and H Eisenberg, Biopolymers 6 (1968) 1077.
273. M Falk, AG Poole and CG Goymour, Can. J. Chem. 48 (1970) 1536.
274. B Wolf and S Hanlon, Biochemistry 14 (1975) 1661.
275. S Pundak and H Eisenberg, Eur. J. Biochem. 118 (1981) 463.
276. POP Ts'o, IS Melvin and AC Olson, J. Am. Chem. Soc. 85 (1963) 1289.
277. POP Ts'o in *Basic Principles in Nucleic Acid Chemistry*, POP Ts'o (Ed), Academic Press, New York, 1974, Vol. 1, pp. 237–318.
278. NI Nakano and SJ Igarashi, Biochemistry 9 (1970) 577.
279. CJ Alden and SH Kim, J. Mol. Biol. 132 (1979) 411.
280. D Pörschle, Mol. Biol. Biochem. Biophys. 24 (1977) 191.
281. VV Fillimonov and PL Privalov, J. Mol. Biol. 122 (1978) 465.
282. J Marmur and P Doty, J. Mol. Biol. 5 (1962) 109.
283. O Gotoh and Y Takashina, Biopolymers 20 (1981) 1033.
284. P Anderson and W Bauer, Biochemistry 17 (1978) 594.
285. A Chan, R Kilkuskie and S Hanlon, Biochemistry 18 (1979) 84.
286. I Sissoëff, J Grisvard and E Guillé, Prog. Biophys. Molec. Biol. 31 (1976) 165.
287. J Brahms and WFHM Mommaets, J. Mol. Biol. 10 (1964) 73.
288. DM Gray, SP Edmondson, D Lang and M Vaughan, Nucleic Acids Res. 6 (1979) 2089.
289. SB Zimmerman and BH Pheiffer, J. Mol. Biol. 135 (1979) 1023.
290. HM Wu, N Dattagupta and DM Crothers, Proc. Natl. Acad. Sci. USA 78 (1981) 6808.
291. FW Pohl and TM Jovin, J. Mol. Biol. 67 (1972) 375.
292. LJ Peck, A Nordheim, A Rich and JC Wang, Proc. Natl. Acad. Sci. USA 79 (1982) 4560.
293. S Brahms, J Vergne, JG Brahms, JG Capua, ED Bucher and T Koller, J. Mol. Biol. 162 (1982) 473.
294. LJ Peck and JC Wang, Proc. Natl. Acad. Sci. USA 80 (1983) 6206.
295. CK Singleton, MW Kilpatrick and RD Wells, J. Biol. Chem. 259 (1984) 1963.
296. M Behe and G Felsenfeld, Proc. Natl. Acad. Sci. USA 78 (1981) 1619.
297. M Behe, S Zimmerman and G Felsenfeld, Nature (London) 293 (1981) 233.
298. S Fujii, AHJ Wang, G. van der Maarel, JH van Boom and A Rich, Nucleic Acid. Res. 10 (1982) 7879.
299. MM Minyat, VI Ivanov, AM Kritzyn, LE Michenkova and AK Schylokina, J. Mol. Biol. 128 (1978) 397.

Bioelectrochemistry: General Introduction
ed. by S. R. Caplan, I. R. Miller and G. Milazzo†
© 1995 Birkhäuser Verlag Basel/Switzerland

CHAPTER 4
Electrochemistry of colloidal systems: Double layer phenomena

Shinpei Ohki[1] and Hiroyuki Ohshima[2]

[1]*Department of Biophysical Sciences, State University of New York at Buffalo, Buffalo, New York, U.S.A.*
[2]*Faculty of Pharmaceutical Sciences, Science University of Tokyo, Tokyo, Japan*

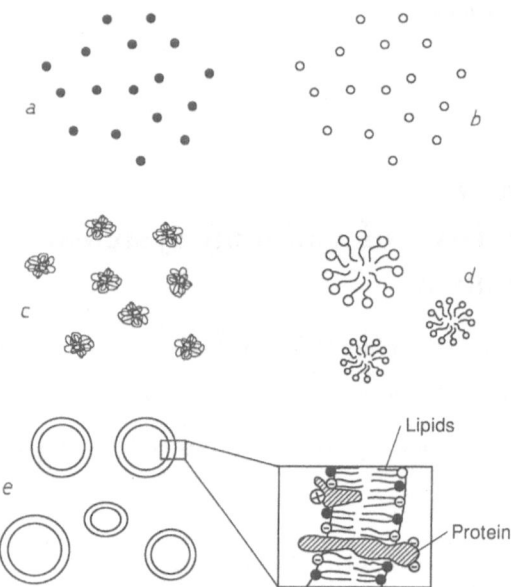

Fig. 1. Various biocolloid suspensions: a) oil-in-water colloids; b) water-in-oil colloids; c) globular polymers; d) miscelles; e) vesicles or cells.

1. The electrical double layer of biocolloidal surfaces

1.1. Biocolloidal interfaces [1]

1.1.1. Biocolloids: Biological colloidal systems may be grouped into three general classifications: 1) colloidal dispersion – microscopic mixtures of two material phases (e.g. an oil-in-water or water-in-oil emulsion, see Fig. 1a,b) 2) solutions of macromolecules (e.g. solutions of globular proteins, see Fig. 1c) and 3) association colloids-dispersion of biomolecular aggregations (e.g. micelles, see Fig. 1d) or biomolecular assemblies (e.g. lipid membrane vesicles, see Fig. 1e). A characteristic feature of colloidal dispersion is the large area to volume ratio of the particle dispersed, and for the definite surface of separation existing between the particles and the medium. Simple colloid dispersions are, therefore, two-phase systems: the dispersed phase and the dispersed medium. At the interface between the dispersed phase and the dispersed medium, characteristic surface properties, such as adsorption and electric double layer effects, are evident and play important parts in determining the physico-chemical properties of the system.

1.1.2. Lyophilic and lyophobic systems: The terms lyophilic (liquid-liking) and lyophobic (liquid-disliking) are often used to describe the tendency of a surface or functional group of the colloid particles to become wetted or solvated. If the liquid medium is aqueous, as most biological colloidal systems are, the terms hydrophilic (water-liking) and hydrophobic (water-disliking) are used. Since biocolloids are discussed in this chapter, the terms hydrophilic and hydrophobic will be used to describe the nature of colloid particles. Most substances (molecules) used to form biocolloids are partly hydrophilic (e.g. peptide linkages, amino, phosphate and carboxyl groups) and partly hydrophobic (e.g. hydrocarbons). These molecules, which consist of two different parts (hydrophilic and hydrophobic) with respect to water associations, are often called amphipathic molecules.

1.1.3. Interfacial charges: The hydrophilic part of the biomolecules is likely to gather at the interface of biocolloids and the hydrophobic part

Table 1. Some acids and their conjugate bases [Refs. 2 and 61]

Acid	Conjugate base	Pk$_a$ in water at 25°C
H_3PO_4	$H_3PO_4^-$	2
$H_3PO_3^-$	HPO_4^{--}	7
HPO_4^{--}	PO_4^{---}	12
H_3O^+	H_2O	—
H_2O	OH^-	15.7
CH_3COOH	CH_3COO^-	4.75
NH_4^+	NH_3	9.3
$CH_3NH_3^+$	CH_3NH_2	10.7
$HOOCCH_2COOH$	$HOOCCH_2COO^-$	2.8
$HOOCCH_2COO^-$	$^-OOCCH^2COO^-$	5.7
$^+H_3N(CH_2)_2NH_3^+$	$^+H_3N(CH_2)_2NH_2$	7.0
$^+H_3N(CH_2)_2NH_2$	$H_2N(CH_2)_2NH_2$	10.0
$^+H_3NCH_2COOH$	$^+H_3NCH_2COO^-$	2.3
$^+H_3NCH_2COO^-$	$H_2NCH_2COO^-$	9.7

Imidazolium ion   Imidazole   7.0

Guanidinium ion   Guanidine   14 (approx)

Phenol   Phenolate ion   10.0

tends to be removed from the interfacial region and gather into the colloidal dispersion phase. The system tends to stay at an energetically stable state. Moreover, the hydrophilic groups of biocolloidal particles often possess dissociable groups in aqueous solutions (see Table 1). In aqueous solutions, the degree of dissociation of the functional polar groups depends on hydrogen ion concentration, and other factors. The degree of dissociation of most functional groups can be expressed in terms of acid-base equilibrium.

According to Brönsted and Lowrey, an acid is a substance which is capable of yielding protons and a base is able to accept protons,

$$A \rightleftarrows H^+ + B^- \quad \text{or} \quad (A^+ \rightleftarrows H^+ + B) \tag{1}$$

where A (A^+) is an acid and B^- (B) is its conjugate base. Some examples of acids and their conjugate bases are listed in Table 1. The equilibrium constant, K_a, for the above reaction (Eqn. 1) is expressed in terms of the activities of H^+, acid, and base:

$$K_a = \frac{(H^+)(B^-)}{(A)} = \frac{(H^+)(\text{base})}{(\text{acid})}. \tag{2}$$

In the case of a base (e.g. a primary amine RNH_2) in water, acid-base equilibrium may be expressed in terms of a base dissociation constant:

$$RNH_2 + H_2O \rightleftarrows RNH_3OH \rightleftarrows RNH_3^+ + OH^-, \tag{3}$$

where the intermediate form, RNH_3OH, is a hydrogen-bonded complex of RNH_2 and H_2O, and concentration can be set to equal to the total concentration of $(RNH_2) + RNH_3OH = $ total conjugate base.

The equilibrium constant (a base dissociation constant), K_b, is then expressed by

$$K_b = \frac{(RNH_3^+)(OH^-)}{(RNH_3OH)} = \frac{(\text{conjugate acid})(OH^-)}{(\text{total conjugate base})}. \tag{4}$$

On the other hand, this acid-base equilibrium can be expressed in terms of an "acid dissociation constant", K_a. Since Eqn. 3 can be written as

$$RNH_3^+ \rightleftarrows RNH_2 + H^+ \tag{5}$$

Since the total conjugate base is expressed by RNH_2, K_a is:

$$K_a = \frac{(R \cdot NH_2)(H^+)}{(RNH_3^+)} = \frac{(\text{total conjugate base})(H^+)}{(\text{conjugate acid})}. \tag{6}$$

Therefore, a relation between K_a and K_b can be obtained:

$$K_a K_b = (H^+)(OH^-) = K_w \tag{7}$$

where K_w is the ion product constant for water: $K_w = (H^+)(OH^-) = 1.0 \times 10^{-14}$ at 25°C. Then from Eqn. (7), the following relation is

Table 2. pK_a values for certain carboxylic acids, phosphoric acids, ammonium ions, amino acids, and carbonic acid (all data for 25°C) [Refs. 2 and 61]

Substance	pK_a	Substance	pK_a
Water (K_w)	14.0	β-Alanine, pK_1	3.6
Carboxylic acids		β-Alanine, pK_2	10.2
Formic acid	3.8	γ-Aminobutyric acid, pK_1	4.0
Acetic acid	4.8	γ-Aminobutyric acid, pK_2	10.6
Propionic acid	4.9	Glycylglycine, pK_1	3.1
Glycolic acid	3.8	Glycylglycine, pK_2	8.3
Lactic acid	3.9	Aspartic acid, pK_1	2.0
Succinic acid, pK_1	4.2	Aspartic acid, pK_2	4.0
Succinic acid, pK_2	5.6	Aspartic acid, pK_3	10.0
Benzoic acid	4.2	Carbonic acid, pK_1	6.4
Phosphoric acid and derivatives		Carbonic acid, pK_2	10.3
Phosphoric acid, pK_1	2.1	Side-chain carboxyl	
Phosphoric acid, pK_2	7.2	Serum albumin	4.0
Glycerol 2-phosphoric acid, pK_1	1.3	Ovalbumin	4.3
Glycerol 2-phosphoric acid, pK_2	6.7	Insulin	4.7
Glucose 1-phosphoric acid, pK_2	6.5	Imidazole	
Ammonium ion and derivatives		Insulin	6.4
Ammonium ion	9.2	Ribonuclease	6.5
Methylammonium ion	10.6	Lysozyme	6.8
Dimethlyammonium ion	10.8	Serum albumin	6.9
Trimethylammonium ion	9.8	Phenolic	
Ethanolammonium ion	9.4	Insulin	9.6
Tris(hydroxymethyl) aminomethane	8.1	Papain (11 of 17 groups)	9.8
Amino acids		Serum albumin	10.4
Glycine, pK_1 (COOH)	2.4	Lysozyme	10.8
Glycine, pK_2 (NH_3)	9.8	Side-chain amino	
α-Alanine, pK_1 (COOH)	2.3	Serum albumin	9.8
α-Alanine, pK_2 (NH_3)	9.9	β-Lactoglobulin	9.9
α-Amino-n-butyric acid, pK_1	2.3	Ribonuclease	10.2
α-Amino-n-butyric acid, pK_2	9.8	Lysozyme	10.4
α-Aminoisobutyric acid, pK_1	2.4	Guanidyl	
α-Aminoisobutyric acid, pK_2	10.2	Insulin	11.9
Serine, pK_1	2.2	α-Corticotropin	12.0
Serine, pK_2	9.2	Ferriheme water	
Threonine, pK_1	2.1	Myoglobin	8.9
Threonine, pK_2	9.1	Hemoglobin	8.0
Hydroxyproline, pK_1	1.8		
Hydroxyproline, pK_2	9.7		
Proline, pK_1	2.0		
Proline, pK_2	10.6		

obtained:

$$pK_b = pK_w - pK_a \qquad (8)$$

where $pK = -\log K$.

Therefore, all acid-base equilibria can be formulated in terms of the acid dissociation constant, K_a.

Depending on mainly pH and also to a lesser extent, the presence of other ions and other environmental conditions, the dissociable groups

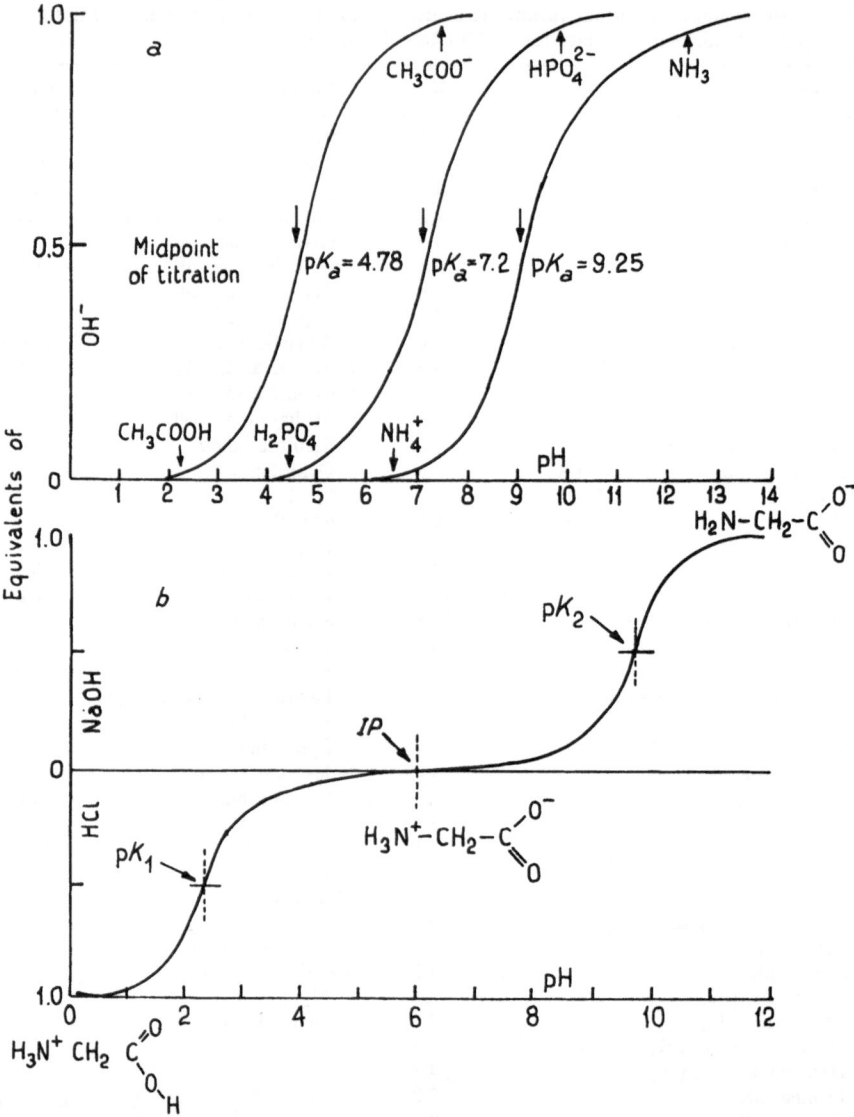

Fig. 2. Acid-base dissociation: a) acid-base titration curves for biologically important dissociable molecular groups. Arrows indicate each pK_a; b) acid-base titration curve of glycine, IP:
isoelectric point.

exist in either electrically neutral, positive, or negative forms in an
aqueous solution. Some of the pK_a values for various amino acids are
given in Table 2, and the approximate behavior of typical dissociation
groups (COOH, PO_4, NH_3) with respect to pH of the aqueous phase is
shown in Fig. 2. The interfacial regions of biocolloids which contain
several such functional dissociation groups, therefore, possess surface
charges depending on the environmental conditions. The over-all surface

Table 3. Protein and lipid-membrane titration at 25°C

Proteins [61]	Isoionic pH	Range of pH for instantaneous and reversible titration	Reversible major conformational change within this range
Conalbumin	6.8	4.2–11.2	Yes
Hemoglobin	7.0	4.4–11.5	No
Insulin	5.6	2–12	No
β-Lactoglobulin	5.3	1.5–9.7	Yes
Lysozyme	(11.1)?	2–12	No
Myoglobin	7.5	4.5–11.5	No
Pepsin	2	5.6	No
Ribonuclease	9.6	2–11.5	No
Serum albumin	5.4	2–12	Yes
Lipid membranes [47]	Isoelectric pH (in 100 mM NaCl)		
Phosphatidylcholine	2–11	2–11	No change
Phosphatidylethanolamine	2–9	2–11	No change
Phosphatidylserine	1.5	1.5–11	No change

charges of biocolloidal surfaces vary with respect to pH of the solution. Isoelectric points, for some of globular proteins and lipid membrane surfaces, are given in Table 3. The pK_a values are affected by several factors: molecular groups and electrostatic charges in the molecule and dipole moments of the molecular groups, and also temperature. For details of these factors refer to reference [2].

1.1.3.1. Effects of ion binding on pK_a. The determination of pK_a values for biopolymers generally is carried out for a particular situation where these molecules are in a given ionic environment. Since there may be different binding affinities for different ions with a given biopolymer, the observed pK_a value must correspond to an "apparent" dissociation constant, which may differ from the intrinsic dissociation constant K_a. Such an "apparent" dissociation constant with respect to H^+ would therefore vary with different ionic environments, such as ionic species and ionic strength, as well as the nature of the surface charge of the biomolecules.

The dissociations of functional groups of the biomolecules are expressed by:

$$AH \rightleftarrows A^- + H^+ \tag{9}$$

$$AM \rightleftarrows A^- + M^+ \tag{10}$$

respectively, where Eqn. (10) refers to the binding of a certain metal cation to an anionic site of the functional group. Therefore, the overall dissociation reaction may be expressed by

$$AH + AM \rightleftarrows A^- + H^+ + M^+. \tag{11}$$

The overall dissociation constant of the above reaction would be

$$K = \frac{(A^-)(H^+)(M^+)}{(AH)(AM)}. \tag{12}$$

However, when the titration of the dissociation reaction is monitored only as a function of hydrogen ion concentration, the apparent dissociation constant may be given:

$$K^{app} = \frac{(A^-)(H^+)}{(AH + AM)}. \tag{13}$$

By defining the acid dissociation constant and metal ion (M^+) complex dissociation constant, respectively:

$$K_a = \frac{(A^-)(H^+)}{(AH)},$$

$$K_M = \frac{(A^-)(M^+)}{(AM)}, \tag{14}$$

the apparent pK_a value obtained from Eqn. (13) is:

$$pK_a^{app} = pK_a + \log\left(1 + \frac{(M^+)}{(H^+)}\frac{K_a}{K_M}\right). \tag{15}$$

In the case where there is no binding of metal ion (M^+) to the anionic site ($K_M \rightarrow \infty$), the apparent pK_a reduces to the intrinsic pK_a.

1.1.3.2. Effect of surface charge and surface dielectric constant on pK_a. The apparent pK_a of a dissociable group at the biosurface differs from the intrinsic pK_a of the group in bulk solution owing to the electrostatic enhancement of the hydrogen ion concentration (activity) at the interface, caused by the surface charge, and also to the shift in the acid-base equilibrium arising from the different local polarity at the interface [3].

$$pK_a' = pK_a + \Delta pK_a^{el} + |\Delta pK_a^p|, \tag{16}$$

where ΔpK_a^{el} is the electrostatic-induced shift and ΔpK_a^p is the polarity-induced shift. The sign of the polarity-induced shift depends on the change in the total number of charges on protonation. For the dissociation of a molecular acid, the pK_a is increased as a result of the lower dielectric constant at the interface than in the bulk solution, whereas for the dissociation of a cationic acid the situation is reversed. The electrostatic shift in pK_a is

$$\Delta pK_a^{el} = -e\psi(0)/2.3kT, \tag{17}$$

where $\psi(0)$ is the surface potential at the surface calculated from the diffuse double layer theory. (The notations $\psi(0)$ and ψ_0 are used interchangeably in this Chapter.) The apparent pK_a' is, therefore,

$$pK_a' = pK_a - \frac{e\psi(0)}{2.3kT} + |\Delta pK_a^p|. \tag{18}$$

Since $\psi(0)$ is a function not only of the surface charge, but also of ionic concentrations in the bulk solution (see Eqns. 32 and 34) the apparent pK' depends also on the ionic concentrations of the bulk solution. Also, if ions other than H^+ can bind to the dissociable groups at the biosurface, an additional term $\Delta pK_a^{binding}$ accounting for this factor should be included in the expression for the apparent dissociation constant (4).

$$pK_a' = pK_a - \frac{e\psi(0)}{2.3kT} + |\Delta pK_a^p| + \Delta pK_a^{binding}. \tag{19}$$

The last term has been discussed in the previous section.

1.2. The diffuse double layer [5, 6]

When biocolloidal particles are suspended in an electrolyte solution, since the polar molecular groups are located at the aqueous interface and some of them are ionized due to the environmental condition as described above, unequal distributions of positive and negative ions due to the surface charge occur in the solution near the surface of the colloidal particles as a function of distance from the surface (Fig. 3a and b). The distribution of ions is related to the electrical potential, the magnitude of which generally decreases with distance from the surface (diffuse potential). This diffuse electrical potential is historically called the "electrical double layer potential". The electrical double layer can be regarded generally as consisting of two regions: an inner region, which may include adsorbed ions, and a diffuse region in which ions are distributed according to the influence of electrical forces and random thermal motion. The diffuse part of the double layer will be considered first. Quantitative and precise treatment of the electric double layers presents an extremely difficult and in some respects unsolved problem.

1.2.1. Double layer potential for a charged flat surface: The simplest quantitative treatment of the diffuse part of the double layer is given by Gouy (1910) [7], and Chapman (1913) [8], and is based on the following assumptions:

(1) The surface is assumed to be flat, of infinite extent and uniformly charged.
(2) The ions in the diffuse part of the double layer are assumed to be point charges distributed according to the Boltzmann distribution.

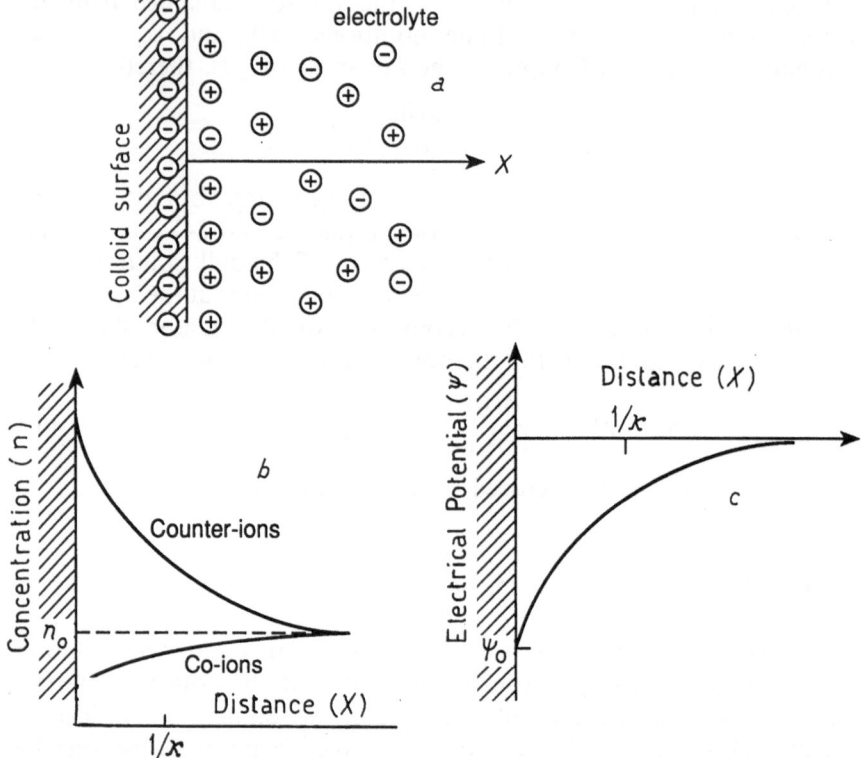

Fig. 3. Schematic diagrams of concentration and electrical potential profiles for a planar charged interface. The shaded and non-shaded areas refer to the hydrocarbon and aqueous electrolyte phases, respectively.

(3) The solvent is assumed to influence the double layer only through its dielectric constant, which is assumed to have the same value throughout the diffuse part.

The electrical potential at a distance x from the surface in an electrolyte solution is denoted by $\psi(x)$ (see Fig. 3). Taking the surface to be negatively charged and applying the Boltzmann distribution for the jth ionic species, the number density of the jth ion at a position x is

$$n_{j(x)} = n_{jo} \exp\left[-\frac{z_j e\psi(x)}{kT}\right] \tag{20}$$

where z_j and n_{jo} are the valency and the number density in the bulk solution of the jth ionic species, respectively. The net volume charge density ρ, at a point where the potential is $\psi(x)$, is given by

$$\rho = \sum_j \left[z_j e n_{jo}\left(\exp -\frac{z_j e\psi(x)}{kT}\right)\right]. \tag{21}$$

For the case of a symmetrical electrolyte, Eqn. (21) becomes

$$\rho = -2zen_o \sinh(ze\psi/kT) \tag{22}$$

where z is the valence of the electrolyte. The quantity ρ is related to the potential by the Poisson's equation, which takes the following form for a flat double layer:

$$\frac{\partial^2\psi}{\partial x^2} = -4\pi\rho/\varepsilon, \quad \text{(C.G.S. unit)} \tag{23}$$

where ε, the dielectric constant of the medium, can be replaced by $4\pi\varepsilon_o\varepsilon$ in SI units. The quantity ε_o is the permittivity of a vacuum which is equal to $8.854 \times 10^{-12} \, kg^{-1} \, m^{-3} \, s^2 \, A^2$ (A: coulomb). Combination of Eqns. (21) and (23) gives the Poisson-Boltzmann equation:

$$\frac{\partial^2\psi}{\partial x^2} = -4\pi\rho/\varepsilon = -\frac{4\pi}{\varepsilon} \sum_j z_j en_{jo}[\exp(-z_j e\psi/kT)] \tag{24}$$

$$= (8\pi zen_o/\varepsilon) \sinh(ze\psi/kT). \tag{25}$$

The solution of Eqn. (25), with use of the boundary conditions ($\psi = \psi_o$ when $x = 0$, and $\psi = d\psi/dx = 0$ when $x = \infty$), can be written in the following form [9].

$$\psi = (2kT/ze) \ln\left(\frac{1 + \alpha \exp[-\kappa x]}{1 - \alpha \exp[-\kappa x]}\right) \tag{26}$$

where

$$\alpha = \frac{\exp[ze\psi_o/2kT] - 1}{\exp[ze\psi_o/2kT] + 1} \tag{27}$$

and

$$\kappa = (8\pi z^2 e^2 n_o/\varepsilon kT)^{1/2} = \left(\frac{8\pi e^2 z^2 CN_A}{l\varepsilon kT}\right)^{1/2}, \tag{28}$$

where N_A is the Avogadro's number and C is the concentration of electrolyte (in mole/liter), κ is the Debye constant. If $ze\psi_o/2kT \ll 1$, the Debye–Hückel approximation

$$\exp[-ze\psi_o/2kT] \cong 1 - (ze\psi_o/2kT) \tag{29}$$

can be made, and Eqn. (26) is simplified to

$$\psi = \psi_o \exp(-\kappa x) \tag{30}$$

which shows that at low potential the potential decreases exponentially with distance from the charged surface. When the Debye–Hückel approximation is not applicable, the potential is predicted to decrease at a greater than exponential rate.

The potential ψ_o can be related to the charge density σ at the surface by means of the integrated form of Eqn. (25) with respect to x, together with another boundary condition at the surface,

$$\frac{\partial \psi}{\partial x}\bigg|_{x=0} = -(4\pi\sigma/\varepsilon). \tag{31}$$

The resulting expression is

$$\sigma = (2n_o \varepsilon kT/\pi)^{1/2} \sinh(ze\psi_o/2kT), \tag{32}$$

which at low surface potential reduces to

$$\sigma = \varepsilon \kappa \psi_o / 4\pi. \tag{33}$$

The surface potential ψ_o, therefore, depends on both the surface charge density σ and the ionic composition of the medium (through κ). The expression in Eqn. (32) is for symmetrical electrolytes. A general expression for the surface potential and surface charge density for any electrolyte can be obtained from Eqns. (24) and (31):

$$\sigma = \left(\frac{\varepsilon kT}{2\pi}\right)^{1/2} \left\{\sum_j n_{jo}(e^{-z_j e\psi_o/kT} - 1)\right\}^{1/2}$$

$$= \frac{1}{273} \left\{\sum_j C_j(e^{-z_j e\psi_o/kT} - 1)\right\}^{1/2} \quad \text{(at 24°C)}. \tag{34}$$

For the uni-univalent and one bivalent electrolyte case, the electrical double layer potential of a smeared charge surface has been solved by Abraham–Shrauner [10] for 2-1-1 electrolytes (one bivalent cation, one univalent cation, and one univalent anion). The electrical potential $\psi(x)$ at a distance x from the membrane surface is expressed by the following equation:

$$X + D = \int_x^\infty \frac{dy}{\sum_j x_j(e^{-z_j y} - 1)^{1/2}}, \tag{35}$$

where $y = e\psi/kT$, $X_j = n_j/n$, $X = x/\lambda_D$, D is an integral constant, and $\lambda_D^2 = kT\varepsilon/2e^2 n$ (λ_D is proportional to the Debye length ($1/\kappa$)). Here, n_j is the number density of the jth species, n is the number density of the monovalent cation, e is the unit positive charge, and ε is the permittivity of the solution (MKS units).

1.2.2. Double layer potential for a spherical surface: For colloidal systems, the surfaces of particles are not flat but curved. The simplest form of such a particle is a sphere or a spherical shell. The Poisson–Boltzmann equation for a spherical interface is

$$\nabla^2 \psi = \frac{1}{r^2} \frac{\partial}{\partial r} \left(r^2 \frac{\partial \psi}{\partial x}\right) = (8\pi z e\psi_o/\varepsilon) \sinh(ze\psi/kT), \tag{36}$$

where ∇ is the nabla operator, and r is the distance from the center of the sphere (see Fig. 4). This expression cannot be integrated analytically

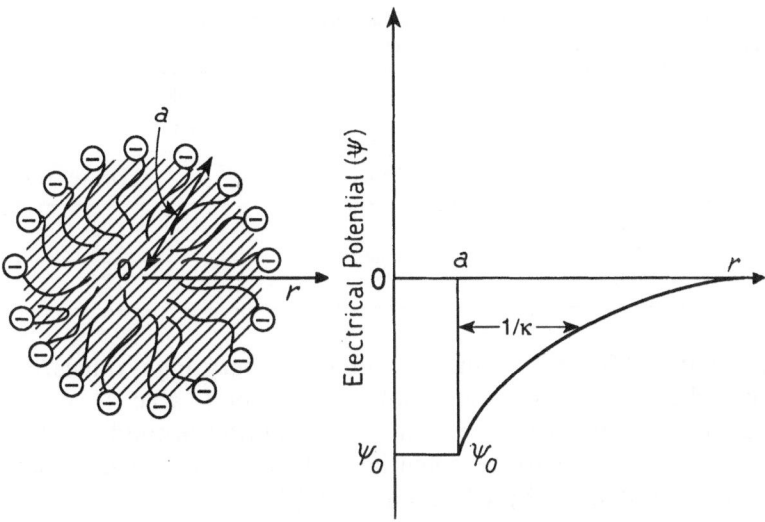

Fig. 4. Diagrams similar to Fig. 3, except for a spherical charge interface. Note that a refers to the radius of the spherical colloid.

without approximation. If the Debye–Hückel approximation is made, the equation is reduced to

$$\frac{1}{r^2}\frac{\partial}{\partial r}\left(r^2\frac{\partial\psi}{\partial x}\right) = \kappa^2\psi \tag{37}$$

which, on integration with the boundary conditions $\psi = \psi_o$ at $r = a$, and $\psi = d\psi/dr = 0$ at $r = \infty$, becomes

$$\psi = \psi_o\frac{a}{r}\exp[-\kappa(r-a)]. \tag{38}$$

The Debye–Hückel approximation ($\psi \ll 25$ mV) is often not a good one in the treatment of colloid and biosurface phenomena. Unapproximated solutions of equation (36) have been obtained numerically.

For small particles, however, the factor $1/r$ causes the electrical potential to fall off more rapidly than the purely exponential expression obtained for the flat double layer. If indeed the particle is small in comparison to the thickness of the double layer ($1/\kappa$ = Debye length), the charge in the surrounding ionic atmosphere going from the particle surface to the bulk of the solution must be distributed among spherical shells of increasing volume. This explains the characteristic difference between spherical and flat double layers, and shows that the equations for the spherical double layer are especially important for $\kappa a \ll 1$. Müller [11] calculated numerically the electrical potential around spherical particles for large values of the potential. Comparison of the results of Müller with those corresponding to the Debye–Hückel equations shows that the differences are not very large, though it should be

Table 4. Illustration of the difference between the approximation of Debye and Hückel and the theory of Müller for spherical particles [Ref. 11]

κr	$\kappa(r-a)$	\multicolumn{2}{c}{$\dfrac{ze\psi}{kT}$}	
		Müller [11]	Debye–Hückel
0.2	0.0	2.83	2.78
0.5	0.3	0.82	0.82
1.0	0.8	0.25	0.25

considered that Müller's results do not go to very high values of the potential where the deviations should be greater (Table 4). Accurate analytical expressions for the potential distribution around a spherical particle as well as the surface potential/surface charge density relationships have been derived by Ohshima et al. [31].

1.3. The inner part of the double layer

The treatment of the diffuse double layer outlined in the previous section is based on the assumption of point charges in the electrolyte medium. The finite size of the ions will, however, limit the inner boundary of the diffuse part of the double layer, since the center of an ion can only approach the surface to within its hydrated radius without becoming specially adsorbed. Stern (1924) [12] proposed a model in which the double layer is divided into two parts separated by a plane (the Stern plane) located at about a hydrated ion radius from the surface (δ), and also considered the possibility of specific ion adsorption at the inner Helmholtz plane (β) (Fig. 5).

Specifically adsorbed ions are those which are attached to the surface by electrostatic and/or van der Waals forces strongly enough to overcome thermal agitation. They may be dehydrated partially. The centers of any specifically adsorbed ions are located in the Stern layer. Ions with centers located beyond the Stern plane form the diffuse part of the double layer, for which the Gouy–Chapman treatment outlined in the previous section, with ψ_o replaced with ψ_δ, is considered to be applicable.

When specific adsorption takes place, counter-ion adsorption usually predominates over co-ion adsorption and a typical double layer situation would be depicted in Fig. 5a. However, it is possible that the adsorption of surface active co-ions could create a situation in which ψ_δ has the same sign as ψ_o and is greater in magnitude (Fig. 5b), or with the adsorption of polyvalent or surface-active counter-ions, even the reversal of charges may take place within the Stern layer, i.e. ψ_o and ψ_δ have opposite signs (Fig. 5c).

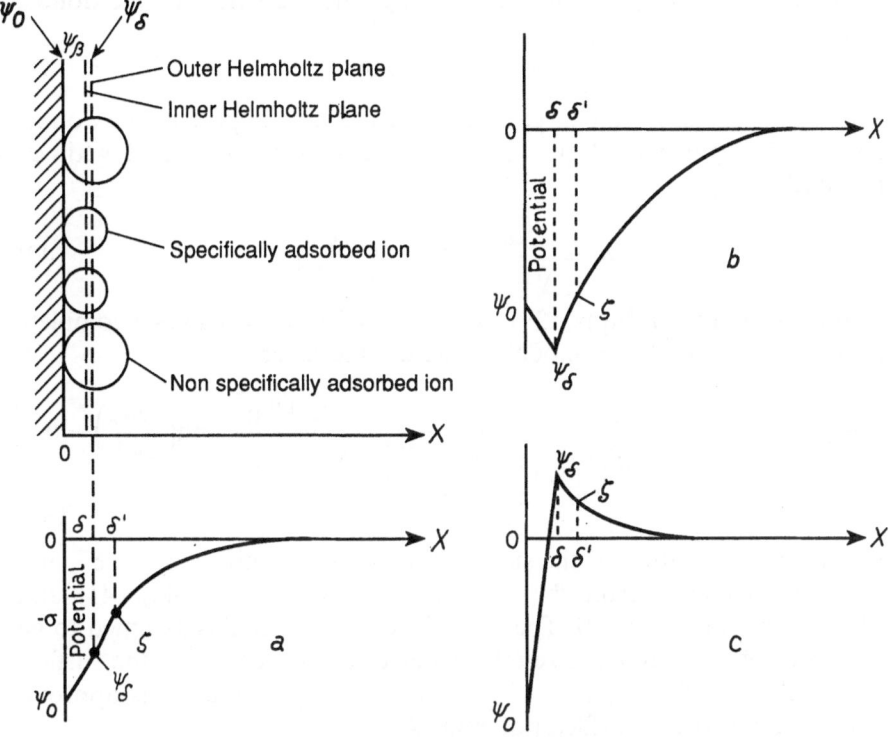

Fig. 5. A schematic diagram of the Stern adsorption layer (top left) and the average potential profile of the Stern layer and Gouy–Chapman diffuse double layer for the three different types ((a), (b), and (c)) described in the text.

Stern assumed that a Langmuir-type adsorption isotherm could be used to describe the equilibrium between ions adsorbed in the Stern layer and those in the diffuse part of the double layer. Considering only the adsorption of counter-ions, the surface charge density σ_1, of the Stern layer due to adsorbed ions is given by the expression

$$\sigma_1 = -\sigma \bigg/ \left\{1 + \left(\frac{N_A}{n_o V_m}\right) \exp\left[\frac{ze\psi_\delta + \phi}{kT}\right]\right\} \tag{39}$$

where σ is the initial surface charge density or proportional to the total number of adsorption sites of the surface, N_A is Avogadro's number, and V_m is the molar volume of the solvent. The adsorption energy is divided into electrical ($ze\psi_\delta$) and van der Waals (ϕ) terms.

Treating the Stern layer as a molecular condenser of thickness δ and with a dielectric constant ε',

$$\sigma = \frac{\varepsilon'}{4\pi\delta} (\psi_o - \psi_\delta), \tag{40}$$

where σ is the surface charge density.

For overall electrical neutrality throughout the whole of the double layer,

$$\sigma = -(\sigma_1 + \sigma_2) \tag{41}$$

where σ_2 is the surface charge density of the diffuse part of the double layer and is given by Eqn. (32) with the sign reversed and with ψ_o replaced by ψ_δ.

$$\sigma_2 = \sqrt{\frac{2\varepsilon kTn_o}{\pi}} \sinh\left(\frac{ze\psi_\delta}{2kT}\right). \tag{42}$$

Substitution from Eqns. (39), (40) and (42) into (41) gives a complete expression for the Stern model of the double layer,

$$\frac{\varepsilon'}{\delta}(\psi_o - \psi_\delta) - \frac{\sigma}{1 + \dfrac{N_A}{n_o V_m}\exp\left[\dfrac{ze\psi_\delta + \phi}{kT}\right]} - \sqrt{\frac{2\varepsilon kTn_o}{\pi}}\sinh\left(\frac{ze\psi_\delta}{2kT}\right) = 0. \tag{43}$$

In the above equation, if the surface potential ψ_o is given, σ, σ_1, σ_2, and ψ_δ can be calculated from the combination of Eqns. (39), (40), (41), and (42) for the case of $\phi = 0$. The Stern layer's distance δ is assumed to be the order of a molecular size (the order of a few Å). When the surface potential is small compared to kT/e, the Stern potential ψ_δ is approximately equal to the surface potential ψ_o.

The above mentioned adsorption is similar to that for a Langmuir type adsorption where the adsorbed molecules are not mobile. The modification required to take into account the mobility of adsorbed molecules has been made by several authors.

1.4. Surface potentials

A static electrical polarization observed at an aqueous interface of amphipathic substances is often called the "surface potential", V, which is due to the contribution of electric polarization of the interfacial molecules and also that of a partial separation of positive and negative ions of the electrolyte brought about by the interface fixed electric charges, and dipole, dielectric constant, and van der Waals forces, including ion adsorption discussed later. It should be noted that the "surface potential" has a slightly different meaning from the surface potential ψ_o (or $\psi(0)$) described before.

Such a potential, "surface potential", V, was first measured at the air-aqueous interface [13]. An accurate theoretical prediction of the measured surface potential in the above system is obviously difficult because of the uncertainty involved in assessing the surface electric dipole moment or polarization and an accurate dielectric constant of the electrolyte solution near the interface of the film. Also a possible breakdown in the Poisson–Boltzmann equation causes some difficulties since the ionic strengths needed to produce accurately measurable potential changes are large. However, the surface potential changes ΔV are interpreted approximately by means of the equation (also see Fig. 6) [14, 15]:

$$\Delta V = \psi_D + \psi_o, \qquad (44)$$

where ψ_D is the effective dipole change normal to the interface created by the addition of one film (monomolecular layer) and counter-ions, and ψ_o is the double layer potential at the interface relative to that far away in the bulk. The potential ψ_o can be derived by solving the Poisson–Boltzmann equation for a certain ideal case, which has been discussed in the previous section.

The potential change due to the effective dipole normal to the interface can be divided into three parts:

$$\psi_D = 4\pi n_1 \mu_1 + 4\pi n_2 \mu_2 + 4\pi n_3 \mu_3, \qquad (45)$$

where μ_1 is the dipole moment or polarization of the film (monolayer) molecule at the air-film interface, μ_2 is the dipole moment or polarization of the bulk component of the film molecule, μ_3 is the dipole

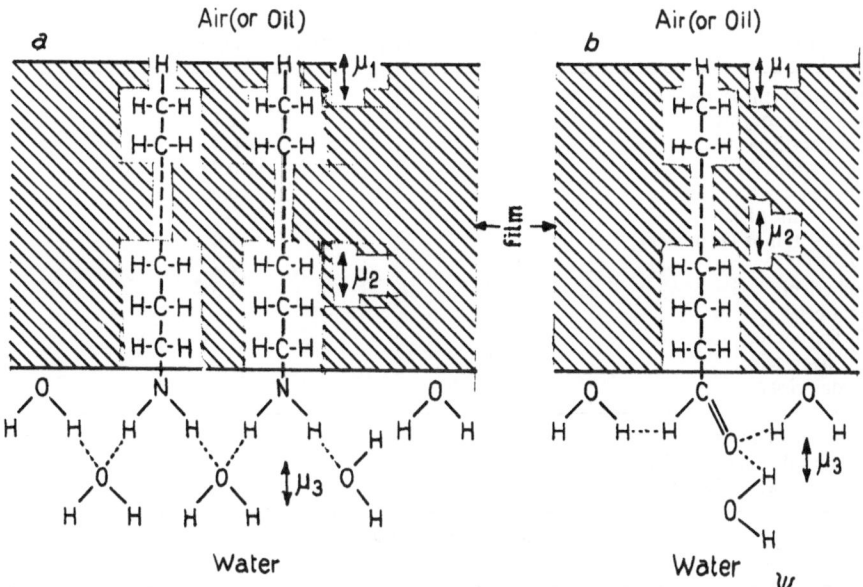

Fig. 6. Components, ΔV_1, ΔV_2, ΔV_3 and ψ of the "surface potentials" ΔV for electrically neutral (a) and charged (b) monolayers, respectively, formed at the air/water (or oil) interface. $\Delta V_i = 4\pi n \mu_i$ are the double layer potential due to molecular dipoles μ_i, and ψ is the double layer potential due to the fixed surface charges and electrolyte solution. $\Delta V = \Delta V_1 + \Delta V_2 + \Delta V_3 + \psi$.

moment (polarization) of the hydrophilic polar group/aqueous molecular (H_2O + ions) complex at the film-water interface, and n is the number of molecules per unit area. The value of μ_1 is the quantity which can be measured (Table 5); the value of μ_2 is usually negligibly small or small but constant, since the membrane interior is composed of hydrocarbon chain substances and as a whole does not possess a permanent dipole, or, if any, its magnitude is small. The value of μ_3 can be evaluated by subtracting the contribution of μ_1, μ_2, and ψ_o from the total surface potential change ΔV, or by summing up the normal component of each molecular bond dipole of the molecular complex. However, the theoretically detailed molecular orientation of the polar groups, H_2O and ions, is difficult to determine. The orientation of each polar segment (bond polarization) depends upon the structure of the molecular complex and the molecular packing, and also depends upon the type of interaction with H_2O molecules and ions from the electrolyte solution. Spectroscopic studies – Raman, infrared, and magnetic resonance (ESR and NMR) – may provide useful information about detailed molecular structure and packing. The surface potential due to a film formed at an oil-aqueous interface can also be analyzed in a manner similar to that above.

Table 5. Electrophoretic data for various calls[a]

Cell type	Electrophoretic mobility (μm sec^{-1} V^{-1} cm) (25°C)	σ electronic charge cm$^{-2} \times 10^{-13}$
Human erythrocytes	-1.16[b]	0.73
Dog erythrocytes	-1.28	1.1
Toad erythrocytes	-0.81	0.5
Pig erythrocytes	-0.88	0.75
Chicken erythrocytes	-0.82	0.69
Epithelium (Toad)	-1.13	—
Kidney tumor (Hamster)	-1.15	1.0
SV-3T3	-1.72	2.2
Escherichia coli	-1.12	1.31
Hela cells	-0.98	—
Ehrlich ascites	-1.07	1.31
Rat liver cells	-1.01	—
Sendai virus	-1.02[c]	—

[a]References 18 and 57.
[b]Standard saline, pH 7.2, 25°C.
[c]0.15 M NaCl, pH 7.4, 24°C.

Lipid vesicles	Electrophoretic mobility[d] (μm sec^{-1} V^{-1} cm)	Surface charge
Multilamellar phosphatidylserine vesicle	-4.4	$-e/200$ Å^2 for $R_{Na} = 0.76$ M^{-1}
Multilamellar phosphatidic acid vesicle	-5.0	$-e/175$ Å^2 for $R_{Na} = 0.38$ M^{-1}

[d]0.1 M NaCl, pH 6.0, 24°C.

As for the calculation of the double layer potential, ψ, there are several assumptions [14–16] used in calculating them theoretically which do not correspond to the real systems, as briefly mentioned in the above. They are:

a) Point charge assumption. It is well known that if high but possible values of the potential ψ and bulk ion concentrations are substituted into the Boltzmann equation (Eqns. 20 or 59), impossibly high surface concentrations of ions are often predicted. This problem arises from the assumption of point charges instead of ions of finite volume. A number of authors have attempted to introduce the ionic volume correction into double layer theory by rewriting the Boltzmann equation with some form of space restriction factor [17, 18].

For a long molecule bearing fixed charges at various points on the molecule, it is apparent that the Gouy–Chapman double layer theory also may not apply. Carnie and McLaughlin (1983) [19] extended the Gouy–Chapman theory to the case where bivalent cations are rod-like or string-like ions. They derived a modified Poisson–Boltzmann equation describing the potential distribution near a flat membrane surface, which is negatively charged, for the case of no ion adsorption onto the surface. Ohshima and Ohki (1991) [20] recently have provided a simple method to obtain an approximate solution for the modified Poisson–Boltzmann equation derived by Carnie and McLaughlin [19] for the case of rod-like bivalent cations. The approximate formula yields the surface potential $\psi(x)$ at $x < 5$ Å which deviates from that of the exact formula within errors of less than 3%. It was found that the rod-like bivalent cations behave like univalent cations having an effective concentration C^{eff}.

Effective charges: the concentration of either multivalent ions or relatively large molecules having multiple charges in the diffuse double layer is described by the Boltzmann distribution:

$$n(x) = n(\infty) \exp(z_f e((\psi(x) - \psi(\infty))/kT) \qquad (46)$$

where $n(\infty)$ and $\psi(\infty)$ are the concentration of molecules and the electrical potential at a reference state. The experimental results obtained by use of the above equation often fit well with use of effective charge z_f which is about a half of the intrinsic valency: $z_f = z/2$. This may be due to the difficulty of the point charge approximation for a finite size of these ions or molecules. However, once these charged ions or molecules are adsorbed onto the colloidal surface, the full charge of such ions or molecules seems to contribute to the total surface charge density [21].

b) Uniform dielectric constant assumption. The Poisson equation contains a term for the continuous dielectric constant of the medium in which the space charge is located. This simple assumption is usually used in calculation of the diffuse double layer potential: the dielectric constant is the same as that in the aqueous solution. However, it is expected that the dielectric constant would vary from a high value in the bulk aqueous solution to a lower value in the vicinity of the membrane surface. The accumulation of ions in the diffuse double layer should influence the dielectric constant of the medium and also the molecular adsorption, including water molecules as well as chemically bound ions at the membrane surface which contribute to the change in the dielectric constant [22, 23]. There are several semi-empirical formulas used to calculate the dielectric constant near the membrane surface. The complete theoretical analysis of the behavior of the dielectric constant near the membrane surface is not yet well established, although there have been a number of attempts to elucidate this problem [22–24]. Some theoretical studies of the electrical potential near the membrane surface using the Gouy–Chapman–Stern theory were successfully made to explain the experimental results, assuming a water dielectric constant $\varepsilon = 78.4$ up to the membrane solution interface [25]. However, the theory cannot be used when the fixed charges are buried within the low dielectric interior of the membrane [26, 27].

c) Non-ion polarization assumption. When an ion is brought from the bulk phase to the diffuse double layer by an electric field, the ion becomes polarized. This polarization contributes an extra energy term to the Boltzmann equation [17, 23].

d) The self-atmosphere effect of the counterions [28–30]. In the double layer as in the bulk phase, there may be a finite free energy of interaction between neighboring ions. Loeb [28] and Williams [29] have both considered this correction to the Poisson–Boltzmann equation. By comparison with the uncorrected Poisson–Boltzmann result, the self-atmosphere effect reduces the potential, and the degree of the reduction becomes less as x increases.

e) Smeared charge assumption for surface charges. The surface charges of the biomembrane surface are obviously discrete rather than uniformly distributed. The effect of discrete membrane surface charges has been studied by a number of investigators [25, 26, 32–38]. Some recent developments in this area will be discussed in a later section.

f) Other assumptions. Levine [39] pointed out that the electrostriction term should be corrected for the Poisson–Boltzmann equation. However, the magnitude of this term depends on ψ^2 and is still negligibly small even at the highest potentials normally encountered in diffuse double layers. Also, as was pointed out above, when the dielectric medium is not uniform, the existence of charge in the medium gives rise to a complex electrical potential treated as an "image potential" effect.

This would affect the ion distribution near the membrane which in turn affects the diffuse double layer potential. Other noncoulombic interaction factors between the counter-ions and the membrane surface should be considered as a correction to the Poisson–Boltzmann equation.

1.5. Electrokinetic phenomena

Electrokinetic phenomena are the general description applied to electrical phenomena which arise when the mobile part of the electric double layer moves in the direction opposite to that of a charged surface. When an electric field is applied tangentially along a charged surface, a force is exerted on both parts of the electric double layer. The charged surface tends to move in the appropriate direction to the electric field (electrophoresis), while the ions in the mobile part of the double layer show a net migration in the opposite direction of the charged surfaces, carrying solvent along with them, thus causing its flow. The movement of electrolyte solution relative to a stationary charged surface by an applied electric field is electro-osmotic flow. The pressure necessary to counterbalance the flow is termed the electro-osmotic pressure. Conversely, when the charged surface and the diffuse part of the double layer are made to move by the external force (e.g. hydrostatic pressure) relative to each other, an electric field is created along the direction of the tangent to the charged surface – the streaming potential [6, 14].

There are at least four electrokinetic phenomena (electrophoresis, electro-osmosis, streaming potential, and sedimentation potential), all of which involve both the theory of the electric double layer and that of liquid flow. Among them, electrophoresis has the greatest practical applicability to the study of the suspension of biocolloids and biological cells. In this section, the relation between electrophoretic mobility and its related electrokinetic potential will be discussed.

a) ζ-potential: When a particle surrounded by an electric double layer is subjected to an applied electric field, the Stern layer and a part of the diffuse double layer move with the particle. The electrical potential at the plane of shear between the bound and free parts of the double layer is called the zeta potential (ζ). It is considered that the shear plane is usually located at a small distance further out from the surface than the Stern plane and that ζ is generally marginally smaller in magnitude than the Stern potential (ψ_δ).

If we consider a spherical particle of radius a and electric charge Q, moving with a uniform velocity u in a field of unit electrical potential gradient $\bar{E}$, the electrical force on the particle, $Q\bar{E} = Q$, will be balanced by the viscous resistance which is assumed to be expressed by Stokes' law:

$$Q = 6\pi a\eta u, \qquad (47)$$

where η is the viscosity of the medium.

If it is assumed that the particle radius a includes a part of the double layer which moves with the particle, then the zeta potential ζ can be expressed as the work done in bringing a unit positive charge from infinity to a distance a from the center of the particle. The force, F, on the unit positive charge at any point r from the center of the particle is:

$$F = Q/\varepsilon r^2, \quad \text{(C.G.S. units).} \tag{48}$$

Therefore,

$$\zeta = -\int_\infty^a F\, dr = -\frac{Q}{\varepsilon} \int_\infty^a \frac{dr}{r^2} = \frac{Q}{\varepsilon a}. \tag{49}$$

Eliminating a between Eqns. (47) and (49), the mobility u of the particle is expressed in terms of ζ, ε, and η:

$$u = \frac{\varepsilon\zeta}{6\pi\eta}. \tag{50}$$

An alternative approach to this problem is to regard the double layer as a parallel plate condenser in which one plate is the particle surface and the other plate is a plane of counter-ions at a potential ζ located at a distance δ from the surface and moving with a velocity u relative to the particle surface. If the surface charge density is σ, the electrical force per unit area of the particle plate in a field of unit potential gradient will be σ, and this force will be balanced by the viscous resistance, which for an assumed Newtonian flow, leads to the equation:

$$\sigma = \eta \frac{u}{\delta}. \tag{51}$$

The capacitance (C) per unit area of the double layer will be given by:

$$C = \frac{\varepsilon}{4\pi\delta} = \frac{\sigma}{\zeta}. \tag{52}$$

Eliminating σ and a from Eqns. (51) and (52), the mobility is

$$u = \frac{\varepsilon\zeta}{4\pi\eta}. \tag{53}$$

Equation (50) differs by a factor of 2/3 from Eqn. (53). The latter equation was first derived by Smoluchowski [40] and the former by Hückel [41]. The latter equation is used for values of κa greater than 10^3, while the former for values of κa smaller than 1. Both equations are limiting forms of a more general expression which was derived by Henry [42].

$$u = \frac{\varepsilon\zeta}{6\pi\eta} f(\kappa a). \tag{54}$$

The equations discussed above are all first-order approximations.

Actually the movement of particles causes a deformation of the diffuse double layer (relaxation effect), which alters the above equation.

However, by using the above simplified equation (53) a number of studies on biological cell [43–45] and lipid vesicle [46, 47, 85, 88, 90] surfaces in regard to the ζ potential and the degree of their surface charges have been made (Table 6). However, in view of the complicated molecular structures of biomembrane surfaces, the use of these simplified equations and the interpretation of the results from such analyses should be carried out with great care. Further studies of the theoretical analysis of cell electrophoresis are discussed in a later section (2.4) of this Chapter.

2. Effects of the double layer potential on biological membrane systems

2.1. Surface potential

2.1.1. Measurements of surface potential: In spite of the many difficulties mentioned in the previous section with the double layer theory, there have been many attempts to measure surface potential in order to correlate the surface potential of cells and their functions. These methods include:

1) The measurement of surface potential of amphipathic monomolecular films formed at either the air/water or the oil/water interface has been done by use of air-ionizing electrode or vibrating electrode methods [46–52, 4], measuring the potential drops across the monolayer at the interface as mentioned before. Since the measurable quantity as the surface potential is $\Delta V = \psi_D + \psi$ (Eqn. 44), if such quantities are measured at two different subphase electrolyte solutions, the difference between the two quantities will be equal to the difference between the corresponding double layer potentials, provided that the polarization potential term ψ_D is the same.

$$\Delta(\Delta V) = \Delta V(1) - \Delta V(2)$$

$$= \psi_D(1) + \psi(1) - \psi_D(2) - \psi(2) \equiv \psi(1) - \psi(2). \qquad (55)$$

2) Measurements of membrane conductance due to the permeation of small amounts of ions. The membrane conductance G may be written as

$$G = \sum_j (Z_j F)^2 U_j(C_j)/h$$

$$= \sum_j (Z_j F)^2 (U_j/h) K_j [C_j] \exp[-Z_j e(\psi_\infty - \psi_m)/kT] \qquad (56)$$

where (C_j) and $[C_j]$ are the concentrations of the jth ionic species in the membrane and in the bulk phases, respectively, U_j its mobility, K_j its

Table 6. Variations in μ_1 for various films at the air-water and oil-water interfaces[a]

Film	ΔV (mV)	$\Delta(\Delta V)$ (mV)	$\Delta\mu_1$ (vertical component of dipole differences in w-bonds) (mD)[b]	Differences in μ for w-bond and C–H bond, from bulk measurements (mD)
Myristic acid with carboxyl group ionized (25 Å²). Air-water	−50			
Perfluorododecanoic acid with carboxyl group ionized (25 Å²). Air-water	−950	−900	−600	−1800
Stearic acid (pH 8.2). Air-water	0			
w-Trifluorostearic acid (pH 8.2). Air-water	−1190	−1190	−800	−1800
Hexadecanoic acid on 1 N NaOH (at 66 Å²). Air-water	−28			
w-Bromohexadecanoic acid on 1 N NaOH (at 66 Å²). Air-water	−28	0	0	−1900
w-Bromohexadecanoic acid on 1 N NaOH (at 66 Å²). Oil-water	−160	−132	−230	−1900
Hexadecanoic acid (pH 4). Air-water (20 Å²)	+390			
w-Bromohexadecanoic acid (pH 4). Air-water (20 Å²)	−870	−1260	−660	−1900

[a]Reference 14.
[b]D = Debye unit = 10 esu cm.

partition coefficient between the membrane/aqueous phases, h the thickness of the membrane, and ψ_∞ and ψ_m are the electrical potential in the bulk and at the membrane surface, respectively; $\psi_\infty - \psi_m$ corresponds to the surface potential of the membrane. Therefore, by measuring membrane conductances under different ionic conditions, we can relate the difference in conductances to the difference in surface potential of the membrane under the two conditions. Then, by comparing a similar experiment made using a membrane that does not have a surface potential with a membrane that does have a surface potential, the absolute value of the surface potential can be determined [53, 54] (Fig. 7).

3) The distribution of paramagnetic amphiphiles between the membrane and bulk phases measured by electron spin resonance spectroscopy (EPR) [55]: The electrical surface potential can be deduced from the distribution of the charged paramagnetic amphiphile (e.g. alkylammonium nitroxide) from its EPR spectrum. The amplitude of the high-field resonance of the EPR spectrum is a function of the aqueous concentration of spin-label probe. The membrane/aqueous partition coefficient of the probe (θ) is determined from the amplitude of the high field resonance and the total spin concentration. A plot of θ^{-1} vs V/m, where V/m is the total volume per unit mass of colloidal

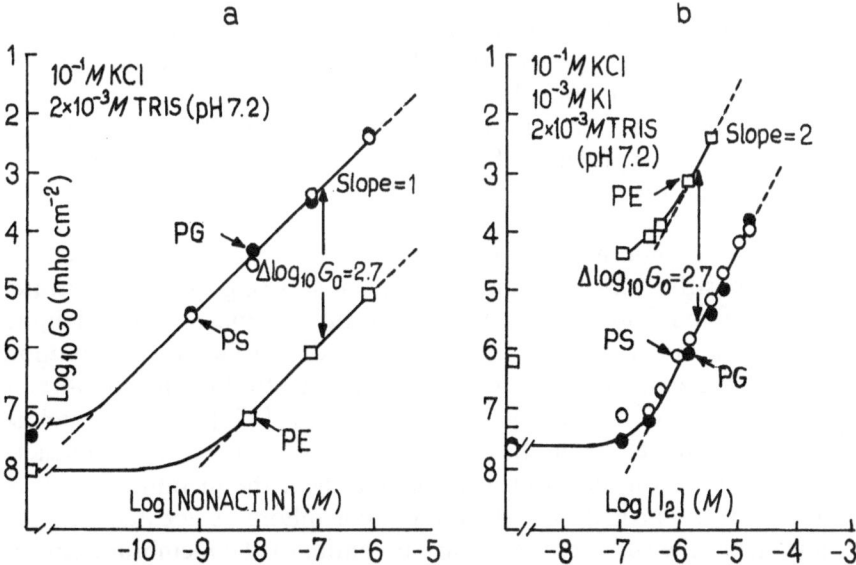

Fig. 7. (a) The effect of the antibiotic nonactin on the conductance of a neutral (phosphatidylethanolamine, PE) and two negatively charged (phosphadylserine, PS and phosphatidylglycerol, PG) lipid bilayer membranes. The permeant species is the positively charged nonactin-K^+ complex. (b) The effect of the neutral iodine molecule on the conductance of bilayers formed from the same lipids. The permeant species is the negatively charged I^- complex. The change of 2.7 log units in the conductance implies that the surface potential at the surface of the negatively charged membrane is $-158\,mV$ (from McLaughlin et al. [53]).

molecules in the sample, yields a straight line relationship with a slope χ [56]. The surface potential is calculated using the following equation:

$$\psi_o = -(kT/ze) \ln(\chi/\chi_o) \qquad (57)$$

where χ_o is the slope of the binding curve for the uncharged surface of colloidal particles and z is the valence of the probe.

4) The measurements of the electrophoretic mobility of liposomes, [46, 47, 85, 88, 90] cells, and organelles [43–45, 57]. By this method, one can observe the ζ-potential of the cell surface. It will be seen later (electrokinetic potential) that this potential (ζ) can be related to the surface potential, ψ_o (or $\psi(0)$).

5) Another method to measure cell surface potential is the use of fluorescence spectroscopy. There are at least three different such methods: 1) fluorescent chromophores (e.g. coumarin) attached to a long hydrocarbon chain are introduced into the membrane surface as probes and utilizing the variation of fluorescence intensity of these probes as a function of H^+ concentration at the membrane surface, the magnitude of the surface potential can be determined [3, 58, 59]. This is done by comparison with that obtained for an uncharged membrane, since the concentration of H^+ is determined by the magnitude of surface potential according to the Boltzmann distribution. 2) Another is the use of univalent ionic fluorophores [38, 85] (e.g. univalent cation 2-p-tolu-idinylnapthalene-6-sulfonate (TNS) derivative), which are adsorbed on the membrane surface due to their hydrophobic parts with a certain partition between the membrane and aqueous phases. The surface concentration of the fluorophore is decided by the Boltzmann factor through the surface potential. Since the adsorbed fluorophore enhances its fluorescence, the adsorbed number of fluorophores and in turn the magnitude and the sign of the surface potential can be estimated from the extent of the fluorescence signal. 3) Another type of measurement is the use of positively charged univalent quenchers (e.g. thallium Nitrate, tempanmine) on fluorophores (e.g. anthraniloyl) located at a definite position on a molecule which extends from the surface to the solution [25]. The concentration of quencher adjacent to the fluorophore is obtained by use of the Stern–Volmer equation [176] and the fluorescence measurements. From the concentration of the univalent quenchers at a certain location, the electrical potential near the membrane surface is determined. Particularly when the fluorophore is located at zero distance from the surface, the surface potential of the membrane can be determined by this measurement.

2.1.2. Ion concentration near the interfacial regions: As seen in the previous section, a fixed charged surface creates a self-consistent electrical potential and ion distribution near the surface according to the Poisson–Boltzmann equation. There will be accumulation of counter-ions and depletion of co-ions near the membrane surface which are expressed by

$$C_j(x) = C_j^0 \exp\left(-\frac{z_j e \psi(x)}{kT}\right), \tag{58}$$

where C_j^0 is the bulk concentration of the jth ionic species. Especially the ion concentration at the surface is given as

$$C_j^{surf} = C_j^0 \exp\left(-\frac{z_j e \psi(0)}{kT}\right). \tag{59}$$

The surface concentrations of univalent and bivalent cations for various surface potential values are shown in Table 7. Most biocolloid and biomembrane surfaces are as a whole negatively charged and therefore positive ions are usually concentrated at the surface with respect to the bulk solution. This gives various effects on the biocolloid surface properties: 1) pH of the surface is usually lower ($1 \sim 2$ pH units) than that of the bulk solution which, of course, depends on the surface charge. The lower pH at the surface will affect the apparent dissociation of the dissociable polar groups at the surface, which has been pointed out in the earlier section (1.1). Also, since counter-ion concentrations are enhanced at the surface by the double layer potential, the adsorption of ion or binding of ion will be affected. This will be discussed in the following section. However, when ions have multi-valence, it seems to be necessary to use an effective valence for the ions in the Boltzmann factor, which is smaller in value than their inherent values (e.g. $z_{eff} = z/2$). This is due to semi-empirical assessment but not well explained theoretically. The reduction of the valence is considered to be due to the use of a point charge assumption for the ion which has a certain volume as well as the distribution of charges within the ion. 2) Since the ion concentration at the surface is different from the bulk, the transport of ions across the membrane may differ between charged and uncharged membranes in certain situations.

The adsorption of charged molecules to membrane surfaces is due to at least two factors: electrostatic and non-electrostatic interaction between the molecule and the membrane. Unlike ions, non-electrostatic interaction (e.g. hydrophobic interaction etc.) plays a major role for some types of molecular adsorption. However, even in such cases, the effect of surface concentration of the molecule, which is subject to the Boltzmann distribution by the surface potential, becomes significant for molecular adsorption onto the membrane surface. In this case also, the use of an effective valence of the molecule in the Boltzmann factor is needed, and the value should be different for each molecule with respect to its molecular configuration and charge distribution.

Also, when macromolecules like proteins are adsorbed on the charged surface, since the pH of the solution at the surface is different from the

Table 7. Enhancement of ion concentration at charged membrane surfaces (25°C)

	H+			M+		M-		M²⁺	
	$C = C_o = 0$	$C = C_o \exp(-e\psi_o/kT)$		$C = C_o \exp(-e\psi_o/kT)$		$C = C_o \exp(-e\psi_o/kT)$		$C = C_o \exp(-2e\psi_o/kT)$	
C_o	$\psi_o = 0$	$\psi_o = -20\,mV$	$\psi_o = -60\,mV$	$\psi_o = -20\,mV$	$\psi_o = -60\,mV$	$\psi_o = -20\,mV$	$\psi_o = -60\,mV$	$\psi_o = -20\,mV$	$\psi_o = -60\,mV$
M	pH	pH	pH	(M)	(M)	(M)	(M)	(M)	(M)
10^{-6}	6	5.66	4.96	2.2×10^{-6}	1.1×10^{-5}	0.4×10^{-6}	0.9×10^{-8}	4.0×10^{-6}	1.2×10^{-4}
10^{-5}	5	4.66	3.96	2.2×10^{-5}	1.1×10^{-4}	0.4×10^{-5}	0.9×10^{-7}	4.0×10^{-5}	1.2×10^{-3}
10^{-4}	4	3.66	2.96	2.2×10^{-4}	1.1×10^{-3}	0.4×10^{-4}	0.9×10^{-6}	4.0×10^{-4}	1.2×10^{-2}
10^{-3}	3	2.66	1.96	2.2×10^{-3}	1.1×10^{-2}	0.4×10^{-3}	0.9×10^{-5}	4.9×10^{-3}	1.2×10^{-1}
10^{-2}	2	1.66	0.96	2.2×10^{-2}	1.1×10^{-1}	0.4×10^{-2}	0.9×10^{-4}	4.0×10^{-2}	1.2
10^{-1}	1	0.66		2.2×10^{-1}	1.1	0.4×10^{-1}	0.9×10^{-3}	4.0×10^{-1}	12.0

bulk, the conformation of the macromolecules may be altered upon adhesion according to the change in pH at the surface. This type of alteration is possible when two cells adhere to each other at a close distance; and if the protein on one cell surface experiences a more acidic surface of the other cell, resulting in conformation of the protein becoming more hydrophobic according to its pK_a value [60, 61]. This alteration to a more hydrophobic nature in the protein would make the protein interact strongly with the membrane core, and the part of the protein may penetrate into the other membrane and consequently may bridge and join together two membrane cores resulting in membrane fusion [62]. This situation may be considered to occur in the fusion processes of viruses with their host cells, leading to infection of the host cells by the viruses [63].

2.1.3. Contribution of surface potential on transmembrane potential: The transmembrane potential, E, is expressed as a difference between the two electrical potentials, φ_i and φ_o ($E = \varphi_i - \varphi_o$) of the two bulk phases separated by a membrane (see Fig. 8). Normally, the transmembrane potential is expressed only in terms of the diffusion potential across the membrane according to Goldman [64], Hodgkin, and Katz [65]. Although the contribution of the membrane boundary potential

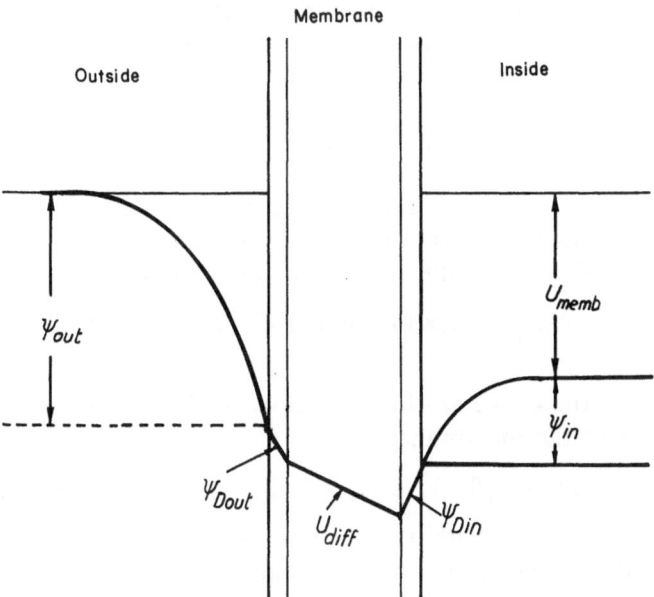

Fig. 8. A schematic transmembrane potential E_{memb} (from Ohki [72]): ψ_{out} = outer surface potential, ψ_{in} = inner surface potential, ψ_{Dout}, ψ_{Din} = polarization potentials due to the membrane molecular dipoles: $\psi_{Dout} \neq \psi_{Din}$ for asymmetrical polarization potentials, E_{diff} = diffusion potential.

to the transmembrane potential was studied by several authors ($E = E_o + E_{diff} - E_i$) where E_o and E_i are the boundary potentials expressed by the Donnan equlibrium potential (see Refs. 66, 79), the effect of surface potential on the transmembrane potential has been considered by several authors [67–72]. Recently, Ohki [73] has developed theoretical as well as experimental analyses for such transmembrane components (surface potentials, diffusion potential) of membrane potential for lipid membranes [72] as well as squid axon membranes [73]. According to Ohki [72], the transmembrane potential is expressed by

$$E = \varphi_i - \varphi_o = V_o - V_i + E_D \tag{60}$$

where V_o and V_i are the "surface potentials" at the two membrane interfaces (o and i refer to outside and inside, respectively) and E_D is the diffusion potential in the membrane. The surface potential is expressed by

$$V = \psi + 4\pi n\mu_D \tag{61}$$

where the first term is the Gouy–Chapman double layer potential at the membrane surface, and the second term is the polarization (permanent dipole) potential ($4\pi n\mu_D = \psi_D$) of the molecules of the surface membrane, where n is the number of dipoles per unit area and μ_D is the dipole moment of a membrane molecule (Fig. 8).

In the case of a membrane symmetrical with respect to its molecular constituents, the terms due to the polarization potential of the surface membrane in Eqn. (60) may cancel each other and the equation becomes:

$$E = \psi_o - \psi_i + E_D \tag{62}$$

where ψ_o and ψ_i are the surface potentials at the two membrane surfaces. Then the concentrations of univalent cation $[C^+]_o^s$ and univalent anion $[C^-]_o^s$ at the outer surface of the membrane are given by

$$[C^+]_o^s = [C^+]_o \exp(-e\psi_o/kT) \quad \text{and} \quad [C^-]_o^s = [C^-]_o \exp(+e\psi_o/kT) \tag{63}$$

and, similarly, the concentrations of univalent cation and anion at the inner surface of the membrane are

$$[C^+]_i^s = [C^+]_i \exp(-e\psi_i/kT) \quad \text{and} \quad [C^-]_i^s = [C^-]_i \exp(+\psi_i/kT). \tag{64}$$

The diffusion potential within the membrane may be expressed by the Goldman [64] and Hodgkin and Katz equation [65] as follows:

$$E_D = (RT/F) \ln\{P^+[C^+]_o^s + P^-[C^-]_i^s)/(P^+[C^+]_i^s$$
$$+ P^-[C^-]_o^s)\} \quad \text{(G–H–K Eqn.)} \tag{65}$$

For precision, the concentration [C] should be replaced by the activity of each ionic species, where P is the permeability coefficient of ionic species through the membrane. The permeability coefficient is defined as $P^+ = RTbU^+/h$, where b is the partition coefficient of the ion at the membrane and aqueous phases, U^+ is the mobility of a cation, and h is the thickness of the membrane, where

$$\left(RT/F = \frac{NkT}{Ne} = \frac{kT}{e} \right).$$

As for the diffusion potential, Eqn. (65) can be rewritten with the use of Eqns. (63) and (64):

$$E_D = (RT/F) \ln\{([C^+]_o Y_o P^+ + [C^-]_i Y_i^{-1} P^-)/([C^+]_i Y_i P^+$$
$$+ [C^-]_o Y_o^{-1} P^-)\} \qquad (66)$$

where $Y_o = \exp(-e\psi_o/kT)$ and $Y_i = \exp(-e\psi_i/kT)$. Therefore, the total transmembrane potential can be expressed by:

$$E = \left(\frac{RT}{F} \right) \ln \frac{\sum_j P_j^+ [C_j^+]_o + \sum_j P_j^- [C_j^-]_i Y_i Y_o^{-1}}{\sum_j P_j^+ [C_j^+]_i + \sum_j P_j^- [C_j^-]_o Y_o Y_o^{-1}} + \psi_o - \psi_i. \qquad (67)$$

Although the Goldman–Hodgkin–Katz (G–H–K) [65] Eqn. (65) has been widely used to explain the membrane potential of many biological cells, the equation has been found not to be adequate to account for changes in resting membrane potential of squid axons, when the internal salt solution is altered in a certain manner [74–76], although attempts have been made [77] to explain the observed membrane potential by retaining the G–H–K equation and introducing modified ionic permeability factors. Recently, Ohki (73) has measured the membrane potential of squid axons as a function of the nature of the ionic species (ion substitution) and the ionic strength (dilution) of the extracellular medium, and has analyzed the observed membrane potentials using both the Goldman–Hodgkin–Katz equation and the surface/diffusion potential equation described above. He concluded that the membrane potential theory including the surface potential of the membrane (the surface/diffusion potential theory) is better able to analyze the observed potential of squid axon.

Whenever the membrane surfaces possess fixed charges, contributions from both the surface potential and the diffusion potential should be involved in the observed membrane potential. The extent of their contributions depends upon the surface charge densities, the magnitude of the surface potential, and also the relative permeabilities of cations and anions involved in the transport process. Since membrane surfaces of most biological cells (such as axons, muscles, etc) are highly charged [78, 79], the experimentally observed membrane potential should therefore involve both a surface potential and an ion diffusion potential.

However, many investigators have so far used solely the diffusion potential equation proposed by Hodgkin and Katz to explain observed membrane potentials for such highly charged membranes.

It should be noted, however, that membrane potentials observed in biological membrane systems involve various complex factors or polyvalent ions and their binding with the charge sites of membrane surfaces. As will be discussed in this chapter, individual ions, especially, bivalent or multivalent ions, have their own characteristic binding affinities for membrane molecules at their surfaces, and the degree of binding can vary from membrane to membrane. It has been shown that surface potential is intimately related to the degree of ion binding with the charged groups of membrane surfaces. Furthermore, when an ion binds strongly with the polar groups of the membrane, the orientation of polar groups at the membrane surface and also the dielectric constant of the polar group region of the membrane surface can be altered. Consequently, the net dipole moment of the membrane molecule can be altered, which would change the surface potential V (Eqn. 61). If the membrane molecular dipole is asymmetric with respect to the membrane, another factor of polarization potential will contribute to the transmembrane potential through the difference in surface potentials. It is known that in many biological membranes, the molecular constituents of the outer surface membrane are different from those of inner membrane surfaces [80, 81]. Therefore, the difference in the molecular dipoles between the two sides (outer and inner surfaces) of the membrane would contribute to the overall membrane potential.

In addition, biological membranes are believed to be composed of an inhomogeneous distribution of molecular assemblies, such as specific ionic channels [82], and molecular segregation with respect to the two-dimensional surfaces of the membranes occurs, which is quite a different situation from that in simple phospholipid membranes. In most cases, therefore, the potential difference across the membrane observed by a pair of electrodes placed at far distance from the membrane surfaces on both sides of the membrane may be controlled by the nature of ion permeation through the inhomogeneous molecular assemblies in the membranes.

2.2. Ion binding and adsorption onto surfaces

2.2.1. Site binding model: When the surface charged sites interact with ions in aqueous solution and bind to them (similar to an ionic bond), the net fixed surface charges will be altered. In order to obtain the free net charge, several interaction modes can be used to describe such ion binding reactions coupled with the surface electrical potential:

1. One charged site to one ion binding model
2. Two charged sites to one ion binding (chelation binding) model
3. Complex binding model

For model 1, we assume that one ion binds to one charged site. Then, such binding reactions can be described as:

$$[A^-] + [M^+] \rightleftarrows [AM] \qquad (68)$$

where $[M^+]$ is the concentration of monovalent cation. $[A^-]$ is the surface concentration of surface fixed charge sites, and $\bar{K}_1$ is the association constant for the above binding reaction:

$$\bar{K}_1 = \frac{[AM]}{[A^-][M^+]}. \qquad (69)$$

Therefore, the free net charge density σ is given by:

$$\sigma = \frac{\sigma^{int}}{1 + \bar{K}_1[M^-]^s} = \frac{\sigma^{int}}{1 + \bar{K}_1[M^+]_o \exp(-e\psi(0)/kT)}, \qquad (70)$$

where $[M^+]^s$ and $[M^+]_o$ are the concentrations of univalent cation, M^+, at the membrane surface and in the bulk solution, respectively, $\psi(0)$ is the electrical potential at the membrane surface, and σ^{int} is the surface charge density for the case of no ion-binding. With use of Eqns. (34) and (70) and knowing the measured surface potential ψ_o and the initial surface charge density σ^{int} one can obtain the binding constant for univalent cations $\bar{K}_i$ [53, 83–85]. In this calculation, no binding of anion to the charge site is assumed.

In the case of the presence of multiple univalent cations (such as K^+, Na^+, and H^+), the above formula can be rewritten as follows:

$$\sigma = \frac{\sigma^{int}}{1 + \sum_j \bar{K}_j[M_j^+]^s} = \frac{\sigma^{int}}{1 + \sum_j \bar{K}_j[M_j^+]_o \exp\left(-\frac{e\psi(0)}{kT}\right)} \qquad (71)$$

where

$$\bar{K}_j = \frac{[AM_j]}{[A^-][M_j^+]} \qquad (72)$$

and the suffix j refers to the jth monovalent cation.

In the case of bivalent cations, the one charged site binding model leaves one positive charge of bivalent cation bound to the site, and such a binding reaction can be expressed as:

$$[A^-] + [M^{2+}] \rightleftarrows [AM]^+, \qquad (73)$$

$$\bar{K}_1' = \frac{[AM]^+}{[A^-][M^{2+}]}. \qquad (74)$$

Therefore, the free net charge density σ is given by

$$\sigma = \frac{\sigma^{int}(1 - \bar{K}'_i[M^{2+}]^s)}{1 + \bar{K}'_i[M^{2+}]^s} = \sigma^{int} \cdot \frac{1 - \bar{K}'_i[M^{2+}]_o \exp\left(-\frac{2e\psi_o}{kT}\right)}{1 + \bar{K}'_i[M^{2+}]_o \exp\left(-\frac{2e\psi_o}{kT}\right)}. \quad (75)$$

In such a binding, the surface charge as well as surface potential could alter the sign of the surface charge as well as the potential, when the adsorbed amount of bivalent cation exceeds a certain value (half of the initial binding sites). As a general case, where many univalent and bivalent binding events occur, the surface charge density is expressed as

$$\sigma = \sigma^{int} \cdot \frac{1 - \sum_i \bar{K}'_j[M^{2+}]_{oj} \exp\left(-\frac{2e\psi(0)}{kT}\right)}{1 + \sum_j \bar{K}_j[M^+]_{oj} \exp\left(-\frac{e\psi(0)}{kT}\right) + \sum_j \bar{K}'_j[M^{2+}]_{oj} \exp\left(-\frac{2e\psi(0)}{kT}\right)}. \quad (76)$$

On the other hand, *for model 2*, the free net charge density, σ, is expressed by;

$$\sigma = \frac{\sigma^{int}}{1 + \bar{K}_2[M^{2+}]^s} = \frac{\sigma^{int}}{1 + \bar{K}_2[M^{2+}]_o \exp\left(-\frac{2e\psi(0)}{kT}\right)}, \quad (77)$$

where $\bar{K}_2$ is the association constant of the following binding reaction:

$$[A^{--}] + [M^{2+}] \rightleftarrows [AM], \quad \bar{K}_2 = \frac{[AM]}{[A^{--}][M^{2+}]}, \quad (78)$$

where $[A^{--}] \equiv [A^-]/2$. This corresponds to the case where one bivalent cation binds two nearest neighboring negative charge sites in a chelating manner for which the concentration is considered as half of the total concentration of negative charge sites. The association constant $\bar{K}$ has the dimension of M^{-1} which is the inverse of the dissociation constant described in the earlier section (1.1). By use of the above treatment, several investigations on ion binding to phospholipid membranes have been carried out [83–90] (Table 8).

For model 3 the binding scheme has been considered by Cohen and Cohen [91] in a more rigorous manner, especially for bivalent cation binding to the phospholipid membrane sites. They consider one site for one bivalent cation in addition to two fixed charge sites for one bivalent cation. By use of a statistical average method, various possibilities for nearest neighbor sites of bivalent cations were taken into account. A more realistic situation for ion binding schemes was introduced and gave a new binding constant at high bivalent ion concentrations. However, at a lower concentration of bivalent cations (in the range less than 1 mM), the binding scheme appears to be mainly 2:1 binding, and at a larger

Table 8. Intrinsic ion binding constants $\bar{K}$ (M^{-1}) for various phospholipid membranes

Ion membrane system													Reference
Phosphatidylcholine membrane 1:1 binding scheme													
Mn^{2+} 0.3	Mg^{2+} 1.0	Ca^{2+} 1.0	Ni^{2+} 1.2	Sr^{2+} 2.8	Ba^{2+} 3.6								[115]
Phosphatidylserine membranes 2:1 binding scheme													
La^{3+} 100	Mn^{2+} 49	Ba^{2+} 37	Ca^{2+} 35	Sr^{2+} 25	Mg^{2+} 40	Li^{+} 0.8		K^{+} 0.2					[83]
			Ca^{2+} 30	Sr^{2+} 30		Li^{+} 0.6	Na^{+} 0.6						[87]
			Ca^{2+} 10	Sr^{2+} 10		Li^{+} 0.4	Na^{+} 0.6						[89]
													[4]
Phosphatidylserine membranes 1:1 binding scheme													
Ni^{2+} 40	Co^{2+} 28	Mn^{2+} 25	Ba^{2+} 20	Sr^{2+} 14	Ca^{2+} 12	Mg^{2+} 8	Li^{+} 0.8	Na^{+} 0.6	NH_4^{+} 0.17	K^{+} 0.15	Cs^{+} 0.05	TEA 0.03	[88] [85]
Phosphatidylglycerol membranes 1:1 binding scheme													
Mn^{2+} 11.5	Cs^{2+} 8.5	Ni^{2+} 7.5	Co^{2+} 6.5	Mg^{2+} 6.0	Ba^{2+} 5.5	Sr^{2+} 5.0							[86]

concentration (>50 mM) the binding scheme becomes 1:1 (one site to one bivalent cation) [92].

2.2.2. Ion condensation model: In the ion condensation model the solution of the Poisson–Boltzmann equation for an infinitely long charged cylinder (cylindrical polymer ion) in an electrolyte solution leads to the following important result [93, 94]: the counter-ion concentration at the cylinder surface (i.e., the concentration within the "condensation" layer of the order of one or two molecular diameters from the surface plane) is independent of bulk counter-ion concentration. The counter-ions are assumed to bind to the polymer ion to reduce the effective charge density on the polymer ion surface to a constant value (σ_{crit}). For a given charge density (σ) on the polymer ion ($\sigma > \sigma_{crit}$), once the number of bound ions is large enough to give a net charge density of the critical value, no more ions would bind. Here, a critical value of the effective charge density is one unit charge per $ez/\varepsilon kT$ of length [93, 94], where z is the valence of the counter-ions.

The solution of the Poisson–Boltzmann equation for the case of two parallel charged plates with an intervening solution containing only counter-ions also gave the ion condensation [95, 96]. However, in this case the critical parameter is determined by the product of the surface charge density on the plate (σ) and the distance (d) between the plates:

$$-(\sigma z d)_{crit} = \frac{1.31}{4} (\varepsilon kT/e). \qquad (79)$$

In considering ion binding to a membrane surface, the above-mentioned condensation layer partially neutralizes the initial surface charge density, so that the quantity $f = 1 - \sigma/\sigma_o$ is independent of the concentration of counter-ions in bulk solution: here σ_o is the surface charge density when no counter-ions are present in the condensation layer, σ is the apparent charge density when neutralization of the surface by the counter-ions in the condensation layer is taken into account, and f represents the fraction of surface charge neutralized by condensed ions. Specificity should be taken into account in the ion condensation model, as it might be manifested in competition between two ionic species i and j for the condensation layer, by introducing a relative distribution parameter, K_{ij}:

$$[C_j]^s/[C_i]^s = K_{ij}[C_j]/[C_i] \qquad (80)$$

where $[C_i]^s$ and $[C_j]^s$ are the respective concentrations of the counter-ion species i and j in the condensation layer, and $[C_i]$ and $[C_j]$ are the respective concentrations of ions i and j in bulk solution. If the parameters f_i and f_j are defined as the fractions of surface charge neutralized, respectively, by ion species i and j in the condensation layer (with

$f_i = f_j$), then one obtains the following relation:

$$f_i = f/(1 + K_{ij}[C_j]/[C_i]) \tag{81}$$

where f, the total fraction of surface charge neutralized, is, as before, independent of the concentration of ions in bulk solution.

On the other hand, the binding site model mentioned in the preceding section assumes that counter-ions of type i bind independently to single sites at the membrane surface with an intrinsic binding constant K_j. Then, the fraction of surface charge neutralized by ion species i is:

$$f_i = Y_{io}K_i[C_i]/\left(1 + \sum Y_{jo}K_j[C_j]\right) \tag{82}$$

where $Y_{jo} = \exp(-ez_j\psi(0)/kT)$.

These two models have in common the following features: (1) the discrete nature of the surface and the solution at the interface is ignored (i.e., ignored in terms of a detailed molecular structure picture); thus the negatively charged membrane surface is represented by a uniform surface charge distribution, with density σ at the interfacial plane, and (2) the solvent is treated as a continuous medium, with uniform dielectric constant.

One obvious distinction between these two theoretical models is that the ion condensation theory gives a fractional neutralization (cation concentration in the condensation layer) which is independent of cation concentration in the bulk solution, while the binding site model does not explicitly predict such a result.

By comparing the above equations (Eqns. (81) and (82)) for the two ion adsorption models and investigating adsorption of counter-ions on membrane surfaces, it has been argued as to which theoretical adsorption scheme is a preferable model for ion binding to lipid bilayer membranes as well as biological membranes. A recent study of univalent cation adsorption on a phosphatidylserine vesicle membrane suggests that the ion binding site model is more appropriate for ion binding to negatively charged phospholipid membrane surfaces [97].

2.3. Adsorption of molecules to membrane surfaces

As mentioned in 2.1.2, the adsorption of molecules onto membrane surfaces is influenced not only by the factors of electrostatic and non-electrostatic interaction, but also the distribution of charges and that of the polar and non-polar molecular groups on both membrane and molecular surfaces. In addition to these factors, hydration interaction plays an important role for adsorption of large molecules, which will be described in the later section 5. Here, three examples of molecular adsorption to membrane surfaces will be described: a) local anesthetics, b) small peptides, and c) macromolecules.

a) Adsorption of local anesthetics onto membranes: Local anesthetics are amphipathic molecules which are composed of well defined polar and non-polar portions. The adsorption of local anesthetics is primarily due to partition of the hydrophobic portion of the molecules between the aqueous and membrane phases. However, local anesthetics usually contain a tertiary amine group which could possess a positive charge depending on the pH of the solution. In such a case, the local anesthetic can have an electrostatic interaction with the negatively charged site of the membrane, besides the hydrophobic interaction responsible for adsorption of the anesthetic molecule to the membrane. The adsorption of local anesthetic on neutral or charged membranes can be analyzed by use of the Gouy–Chapman double layer theory and the partition equation ($K = C_m/C_s$), where C_m and C_s are the concentrations of local anesthetic in the membrane and at the membrane surface in the solution, respectively [90].

b) Adsorption of small peptides: Adsorption of small peptides, particularly having basic residues, to membrane surfaces will be considered. Unlike local anesthetics, many such small peptides have both hydrophilic or hydrophobic portions and charged groups which are located within the molecule in a mixed way; therefore they are often not amphipathic molecules. In spite of such geometrical diversity, the binding of such molecules to membranes can be treated by use of the Gouy–Chapman theory to describe the accumulation of the peptide in the electrical diffuse double layer, and the mass action equations in expressing adsorption of the peptide to the membrane surface. The accumulation of peptides can be described by the Boltzmann relation by assuming the peptide to have a point charge:

$$P = P_o \exp(-z_{eff}\psi(0)/kT) \qquad (83)$$

where P_o is the concentration of the peptide in the aqueous solution immediately adjacent to the membrane, and z_{eff} is the effective valence of the peptide.

The electrical double layer potential enhances the binding of the basic peptide to the membrane because it increases its free concentration in the aqueous solution immediately adjacent to the surface. However, adsorption of the peptide from three dimensional space to a two dimensional plane can also produce binding with a sigmoidal dependence on mole fraction of binding sites on the membrane surface even if the peptide has only one binding site. If the peptide has more than one binding site, the reduction of dimensionality further enhances the sigmoidal dependence of binding [98]. In the case of a molecule having two identical and independent binding sites ($n = 2$), by applying the mass action equation, such as the one used for antigen-antibody reactions [177, 178], the fraction of binding sites of the molecule saturated by ligands (membrane binding sites), y, is expressed as

$$y = \frac{\bar{K}[L](1 + \bar{K}[L]_s)}{1 + \bar{K}[L](2 + \bar{K}[L]_s)} \qquad (84)$$

where $\bar{K}$ is the association constant ($\bar{K} = [PL]/2[P][L]$). [L] is the concentration of the membrane binding sites for the first binding site of the molecule and $[L]_s$ is the surface concentration of the membrane binding sites for the second binding site. This is equivalent to the binding of the first ligand (membrane binding site) producing an allosteric transition that increases the association constant $\bar{K}$ for the second site of the molecule, because $\bar{K}$ and $[L]_s$ appear as a product when $\bar{K}[L]_s \gg 1$. This corresponds to a Hill coefficient $\alpha = n = 2$ in the case of cooperative binding of ligands to a macromolecule analyzed by Hill [179] ($y = (\bar{K}[L])^\alpha/(1 + \bar{K}[L])^\alpha$). In addition to this, the inclusion of the electrostatic potential yields

$$y = \frac{\bar{K}[L] \exp(-ze\psi_o/kT)(1 + \bar{K}[L]_s)}{1 + \bar{K}[L] \exp(-ze\psi_o/kT)(2 + \bar{K}[L]_s)}. \qquad (85)$$

The electrostatic potential enhances the apparent cooperativity even more if the region of the macromolecule that binds to the membrane has a cluster of positive charges and most of the negative charges of the molecule are located more than a Debye length from the membrane surface. In such a case, the apparent Hill coefficient is greater than the number of the binding sites of the molecule, i.e. $\alpha > n$.

The above behavior of peptide adsorption has been demonstrated for the adsorption of lysine-5 on a phospholipid membrane containing various amounts of acidic lipids [98]. The effect of another small peptide, melittin, has been studied as a model for lipid-protein interaction systems by many workers. Melittin is a peptide consisting of 26 amino acids. It interacts with membranes and induces leakage and fusion. Recent studies on the interaction of melittin with phospholipid membranes indicate that melittin is adsorbed more than ten times more strongly on a phosphatidylserine (negatively charged) membrane than a phosphatidylcholine (neutrally charged) membrane. However, the phosphatidylcholine membrane is more susceptible to membrane leakage and fusion caused by melittin than the phosphatidylserine membrane. It is deduced that the state of adsorption of melittin on the neutrally charged membrane is different from that on the charged membrane. In the analysis of melittin adsorption, it was shown that the experimental results were analyzed well by using an effective charge of 2.3 for melittin in the solution but using the full fixed charge of 5 when it is adsorbed on the membrane surface [21].

The utilization of the double layer for the adsorption of macromolecules, such as proteins or polymers, onto the membrane surface seems to be ineffective since the size of the molecule is larger than the double layer distance (Debye length) in normal physiological environments. Also, almost all assumptions used for the Gouy–Chapman double layer theory may not hold for the adsorption of macromolecules onto the membrane surface. Major factors for macromolecular adsorption onto the membranes may be due to electrostatic, hydrophobic, and hydration interactions. It depends upon how these functional groups are distributed on the surface of the macromolecule; it also depends on the structure of the macromolecule, which is not unique but dependent on the environmental parameters temperature, pH, and salt concentration and composition. Nevertheless, there are some attempts to analyse such macromolecular adsorption phenomena by using the double layer theory.

2.4. Electrophoresis

When an electric field is applied to colloidal particles suspended in an electrolyte solution, the colloidal particles move according to the magnitude and size of the fixed charges of the particles. The mobility of a particle, u, is generally defined as the velocity of the particle, v, per unit electric field (E):

$$u = v/E. \tag{86}$$

When the term is applied to colloidal particles, the mobility is known specifically as the electrophoretic mobility. In this case, the overall phenomenon is known as electrophoresis and the specific experimental technique of direct microscopic observation of the electrophoretic mobility is called microelectrophoresis. Some examples of cell electrophoretic mobility are listed in Table 5.

For large colloidal particles ($>1\ \mu$m in diameter), including biological cells and lipid vesicles, the observations of particle velocity under a given applied voltage [45, 57] can be made by a visual method using an optical microscope and timer, or alternatively by the aid of electronic devices (rotating prism technique) to measure particle velocity [99]. For small colloidal particles (less than 1 μm) another method using a laser, which depends on the doppler effect of scattered light from the particles, is able to measure the mobility of particles of 1 μm $\sim$ 3 nm in diameter [100].

Although the electrophoretic mobility is a readily measurable quantity, its interpretation is considerably more difficult for colloidal particles than for small ions: e.g. the charge carried by a colloidal particle is not a constant known quantity as is the case for simple ions. The charge distribution affects the mobility of the colloidal particle because it is affected by the electrical double layer around the particle. If the fixed charges are located only at the surface region of a spherical and non-conducting colloidal particle, the situation can be clearly discussed with respect to the potential at the interface as shown in the preceding section. Once the potential profile is solved, another difficulty is to decide at what distance from the surface the potential relates to the migration of the particle. The layer of liquid immediately adjacent to a particle would move with the same velocity as the surface of the particle, and it is not clear what the actual distance is from the surface at which the relative motion sets in between the immobilized liquid layer (relative to the moving surface) and the mobile liquid. Although the precise location of the surface of shear is not known, it is presumably within a couple of molecular diameters of active particle surface for smooth particles (or surfaces). This is called a slipping plane. The potential at the surface of shear is called the zeta potential ζ, which may be fairly close to the Stern potential but smaller in magnitude (Figs. 5 and 9). Once the zeta potential, ζ, is obtained, the surface potential, ψ_o, can be calculated by using Eqs (26), (27) and (28) for an uni-univalent electrolyte system.

The surface of phospholipid membranes is well defined and smooth. For the determination of surface potential for such lipid membranes, a value of the slipping layer of $2 \sim 3$ Å is usually used [85, 90] and produces a consistent result with other experimental observations [85, 90, 92].

On the other hand, the surfaces of biological cells and biocolloids consisting of protein and other macromolecular moieties are usually not well defined surfaces. In particular, most biological cell surfaces such as those of erythrocytes have glycolipid and glycoprotein moieties (glyco-calyx) which extend quite a distance to the bulk solution (nearly

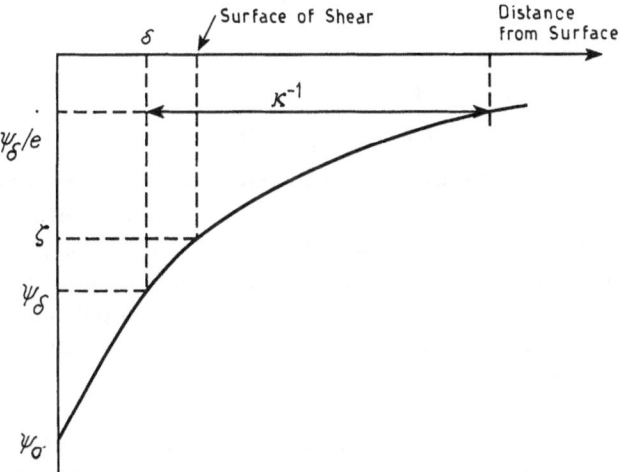

Fig. 9. Components of an electrical double layer potential and their relative spatial dimensions; ψ_o: surface potential, ψ_δ: the potential at the Stern plane, ζ: zeta-potential.

50 Å $\sim$ 100 Å) [101]. Some of the residues have dissociable molecular groups (acetyl-neuraminic acids) here and there in non-systematic manner. Therefore, it is more difficult to determine the point where the proper slipping layer is. Since the fixed charges for most such surfaces are distributed in not only a discrete manner but also over the three dimensional space (over a certain thickness of surface layer), even if the slipping layer is given, the evaluation of surface potential and surface charge density from the measured ζ potential will encounter yet another difficulty which will be discussed later.

2.4.1. *Extreme cases of κR:* For small values of κR where R is the radius of the particle, the zeta potential is expressed in terms of the electrophoretic mobility, u, as shown previously (Eqn. 50)

$$u = \frac{\varepsilon \zeta}{6\pi\eta}, \quad \text{(Hückel)}. \tag{87}$$

This expression is valid for spherical particles when κR is less than about 0.1. This imposes a rather severe restriction on the application of this relation in an aqueous system, since for $R = 10^{-6}$ cm, the corresponding concentration is about 10^{-5} M for a uni-univalent electrolyte.

For larger values of κR, the zeta potential is expressed (Eqn. 53) as

$$u = \frac{\varepsilon \zeta}{4\pi\eta}, \quad \text{(Helmholtz–Smoluchowski)}. \tag{88}$$

This expression is valid for $\kappa R > 100$.

The above two cases are the extreme cases and the general expression of mobility for any value of R can be rewritten as follows:

$$u = C(\kappa R)\varepsilon \zeta / \eta \tag{89}$$

where

$$\frac{1}{6\pi} \leq C(\kappa R) \leq \frac{1}{4\pi}. \tag{90}$$

For larger cells and vesicles (biological cells and large lipid vesicles), the Helmholtz–Smoluchowski equation [40] (Eqn. 88) is often used, and for small biocolloid particles (such as lipid micelles) the Hückel equation [41] (Eqn. 87) should be applicable within a certain limitation as mentioned above. However, many other biocolloidal particles fall in the intermediate range between the two limiting cases. The situation is summarized in Fig. 10.

2.4.2. *Intermediate cases of κR:* The generalized electrophoresis problem for the non-limiting case of κR has been solved for spherical and rod-shaped particles and approximately for random coils. Here the case of spherical particles at low concentration will be described. Because of

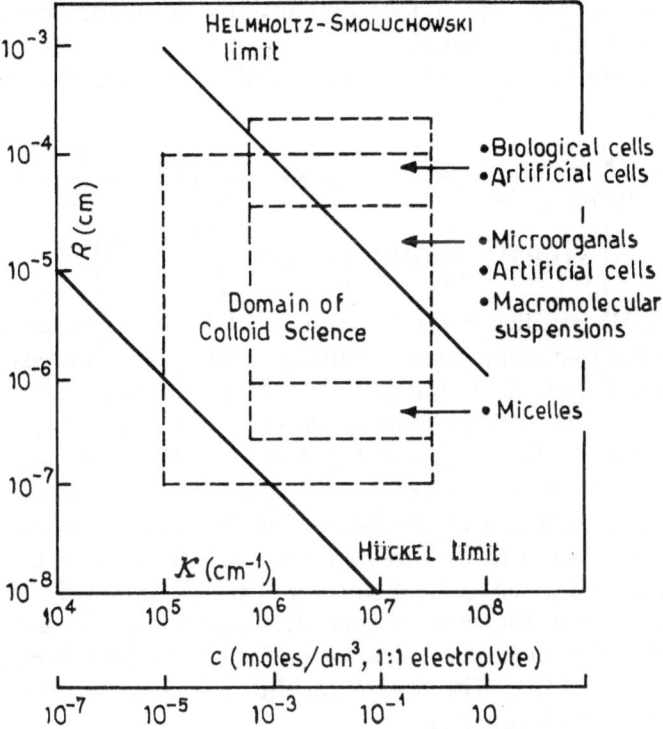

Fig. 10. The domain within which most investigations of aqueous colloidal systems lie in terms of particle radii and 1:1 electrolyte concentration. The diagonal lines indicate the limits of the Hückel and the Helmholtz–Smoluchowski equations. (Modified from Overbeek [102].)

the low concentration, there is no interaction between the colloidal particles. Since the surface of shear coincides more or less with the Stern layer, specific adsorption in the Stern layer may affect the zeta potential, but should be unimportant as long as the relation between u and ζ is concerned.

By assuming that the external field deformed by the presence of the colloidal particle and the field of the double layer are additive, Henry [42] derived the following expression for mobility u:

$$u = \frac{\varepsilon}{4\pi\eta} \left(\zeta + 5R^5 \int_\infty^R \frac{\psi}{r^6} \, dr - 2R^3 \int_\infty^R \frac{\psi}{r^4} \, dr \right), \qquad (91)$$

where r is the radial distance from the center of the particle. Using the Debye–Hückel approximation for ψ (Eqn. 37), which is

$$\psi = \psi(0) \frac{q}{r} \exp[-\kappa(r-a)]. \qquad (92)$$

one obtains

$$\psi = \frac{R\zeta}{r} \exp[-\kappa(r-a)], \qquad (93)$$

where R is the position of the slipping layer from the center of the particle and ψ is its potential.

By combining Eqns. (91) and (93), and integrating the terms, the mobility is

$$u = \frac{\varepsilon\zeta}{6\pi\eta} \left\{ 1 + \frac{1}{16}(\kappa R)^2 - \frac{5}{48}(\kappa R)^3 - \frac{1}{96}(\kappa R)^4 + \frac{1}{96}(\kappa R)^5 \right.$$

$$\left. - \left[\frac{1}{8}(\kappa R)^4 - \frac{1}{96}(\kappa R)^6 \right] \exp(\kappa R) \int_{\infty}^{\kappa R} \frac{e^{-t}\,dt}{t} \right\}. \tag{94}$$

Equation (94) is called Henry's equation [42]. The equation has been derived under two assumptions: 1) the ion atmosphere is undistorted by the external electric field, and 2) the potential is low enough to justify the Debye–Hückel approximation ($e\psi_o/kT < 1$). It can be easily seen that in the limit of $R \to 0$, Eqn. (94) reduces to the Hückel equation, and in the limit of $R \to \infty$ it reduces to the Helmholtz–Smoluchowski equation. The variation of the factor C in the mobility expression (cf. Eqn. 98) with respect to various values of κR is shown in Fig. 11. These different effects are due to the distortion of the electric field surrounding the particles. For the cases where the zeta potentials are not low ($\zeta > 25$ mV), the numerical calculations for $C(\kappa R)$ have been done by Wiersman et al. [103]. The results are shown in Fig. 11, together with the case of Henry's formula.

2.4.3. Retardation and relaxation effects on mobilities: Since the charged particle and its ionic atmosphere move in opposite directions, there will be at least two effects exerted on the movement of colloidal particles: 1) a counter-current effect which pulls back the movement of the particle according to the electric field – the retardation effect, and 2) the

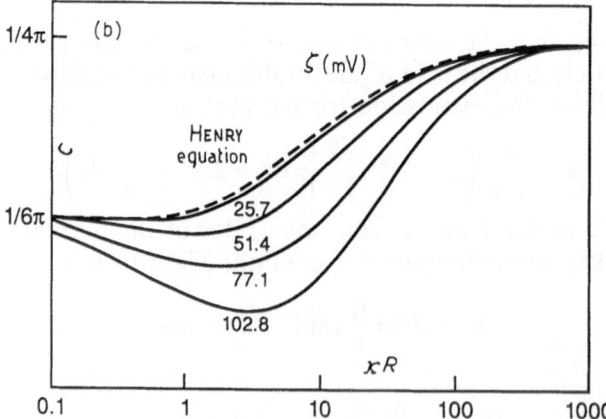

Fig. 11. Variation of the factor $C(\kappa R)$ with respect to κR for various ζ potentials (data from PH Wiersman et al. [103]). – – – Henry's equation with the low potential approximation. The values are in mV for each ζ potential.

deformation of the ionic atmosphere – the relaxation effect. The retardation effect is incorporated in Henry's equation but the relaxation effect is not included.

A number of workers have studied the problem of relaxation [103–105]. The use of computers has greatly assisted this area of research because of the complexity of the mathematics involved. Loeb et al. [104] report the results of some numerical solutions to the mobility problem with consideration of the relaxation effect. Figure 11 summarizes some results from these studies for the case of a 1:1 electrolyte. The various curves correspond to values of the zeta potential equaling 25.7, 51.4, 77.1, and 102.8 mV at 25°C (Fig. 11). It is evident from the figure that the relaxation effect is negligible when $\zeta < 25$ mV regardless of the value of κR, and in the limit of both large and small values of κR regardless of the value of ζ.

A family of curves qualitatively similar in appearance to those shown in Fig. 11 results when $C(\kappa R)$ versus κR is plotted at constant ζ with the valence of the electrolyte taken as the variable parameter. The relaxation effect is found to increase with the valence of the counter-ions [103]. An approximate expression [30] taking into account the relaxation effect has been derived with a relative error less than 1% for $\kappa R > 10$.

In this section, we have considered the relationship between u and ζ under conditions of intermediate κR values, a wide range of ζ values, and a number of ionic valence possibilities. The relationship is seen to be quite complex, except in the Hückel and Helmholtz–Smoluchowski limits. When the particle size-electrolyte concentration conditions are such that one of these limits clearly applies, ζ can be evaluated unambiguously from experimental mobilities. So far, the treatment of electrophoresis of colloidal particles assumes the particle is spherical and the surface is smooth, and the fixed charges are distributed on the two dimensional surface with no thickness. In biocolloidal particles, the fixed charges are distributed in a discrete manner at the interface, and also the distribution is not only on the two-dimensional surface but often within the three dimensional surface layer. The consideration of these factors would complicate the interpretation and analysis of the electrokinetic phenomena and potentials. The double layer potential for the discrete charge surface and three dimensional distribution of the fixed charges will be considered in the next two sections (3 and 4).

3. Discrete charge and charge distribution effects on double layer potential

3.1. Discrete charge on surface

The electrostatic potential due to a charged surface is conventionally described by the Gouy–Chapman theory, which assumes the charges are smeared uniformly over the surface. This assumption is not true for

most of the colloid particles, including biomembranes, and the discrete
character of the charges on the interfacial region must influence the
properties of the diffuse double layer.

Grahame (1958) [106], Levine et al. (1967), [32] and Barlow and
MacDonald (1967) [107] reviewed the early discrete charge theories,
which dealt mainly with the inability of the Gouy–Chapman formalism
to describe the adsorption of anions to metalic surfaces. In addition to
the discrete surface charge distribution, a change in dielectric constant
near the metal-water interface was the subject for investigation to
elucidate anion adsorption on the metal surface. Several other investigators have more recently adapted these discrete charge theories to interpret membrane phenomena [36–38]. Except for Sauve and Ohki 1979
[37], all others used the linearized Poisson–Boltzmann equation to solve
the problem [108, 109, 36]. Nelson and McQuarrie [36] used the
linearized Poisson–Boltzmann equation of the Debye–Hückel theory to
calculate the electrostatic potential due to a two dimensional array of
surface charges fixed in a square lattice. They have shown that indeed
the potential differs greatly near the charged surface from that for the
smeared charge case. Sauve and Ohki [37] have studied the effect of
discrete charge on the electrical potential of a membrane–electrolyte
system having three dielectric constant regions (nonpolar, fixed charge,
and aqueous phases) using the non-linearized form of the Poisson–
Boltzmann equation (Fig. 12). The mathematical description of the
local potential produced by an arbitrary arrangement of polar groups in
contact with an electrolyte solution was derived by use of the Bessel–
Fourier Integral equation. The main conclusions derived from the
theory are: (1) For highly charged surfaces, both smeared and discrete
charge theories result in approximately the same potential. On the other
hand, for sparsely charged surfaces, the discrete charge approach becomes more reasonable. (2) A smeared charge theory is adequate to
describe membrane phenomena which depend upon the space average
value of the surface charge potential. (3) In the systems involving
localized structures of charges such as specific ion channels, a uniform
charge (smeared charge) approach will lead to an underestimate of the
membrane surface charge density. (4) The microenvironment of a
localized structure is an extremely important factor in the general
behavior of the local potential acting at that point. This effect results
from the mutual interaction of the surface charges with each other's
ionic atmospheres, and cannot be taken into account with a discrete
charge theory based on a linearized form of the Poisson–Boltzmann
equation. However, it was shown that with the use of lower surface
charges, both formalisms (linearized and non-linearized) give similar
results. The discrete charge approach should be required in cases of
(1) discussion of the molecular mechanism of ion passage through
ionic channels which have localized structures, and (2) elucidation of

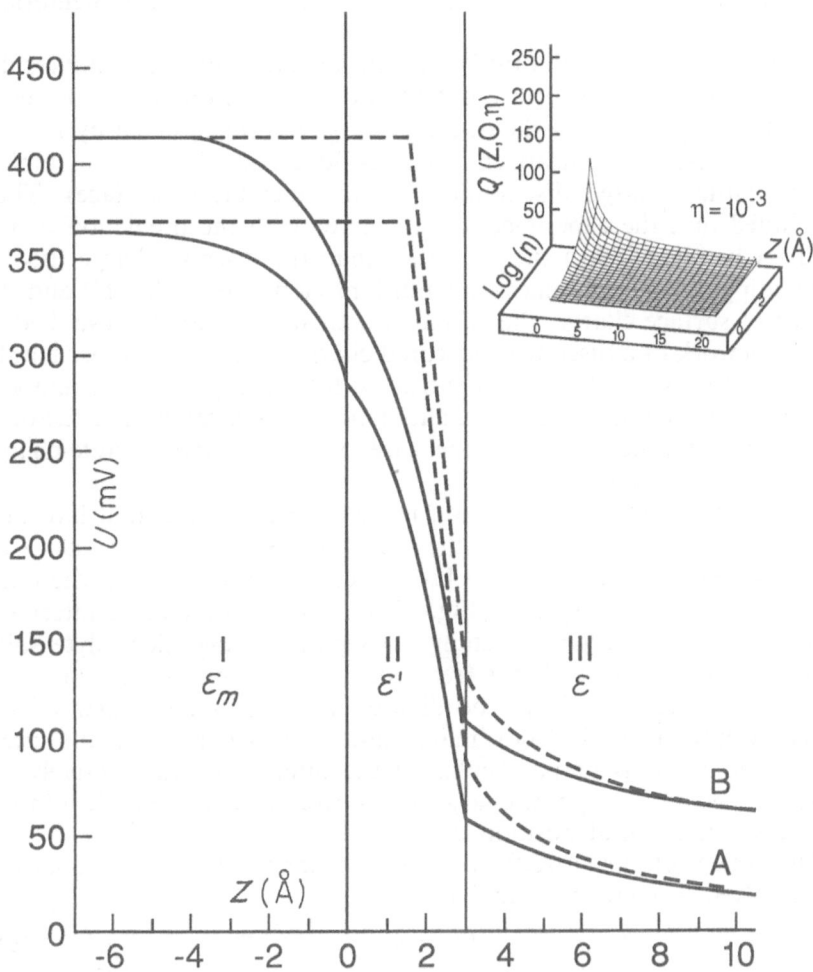

Fig. 12. Comparison between spatially averaged discrete charge (continuous lines) and smeared charge (dashed lines) potentials, where (A) $[M^{2+} M_2^-] = 35$ mM, (B) $[M^{2+} M_2^-] = 1$ mM and $[M^+ M^-] = 100$ mM. A membrane-electrolyte system having three dielectric constant regions (nonpolar ($\varepsilon_m = 2$), fixed charge ($\varepsilon' = 20$), and aqueous ($\varepsilon = 80$)) is represented, and the surface charge density was set to $e/48$ Å^2. The fixed charges were placed at a distance of 1.5 Å from the surface of the membrane on a hexagonal lattice. Q is a quantity related to the electrical potential at a distance from the membrane surface in certain electrolyte solutions. n: the number of univalent negative ions in the bulk aqueous phase per cm^3, $\eta = 2C^{2+}/n$ where C^{2+} is the bulk concentration of divalent cations in number of ions/cm^3. (From Sauve and Ohki [37].)

specific ion binding or adsorption at the membrane binding sites which also have localized structures. The specific nature of ion binding onto the charged membrane surface may be analyzed in terms of this micropotential, the image potential, the size of an ion, and its degree of hydration. Also, the structured water at the membrane surface seems to be an important factor for ion binding onto the membrane surface.

Recently, Winiski et al. (1987) [38] analyzed their experimental results on ion adsorption and surface potential of lipid bilayers in the presence of various ions by use of the discrete charge approach taken by Nelson and McQuarrie (the linearized Poisson–Boltzmann equation assuming square lattice charge distribution on the membrane surface). They concluded that the experimental data agree with the prediction of the Gouy–Chapman–Stern theory rather than the discrete charge theory. Depending on the experimental conditions (ionic strength, pH) and the degree of surface charge, the surface electrostatics may be described in the framework of a discrete or smeared charge distribution approach. In the case of $\bar{a}\kappa < 1$, where $\bar{a}$ is the nearest fixed charge distance and κ is the Debye constant, the Gouy–Chapman approach can be satisfactorily used. On the other hand, in the case of $\bar{a}\kappa > 1$, the discrete-charge approach should be used.

A general theoretical expression for the electrical potential having a discrete charge distribution on the membrane surface in the linearized case has been derived as follows [109]. Let us consider a case where the discrete charge model system (Fig. 13) consists of a dielectric interlayer of thickness a (dielectric constant ε_m) separating two electrolyte solutions of different compositions (reciprocal Debye lengths κ_1, κ_2; relative dielectric constants ε). The coordinate origin is in the center of the dielectric interlayer; the coordinate plane x, y is parallel to the interfaces. The z-axis is perpendicular to the interfaces and positively directed to region 3. A point charge e is located near the interface in the electrolyte solution at position q_i.

The system of (linearized) Poisson–Boltzmann differential equations to be solved in regions 1 and 3 is

1) $$\Delta\psi_1 = \kappa_1^2\psi_1 \tag{95}$$

2) $$\Delta\psi_2 = 0 \tag{96}$$

3) $$\Delta\psi_3 = \kappa_2^2\psi_3 - \frac{4\pi\delta(\dot{q} - \dot{q}_i)}{\varepsilon} \tag{97}$$

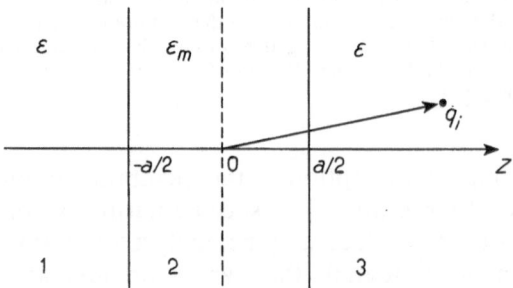

Fig. 13. Coordinate system and physical parameters for the calculation of the electrical potential of a discrete charged membrane. Explanations are given in the text.

with ε the dielectric constant and δ the Dirac's Delta-function. Δ is the Laplace operator, equal to ∇^2. The boundary conditions are

$$\psi_1 = \psi_2 \quad \text{for } z = -a/2 \tag{98}$$

$$\varepsilon \frac{\partial \psi_1}{\partial z} = \varepsilon_m \frac{\partial \psi_2}{\partial z} \quad \text{for } z = -a/2 \tag{99}$$

$$\psi_2 = \psi_3 \quad \text{for } z = a/2 \tag{100}$$

$$\varepsilon_m \frac{\partial \psi_2}{\partial z} = \varepsilon \frac{\partial \psi_3}{\partial z} \quad \text{for } z = a/2. \tag{101}$$

Theoretical approaches have been developed in a series of papers considering similar geometrical and physical problems (36, 37, 110–115). The solution found involves Fourier–Bessel integral expressions. This technique was first introduced to solve the electrical potential due to a fixed charge on a dielectric surface by Oka (1932) [116]. The solution has to be solved finally by a numerical procedure. The resulting fundamental system of equations representing the solution for the potential problem in regions 1 to 3 is expressed by the Bessel function of the zeroth order J_0.

$$\psi_1 = \frac{e}{\varepsilon} \int_0^\infty E(\lambda) \exp(z\sqrt{\lambda^2 + \kappa_1^2}) J_0(\lambda |\vec{\beta} - \vec{\beta}_i|) \, d\lambda, \tag{102}$$

$$\psi_2 = \frac{e}{\varepsilon_m} \int_0^\infty [C(\lambda) \exp(\lambda z) + D(\lambda) \exp(-\lambda z)] J_0(\lambda |\vec{\beta} - \vec{\beta}_i|) \, d\lambda, \tag{103}$$

$$\psi_3 = \frac{e}{\varepsilon} \int_l^\infty \left[\frac{\lambda \exp(-|z - z_i|\sqrt{\lambda^2 + \kappa_2^2})}{\sqrt{\lambda^2 + \kappa_2^2}} \right. $$
$$\left. + V(\lambda) \exp(-z\sqrt{\lambda^2 + \kappa_2^2}) \right] J_0(\lambda |\vec{\beta} - \vec{\beta}_i|) \, d\lambda, \tag{104}$$

with

$$|\vec{\beta} - \vec{\beta}_i| = \sqrt{(x - x_i)^2 + (y - y_i)^2}. \tag{105}$$

The coefficients $E(\lambda)$, $C(\lambda)$, $D(\lambda)$, and $V(\lambda)$ are obtained from the respective algebraic systems of equations by substituting Eqns. (102–104) into Eqns. (95–97) under the boundary conditions of Eqns. (98–101). Ultimately, they are expressed as:

$$E(\lambda) =$$
$$-\frac{2\lambda^2 \xi \exp(a\sqrt{\lambda^2 + \kappa_1^2}/2) \exp(a\lambda) \exp[(a/2 - z_j)\sqrt{\lambda^2 + \kappa_2^2}]([-] + [+])}{(\sqrt{\lambda^2 + \kappa_1^2} - \lambda \xi)(\sqrt{\lambda^2 + \kappa_2^2} \cdot [-] - \lambda \xi \cdot [+])}, \tag{106}$$

$$C(\lambda) = \frac{\lambda \xi \exp[(a/2 - z_j)\sqrt{\lambda^2 + \kappa_2^2}] \exp(-a\lambda/2)([-] - [+])}{\sqrt{\lambda^2 + \kappa_2^2} \cdot [-] - \lambda \xi [+]} \tag{107}$$

$$D(\lambda) = \frac{\lambda \, \exp[(a - z_j)\sqrt{\kappa^2 + \kappa_2^2}] \, \exp(a\lambda/2)([-] + [+])}{\sqrt{\lambda^2 + \kappa_2^2}[-] - \lambda\xi \cdot [+]} \tag{108}$$

$$V(\lambda) = \frac{\lambda \, \exp[(a - z_j)\sqrt{\lambda^2 + \kappa_2^2}](\sqrt{\lambda^2 + \kappa_2^2} \cdot [-] + \lambda\xi[+])}{\sqrt{\lambda^2 + \kappa_2^2}(\sqrt{\lambda^2 + \kappa_2^2} \cdot [-] - \lambda\xi[+])} \tag{109}$$

where

$$\xi = \varepsilon_m/\varepsilon, \tag{110}$$

$$[-] = (\sqrt{\lambda^2 + \kappa_1^2} - \lambda\xi) \, \exp(-\lambda a) - (\sqrt{\lambda_1^2 + \kappa_1^2} + \lambda\xi) \, \exp(\lambda a), \tag{111}$$

$$[+] = (\sqrt{\lambda^2 + \kappa_1^2} - \lambda\xi) \, \exp(-\lambda a) + (\sqrt{\lambda^2 + \kappa_1^2} + \lambda\xi) \, \exp(\lambda a). \tag{112}$$

Then the electrical potential in each region can be solved numerically from Eqns. (102–104).

However, if a few fixed charged sites, which interact with specific ions in the solution, are located around a small area of the surface (e.g. an ionic channel of the membrane), the non-linearized discrete charge approach may be suitable. It might be used for the analysis of the ionic concentration near the channel area (the surface may be curved), and the interaction of the charged sites with the specific ions, even though $\bar{a}\kappa < 1$ may hold at the local area [37].

3.2. Comparison between discrete charge and smeared charge approaches

According to the studies, when the membrane surface has only univalent fixed charge sites the electrical potential adjacent to the membranes can be described quite well by the Gouy–Chapman–Stern theory, which agrees with the potential observed experimentally. The experiments were done using fluorescence spectroscopy [25, 38] and measurement of the force exerted on two interacting membranes [118]. In the fluorescence experiment, the potential was determined by measuring the extent of quenching of the fluorescence probe, which was placed at different distances from the membrane surface in the solution (the fluorescence probe was attached to a certain part of a molecule which extends from the surface). The electrical potential was obtained through the distribution of univalent ionic quenchers in the solution with respect to the extent of fluorescence quenching. Such experimental results show good agreement with those calculated by the Gouy–Chapman theory (Eqn. 26). In the calculation the dielectric constant of water was used for the electrical double layer right up to the distance of the first layer of adsorption. The successful agreement was obtained for the case where the membrane contains only univalent fixed charge sites in univalent ionic solution (Fig. 14) from the actual surface ($x = 0$) to 20 Å in the aqueous solution. Also, the effective potentials arising from the

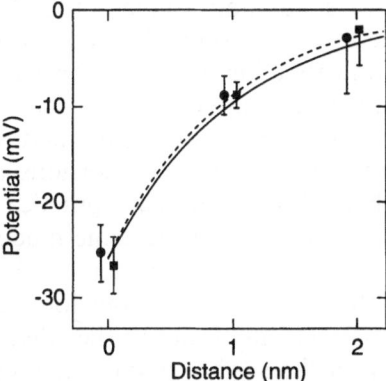

Fig. 14. The profile of the electrical potential in the aqueous phase adjacent to membranes formed from 5:1 PC/PI (circles) and 5:1 PC/PS (squares). The potentials were determined by quenching the fluorescence of probes located 0, 1, or 2 nm from the surface. The aqueous solutions contained 0.1 M KNO_3, buffered to pH 7.4 at 25°C with 5 mM MOPS. The two curves illustrate the predictions of the linear (solid curve) and nonlinear (dashed curve) Gouy–Chapman theory. The surface potential was chosen to fit the experimental points. (From Langer et al. [25].)

predicted counter-ion and co-ion concentrations were obtained experimentally and theoretically when the discrete surface charges and groups were placed at different distances. When the membrane surface charges were composed of univalent fixed charges, the effective potentials for counter-ion and co-ion are the same and agreed with the predicted potential from the Gouy–Chapman theory. However, when the surface charge contained multi-charged sites, the effective potentials were asymmetrical for co-ion and counter-ion distribution; the effective potential sensed by the co-ion was not as large as the effective potential sensed by the counter-ion probe. The results were not analyzed successfully by the Gouy–Chapman theory.

The above finding also agrees with those obtained by using different techniques: measurements of the force and distance between charged bilayers using a surface force apparatus [118], and X-ray diffraction measurements [119, 120] of the distance between charged bilayers upon application of a known force. Both demonstrate that the Gouy–Chapman theory (Eqn. 26) describes the experimental data very well for bilayer separation distances greater than 2 nm (see Fig. 19). Neutron diffraction estimates of the concentration profiles of deuterated tetramethylammonium ions between negatively charged lipid bilayers also support the Gouy–Chapman theory [121]. So far, all available evidence suggests that the Gouy–Chapman theory correctly describes the dependence of potential on distance from a lipid bilayer when the surface charges are of univalent ionic lipids and the aqueous solution contains only univalent ions.

Several Monte Carlo simulation studies [122, 123] on a primitive model electrolyte next to a uniformly charged surface have been performed and demonstrate that ion size and correlation effects are not very important for the surface charge densities characteristic of biological membranes ($\sigma < 1$ electronic charge/nm^2) and salt concentrations lower than 0.2 M, provided that the counter-ions are univalent. The hypernetted chain (HNC) [124] theory gave practically the same results as the Monte Carlo simulations with the same model. In HNC results, changing from continuous to discrete univalent surface charges has only a minimal effect on the calculated average ion density or electrical potential. It was also theoretically found earlier that the average potential beyond a few Å from the surface is the same for both the discrete charge and smeared charge approaches (37). The same is not true, however, when the valency of the surface charges is higher than one. The effect of correlation in the position between the surface group and the electrolyte ions becomes important, leading to a very strong screening of the multivalent surface charge site. As a result, co-ions are not as strongly excluded from the immediate vicinity; the average co-ion density is higher than expected by the average value of the electrostatic potential.

The Gouy–Chapman method is a consistent mean-field theory, where the interaction between the pairs of ions is replaced by the interaction of an ion in the diffuse double layer with the average electrostatic potential. In the discrete charge approach, however, ion pair interaction between the surface and solution charges should be considered rather than the mean potential field. As soon as these explicit interaction terms are included in a theory, a finite size for the ions must also be included. This brings about a very difficult and complicated problem. To make a fully quantitative model of the behavior of electrolytes near surfaces by the discrete charge approach is not feasible at this moment.

3.3. Three dimensionally distributed surface charges

Except for the discrete charge approach, the Gouy–Chapman double layer theory mentioned so far assumed the fixed charges of the colloidal particle interface to be placed on a two dimensional interface. However, the surfaces of most biocolloids including biological cell surfaces have surface layers of a finite thickness where the fixed charges are distributed in three dimensional space. It is known that the glycocalyx of animal cell surfaces carries a fixed surface charge of density about 0.02 Coulomb/m^2 distributed inside a layer of a finite thickness under physiological conditions [125]. For the erythrocyte cells, there exists about 1.5×10^7 electric charges distributed in a surface layer of a thickness of about 60 Å and an area of 140 μm^2 [126, 128].

A number of investigators have studied the electrostatic structure of such cell surfaces, in relation to cell-cell interaction (Parsegian and Gingell 1973 [125]) and to cell electrophoresis (Haydon 1961 [127]; Donath and Pastushenko 1979 [128]; Wunderlick 1982 [129]; and Levine et al. 1983 [130]). Most of the investigators have used the assumption of a continuous distribution of charge to solve the potential behavior.

If the quotient of mean fixed charge distance inside the glycoalyx $\bar{a}$ and the Debye length $1/\kappa$ fulfills the relation $\bar{a}\kappa \leq 1$ the problem of an adequate description of the surface electrostatics can be solved in a less difficult way. The following case is considered: the fixed charges are assumed to be smeared or smoothly distributed over the cell surface layer and are represented by a space fixed charge density function ρ which depends only on the coordinate z perpendicular to the membrane surface. Plane charge distributions of density σ (C/m^2) are assumed at arbitrary positions inside the surface layer (especially at the membrane/glycocalyx transition at $z = 0$ and the glycoalyx/bulk electrolyte solution transition at $z = \delta'$). Representing the fixed charges in this way leads to a drastic reduction of mathematical efforts compared with the discrete charge case. The modeling following this picture started with Parsegian and Gingell (1973) [125]. Thereafter, an extension of this approach was achieved and general expressions for the calculation of the electric potential profiles at single cell surface were derived (Pastushenko and Donath, 1976) [131].

Pastushenko and Donath (1976) [131] derived the following expression for the electric potential profile of a single charged layer by using the method of the Green's function:

$$\psi(z) = \frac{4\pi\sigma}{\varepsilon\kappa} \exp(-\kappa z) + \frac{4\pi \cosh(\kappa z)}{\varepsilon\kappa} \int_0^{\delta} \rho_f(z') \exp(-\kappa z') \, dz'$$

$$-\frac{4\pi}{\varepsilon\kappa} \int_0^z \rho_f(z') \sinh\{\kappa(z-z')\} \, dz', \qquad (113)$$

where σ is the plane charge density at $z = 0$ (membrane surface), $\rho_f(x)$ the density profile of fixed electric charges, and δ the thickness of the surface layer (glycocalyx). In case of a constant density of fixed charges $\rho_f = \rho =$ const. the following potential profile is obtained

$$\psi(z) = \frac{4\pi\rho}{\varepsilon\kappa^2} \{1 - \cosh(\kappa z) \exp(-\kappa\delta)\}, \quad \text{for } z \leq \delta \qquad (114)$$

$$\psi(z) = \frac{4\pi\rho}{\varepsilon\kappa^2} \{\cosh[\kappa(z-\delta)] - \cosh(\kappa z) \exp(-z\delta)\} \quad \text{for } z \geq \delta. \quad (115)$$

However, when the charges are distributed as a function of the distance from the surface, in general the potential equation (Poisson–Boltzmann equation) is not soluble in analytical form. Ohshima and Ohki (1985)

[132, 133] analyzed the potential profile of a membrane surface having a finite thickness of surface charge layer, in the case of a uni-univalent electrolyte containing bivalent cations, by assuming the charges are distributed uniformly in the surface charge layer and using the non-linear Poisson–Boltzmann equation. They showed that the potential distribution depends significantly on the thickness, d, of the surface charge layer when $d < 1/\kappa$ (κ is the Debye constant), as well as on the types of ion binding to the charge sites of the surface charge layer. The conventional method to estimate surface potential and surface charge density from the measured value of cell electrophoretic mobility by assuming zero thickness of the surface charge layer would result in significantly underestimated values for a membrane having non-zero thickness of the surface charge layer in the case of a large κR value. It should be noted that in the case of the ζ potential, depending on the size of the particles, more complicated considerations should be made due to the change in hydrodynamic radius of the particle, etc.

4. Relation between surface potential and Donnan potential

When a membrane possessing fixed charges is in equilibrium with an electrolyte solution, an electric potential difference is generally established between the membrane and the solution.

There have been two entirely different approaches to describe this potential difference. One approach considers the potential difference to be the Donnan equilibrium potential. This approach has been used particularly for studies of ion transport processes through membranes (Toerell, 1953) [66]. In the other approach it is regarded as the surface potential, i.e., the Gouy–Chapman diffuse double layer potential, which is familiar in Colloid Sciences (Verwey and Overbeek, 1948, [134] and others [5, 14]).

When a membrane which is permeable to electrolyte ions and contains fixed negative charge groups at a uniform density N is in equilibrium with a symmetrical electrolyte solution of concentration n and valence z, the Donnan potential ψ_D relative to the external bulk solution is expressed as (Davies and Rideal, 1961 [14]; Ohki, 1965) [135]:

$$\psi_D = -\frac{kT}{ze} \ln\left[\frac{N}{2zn} + \left\{\left(\frac{N}{2zn}\right)^2 + 1\right\}^{1/2}\right] = -\frac{kT}{ze} \text{arc sinh}\left(\frac{N}{2zn}\right), \quad (116)$$

where k is the Boltzmann constant, T the absolute temperature, and e the elementary electric charge. Conventionally, the Donnan potential is considered to be accompanied by a discontinuous potential gap across the membrane surface (Fig. 15a). This is due to the assumption of local electroneutrality, which is used in the classical derivation of Eqn. (116). Replacing this assumption by the Poisson–Boltzmann equation, Mauro

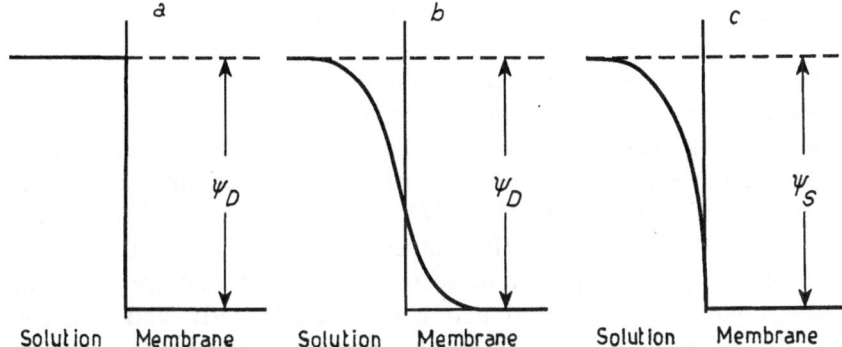

Fig. 15. Schematic representation of the potential distribution near the membrane surface for the proposed membrane models. (a) Classical (discontinuous) Donnan equilibrium potential, (b) continuous Donnan potential (Mauro 1962 [136]), (c) surface potential.

(1962) [136] has shown that the Donnan potential diffuses over distances of order $1/\kappa$ (κ is the Debye–Hückel constant) on both sides of the membrane surface as shown in Fig. 15b, and has a continuous nature across the membrane surface.

If, on the other hand, it is assumed that all the fixed membrane charges are located only at the membrane surface and the electrolyte ions cannot penetrate into the membrane, then the electrical double layer is considered to be formed around the surface (Fig. 15c) and the surface potential relative to the bulk solution takes the following form (Verwey and Overbeek, 1948 [126]; Davies and Rideal, 1961 [14]):

$$\psi_s = \frac{2kT}{ze} \text{ arc sinh}\left[\frac{\sigma}{(2n\varepsilon kT/\pi)^{1/2}}\right] = \psi_o \quad \text{(C.G.S. Unit)} \quad (117)$$

where σ is the surface charge density of the membrane and ε the dielectric constant of the solution. This is quite different from Eqn. (116) described above. Several modifications of the surface potential approach have been attempted by allowing the surface region to be permeable to electrolyte ions in relation to cell-cell interactions [125, 137] and to cell electrophoresis [127–130].

In spite of the above-mentioned improvements or modifications on each of the two different approaches, there remained ambiguity as to the interrelation or transition between the Donnan potential and the surface potential.

Ohshima and Ohki (1985) [138] have shown that these Donnan and surface potential concepts do smoothly make the transition into each other in their potential distributions across a model charge membrane, in which the membrane fixed charges are uniformly distributed through a layer of finite thickness at the membrane surface and the charge layer is permeable to ions.

4.1. Theory (1:1 electrolyte and non-ion binding case)

Let us consider a planar charged membrane which is in equilibrium with a large volume of a uni-univalent electrolyte solution of concentration n. The x-axis is chosen in the direction normal to the membrane surface so that the plane at $x = 0$ coincides with the left boundary between the membrane and the solution (Fig. 16). The membrane is composed of three layers: two identical surface layers of thickness d which contain negatively charged groups at a uniform density $-eN$ and are permeable to electrolyte ions ($0 \leq x \leq d$ and $h + d \leq x \leq h + 2d$), and one core layer ($d \leq x \leq h + d$). Since the membrane considered here is symmetrical with respect to the plane $x = h/2 + d$, we need to consider only the region $-\infty < x \leq h/2 + d$.

We assume that the electric potential $\psi(x)$ at position x in the regions $x < 0$ and $0 < x < d$ (relative to the bulk solution ($x = -\infty$)) satisfies the Poisson–Boltzmann equation,

$$\frac{d^2\psi}{dx^2} = -\frac{4\pi\rho(x)}{\varepsilon} = -\frac{4\pi en}{\varepsilon}(e^{-e\psi/kT} - e^{e\psi/kT}) = \frac{8\pi en}{\varepsilon}\sinh\frac{e\psi}{kT}, \quad x < 0$$

(118)

$$\frac{d^2\psi}{dx^2} = -\frac{4\pi\rho(x)}{\varepsilon'} = -\frac{4\pi en}{\varepsilon'}(e^{-e\psi/kT} - e^{e\psi/kT}) + \frac{4\pi eN}{\varepsilon'}, \quad 0 < x < d \quad (119)$$

where ε_r and ε_r' are the dielectric constants of the solution and of the surface charge layer, respectively. For the region $d < x < h/2 + d$, in

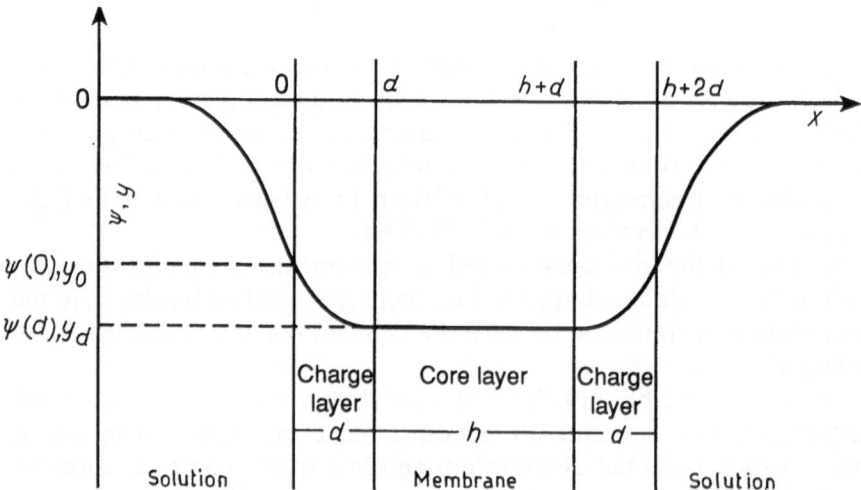

Fig. 16. Schematic representation of the potential distribution $\psi(x)$ across a membrane with two surface charge layers (of thickness d) and one core layer (of thickness h). (From Ohshima and Ohki [138].)

which there are no true charges, we have

$$\frac{d^2\psi}{dx^2} = 0, \quad d < x < h/2 + d. \tag{120}$$

The boundary conditions are:

$\psi(x)$ is continuous at $x = 0$ and $x = d$

$$\varepsilon \frac{d\psi}{dx}\bigg|_{-o} = \varepsilon' \frac{d\psi}{dx}\bigg|_{+o}, \tag{121}$$

$$\varepsilon' \frac{\partial\psi}{\partial x}\bigg|_{d-o} = \varepsilon_m \frac{d\psi}{dx}\bigg|_{d+o} \tag{122}$$

and

$$\psi(x) \to 0, \quad (x \to -\infty), \tag{123}$$

where ε_m is the relative permittivity of the core layer; and because of symmetry of the membrane we have

$$\frac{d\psi}{dx}\bigg|_{h/2+d} = 0. \tag{124}$$

From the symmetry of the system, the boundary condition (122) may be replaced by

$$\frac{d\psi}{dx}\bigg|_{d-o} = 0. \tag{125}$$

The solution of Eqns. (118) and (119) subject to the boundary conditions (121), (122), (124), and (125) completely determines the potential function (x) in our system. By introducing the following dimensionless potential:

$$y \equiv \frac{e\psi}{kT}. \tag{126}$$

Equation (118) can be easily integrated subject to the boundary condition (123) to give

$$y = 4 \,\mathrm{arc\,tanh}\left[\tanh\frac{y_o}{4} e^{-\kappa|x|}\right], \quad x < 0 \tag{127}$$

where κ is the Debye–Hückel parameter of the solution

$$\kappa = (8\pi n e^2/\varepsilon kT)^{1/2} \tag{128}$$

and y_o is defined as

$$y_o = y(0) = \frac{e}{kT}\psi(0). \tag{129}$$

From Eqn. (127), we obtain:

$$\frac{dy}{dx}\bigg|_{-o} = 2\kappa \sinh \frac{y_o}{2}. \tag{130}$$

Integration of Eq. (119) subject to the boundary condition (125) yields

$$\frac{dy}{dx} = -\kappa'\left[2\left(\cosh \frac{y}{y_o} - \cosh \frac{y_d}{y_o}\right) + \frac{N}{n}(y - y_d)\right]^{1/2}, \quad 0 < x < d \tag{131}$$

where κ' is the Debye–Hückel parameter of the surface charge layer

$$\kappa' = (8\pi ne^2/\varepsilon' kT)^{1/2} \tag{132}$$

and y_d is defined by

$$y_d = y(d) = \frac{e}{kT}\psi(d). \tag{133}$$

Using Eqns. (130) and (131) and the boundary condition of Eq. (121) we obtain

$$2 \sinh \frac{y_o}{2} = -\left(\frac{\varepsilon'}{\varepsilon}\right)^{1/2}\left[2(\cosh y_o - \cosh y_d) + \frac{N}{n}(y_o - y_d)\right]^{1/2}. \tag{134}$$

Equation (130) can be integrated further to give

$$\kappa'd = \int_{y_d}^{y_o} \frac{dy}{\left[2(\cosh y - \cosh y_d) + \dfrac{N}{n}(y - y_d)\right]^{1/2}}. \tag{135}$$

Consider the two limiting cases $d \to \infty$ and $d \to 0$:

Case of d → ∞. This is the case where the membrane itself is a semi-infinite charge layer.

When $d \to \infty$, the region $0 < x < d$ where Eqn. (119) is applicable is extended to the region $0 < x < +\infty$, so that

$$\frac{d^2y}{dx^2} = \kappa'^2\left(\sin y\, y + \frac{N}{2n}\right), \quad 0 < x < +\infty, \tag{136}$$

and the total potential difference across the membrane $\psi(d)$ (or y_d) becomes equal to $\psi(+\infty)$ or $(y(\infty))$. Noting that $d^2y/dx^2 = 0$ at $x = +\infty$, from Eqn. (136) we obtain

$$y_\infty = y(\infty) = -\text{arc} \sinh \frac{N}{2n}, \tag{137}$$

which agrees with the Donnan potential Eqn. (116). We note that Eq. (136) does not depend on ε or ε'.

Case of d → 0. This is the case where all of the charged sites are located on the membrane surface. Noting that $y - y_d \ll 1$ $(0 < x < d)$ when $\kappa'd \ll 1$, we put $y = y_d + \Delta y$ $(\Delta y \ll 1)$ in Eqn. (131) and its integrated form Eqn. (135), linearize them with respect to Δy, and solve the differential equation with respect to Δy under the boundary conditions Eqns. (121–123). Then we find

$$\kappa' d = 2 \left[\frac{y_o - y_d}{2 \sinh y_d + (N/n)} \right]^{1/2}, \quad \kappa' d \ll 1 \tag{138}$$

and

$$2 \sinh \frac{y_o}{2} = -\left[\left(\frac{\varepsilon'}{\varepsilon} \right)\left(2 \sinh y_d + \frac{N}{n} \right)(y_o - y_d) \right]^{1/2}, \tag{139}$$

respectively. From Eqns. (138) and (139) we find

$$2 \sinh\left(\frac{y_o}{2} \right) = -\frac{eNd}{(n\varepsilon kT/2\pi)^{1/2}} - \kappa d \sinh y_d. \tag{140}$$

If we take the limit $d \to 0$ in Eqn. (140), keeping the product Nd constant, i.e. keeping the total amount of membrane fixed charges – eNd contained in a unit area of the surface charge layer $(0 < x < d)$ constant, then the second term on the right hand side of Eqn. (140) becomes negligible and we obtain the following limiting result, which is identical to Eqn. (117) (the surface potential).

$$y_o = y_d = 2 \text{ arc } \sinh\left[\frac{\sigma}{(2n\varepsilon kT/4\pi)^{1/2}} \right], \tag{141}$$

where we have defined σ as

$$\sigma = -e \lim_{d \to 0} (Nd)$$

$$(Nd = \text{constant}) \tag{142}$$

which can be interpreted as the surface charge density of the membrane.

4.2. Transition between the Donnan and surface potentials

In order to see the transition between these two limiting cases, we consider here the potential distribution for $\kappa d \leq 1$. In order to perform numerical calculations of $y(x)$ $(0 < x < d)$ in the range of $\kappa d \leq 1$, we expand $y(x)$ $(0 < x < d)$ around $x = d$ in powers of $\kappa'(x - d)$ (the Taylor expansion series).

Since all terms with an odd power of $\kappa'(x - d)$ vanish due to the boundary condition of Eqn. (125), $y(x)$ can be written as

$$y(x) = y_d + \sum_{n=1}^{\infty} A_n [\kappa'(x - d)]^{2n}, \quad 0 < x < d, \tag{143}$$

where the coefficients A_n can be obtained easily from Eqn. (136) and are expressed in terms of N, n and y_d.

Some such expressions for A_n ($n = 1, 2, 3$) are given below:

$$A_1 = \frac{1}{2} B_1$$

$$A_2 = \frac{1}{24} B_1 B_2 \tag{144}$$

$$A_3 = \frac{1}{826} B_1 (3B_1 B_3 + B_2^2)$$

where

$$B_1 = \sinh y_d + (N/2n),$$

$$B_2 = \cosh y_d \tag{145}$$

$$B_3 = \sinh y_d$$

As $x \to 0$, Eqn. (143) becomes

$$y_o = y_d + \sum_{n=1}^{\infty} A_n (\kappa' d)^{2n}. \tag{146}$$

By combining Eqn. (146) with Eqn. (134), it is possible to obtain numerical values of y_o and y_d which in turn determine $y(x)$ (or $\psi(x)$) in the regions $x < 0$ and $0 < x < d$, respectively.

4.3. Numerical results

Figure 17 shows numerical results for $\psi(x)$ when $d = 0$, 5, and 10 Å. Here we have varied d, keeping the product Nd constant, i.e., keeping the total amount of membrane fixed charges $-eNd$ per unit area of the surface charge layer ($0 < x < d$) constant, and using the value $-eNd = 0.190 \, \text{C m}^{-2}$ so as to give $\psi(d) = -120 \, \text{mV}$, as $d \to 0$. The values of the other parameters were $n = 0.1$ M, $T = 298$ K and $\varepsilon'/\varepsilon = 1$ and 0.5, respectively. The value of $1/\kappa'$ then becomes 9.6 Å for $\varepsilon'/\varepsilon = 1$ and 6.8 Å for $\varepsilon'/\varepsilon = 0.5$.

Figure 17 shows a strong dependence of $\psi(x)$ on the thickness d of the surface charge layer. We also see from Fig. 17 that the potential drop within the membrane is sharper for smaller ε', which gives rise to less negative values of $\psi(0)$ and more negative values of $\psi(d)$.

The case of no ion binding to the charge layer has been described above. For the case of ion binding to the charge layer, the fixed charges as well as the profile of surface potential will be affected significantly depending on the affinity of binding of ions to the charged sites in the layer [132, 133].

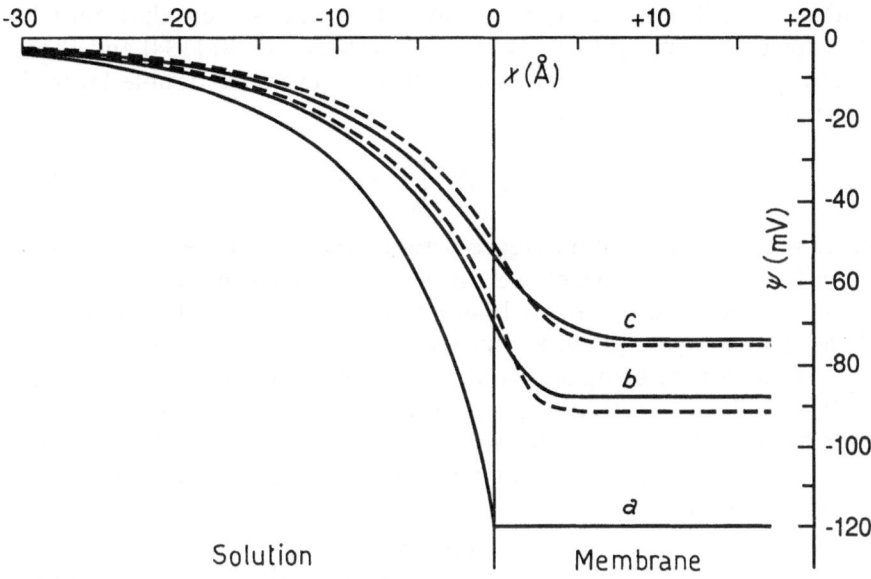

Fig. 17. Potential distribution $\psi(x)$ across a membrane with a surface charge layer of thickness $d = 0$, 5, and 10 Å (curves a, b, and c, respectively). Calculated with $T = 298°K$, $n = 0.1$ M, $-eNd = 0.190$ Cm^{-2}, $\varepsilon'/\varepsilon = 1$ (solid line) and $\varepsilon'/\varepsilon = 0.5$ (broken line). $\varepsilon = 78.5$. (From Ohshima and Ohki [138].)

5. Effect of double layer on the interaction of colloidal particles

When two colloidal particles interact with each other, there exist several interaction forces exerted on the colloidal particles which are functions of the separation distance of the two particles: interaction forces due to two double layers, van der Waals forces, short range forces including hydration forces, etc.

5.1. Electric double layer interaction and DLVO theory

5.1.1. Double layer interaction energy: It is well known that Derjaguin and Landau (1941) [139] and Verwey and Overbeek (1948) [134] (DLVO) have given a quantitative treatment of the interaction of electric double layers for two interacting colloid bodies. According to them, the free energy of a double layer may be expressed as a difference between the surface energy G_s of the system in its equilibrium state and the surface free energy G_s^0 of a standard state in which no double layer is present:

$$G_{\text{double layer}} = G_s(\sigma) - G_s^0(\sigma = 0). \qquad (147)$$

Although this expression presents the basis of the free energy of the double layer, it is not truly expressive of the parameters that must be considered. According to Verwey and Overbeek (1948) [134] and Ikeda (1950) [140], the free energy of a system of reversible double layers is determined by the following:

$$G = \sum dS_k \int_0^{\psi_k} \sigma \, d\psi \qquad (148)$$

where ψ is the potential difference between the two phases separated by the kth interface of which dS_k is a surface element and ψ_k is the equilibrium surface potential. It should be noticed that the free energy of the double layer system is a negative quantity which implies double layer formation as being a spontaneous process. Verwey and Overbeek [134] formulated a second expression for the free energy of the double layer based on the Gouy–Chapman model by considering the reversible work required to discharge all ions in the system while keeping the surface potential constant:

$$G = G(0) + \int_0^1 \frac{d\lambda}{\lambda} \int_v (\rho'\psi')_{\psi_k} \, dv, \qquad (149)$$

where λ is a parameter describing the stage of the discharging process, which varies from 1 to 0 as the charge of the ith ion varies from $z_i e$ to 0, the second term on the right represents the reversible work done in discharging all ions at constant surface potential ψ_k, the quantities ρ' and ψ' represent the space charge density and the potential in the Gouy diffuse layer at any stage λ, and dv is a volume element in the solution. The term $G(0)$ is the free energy of the discharged system at $\lambda = 0$ and at fixed surface potential. By use of Ikeda's basic derivation for $G(0)$, $G(0)$ may be substituted into the above equation to give the free energy per unit area of a system composed of two large parallel plates immersed in solution:

$$G = -\frac{\varepsilon}{8\pi d} (\psi_o - \psi_R)^2 + \int_0^1 \int_0^R \frac{\lambda'\psi'}{\lambda} \, dx \, d\lambda. \qquad (150)$$

This expression may be reduced using the Poisson equation; a detailed analysis may be found in Devereux and de Bruyn (1963) [141].

The equations used are for the free energy per square centimeter of two flat double layers at a fixed distance of separation R. Therefore, the reversible work required to bring the two parallel surfaces together from infinity to a distance R apart is determined by the expression:

$$V = G - G_\infty \qquad (151)$$

where G_∞ is the free energy per square centimeter of surface when the distance of separation is infinite. V determines the free energy of interaction of the double layer system. This interaction gives normally repulsive forces for two interacting bodies.

The electrostatic interaction energy between diffuse double layers based on the linear Debye–Hückel approximation for two parallel plates having constant surface charge are given by Usui [142] in the form

$$V^\sigma = \frac{\varepsilon \kappa}{8\pi} [(\psi_{o1}^2 + \psi_{o2}^2)(\coth(\kappa R) - 1) + 2\psi_{o1}\psi_{o2} \operatorname{cosech}(\kappa R)], \quad (152)$$

where ψ_{o1} and ψ_{o2} are the surface potentials of plane 1 and plane 2 at infinite separation distance, respectively.

The linear approximation and employment of the boundary condition for constant surface potentials yields [143]:

$$V^\psi = \frac{\varepsilon \kappa}{8\pi} [(\psi_{o1}^2 + \psi_{o2}^2)(1 - \coth \kappa R) + 2\psi_{o1}\psi_{o2} \operatorname{cosech} \kappa R]. \quad (153)$$

Under most physiological conditions, the Debye–Hückel length $(1/\kappa)$ is about 10 Å. At distances greater than 2 to $3\kappa R$, the difference between V^σ and V^ψ becomes negligibly small.

The expression for the electrostatic interaction energy between the double layers of two spherical bodies, for the case of constant surface potential with use of the Debye–Hückel approximation, is given by [143]

$$V^\psi = \frac{\varepsilon a_1 a_2 (\psi_{o1}^2 + \psi_{o2}^2)}{4(a_1 + a_2)} \left\{ \frac{2\psi_{o1}\psi_{o2}}{(\psi_{o1}^2 + \psi_{o2}^2)} \ln\left[\frac{1 + \exp(-\kappa R)}{1 - \exp(-\kappa R)}\right] \right.$$
$$\left. + \ln[1 - \exp(-2\kappa R)] \right\}, \quad (154)$$

where a_1 and a_2 are the respective radii of the two bodies 1 and 2.

The expression for the case of constant surface charge, for the same system as the above, is given [144]

$$V^\sigma = V^\psi - \frac{\varepsilon a_1 a_2 (\psi_{o1}^2 + \psi_{o2}^2)}{2(a_1 + a_2)} \{\ln[1 - \exp(-2\kappa R)]\}. \quad (155)$$

When $a_1 = a_2$ and small electric double layers overlap, $\exp(-\kappa R) \ll 1$, both expressions, V^σ and V^ψ, reduce to

$$V = \tfrac{1}{2} \varepsilon a \psi_o^2 \exp(-\kappa R). \quad (156)$$

The expression for the electrostatic interaction energy between the double layers of two spherical bodies, for a sufficiently large interseparation distance and without the Debye–Hückel approximation, was found to be

$$V = \frac{16\varepsilon a_1 a_2 k^2 T^2 \gamma_1 \gamma_2}{(a_1 + a_2) e^2 z^2} \exp(-\kappa R) \quad \text{(C.G.S. units)}. \quad (157)$$

which, for equal spheres, reduces to

$$V = \frac{8\varepsilon a k^2 T^2 \gamma^2}{e^2 z^2} \exp(-\kappa R), \tag{158}$$

where z is the valency of counter-ion and

$$\gamma = \frac{\exp[ze\psi_o/2kT] - 1}{\exp[ze\psi_o/2kT] + 1}. \tag{159}$$

If the Debye–Hückel approximation, $Z\psi_o/kT \ll 1$, is made, Eqn. (158) is reduced to Eqn. (156).

5.1.2. van der Waals interaction energy: The terms due to the attractive interaction forces, van der Waals forces, will not be discussed in detail. These terms consist of the interaction energies due to London dispersion [145], dipole-induced dipole (Debye), and dipole-dipole (Kessom) interaction forces (see Ref. 1). Among them, the London dispersion interaction term is due to the induced dipole-induced dipole interaction between the two bodies (1 and 2). The interaction energy is expressed in terms of second order perturbation theory [145] as

$$W_{12} = -\frac{3}{2}\frac{\hbar}{R_{12}^6}\sum_{ij}\frac{w_{i1}w_{j2}\alpha_{i1}\alpha_{j2}}{w_{i1} + w_{j2}} = -\frac{\lambda_{12}}{R_{12}^6} \cong -\frac{3}{2}\frac{\hbar}{R_{12}^6}\frac{I_{o1}I_{o2}}{I_{o1} + I_{o2}}\alpha_1\alpha_2, \tag{160}$$

where R_{12} is the separation distance between the two molecules 1 and 2. Here $\hbar = h/2\pi$, where h is Planck's constant, $\omega_{i1} = E_{i1} - E_{o1}$, E_{i1} is the ith energy level of molecule 1, and α_i is its polarizability. I_o is the ionization potential of the molecule.

Although this interaction energy is inversely proportional to the 6th power of the separation distance R_{12}, when the interaction energy is summed up over the entire two bodies, it becomes a long-range interaction energy:

$$U_{12} = \iint W_{12}\rho_1\rho_2 \, dV_1 dV_2, \tag{161}$$

where dV and ρ are the volume element and the density of the molecules composing the bodies 1 and 2.

For the case of two interacting spherical particles (bodies) of radii a_1 and a_2 separated by a distance R, the expression for the London dispersion interaction energy U is

$$U = -\frac{A_{12}}{12}\left[\frac{y}{x^2 + xy + x} + \frac{y}{x^2 + xy + y + x} + 2\ln\left(\frac{x^2 + xy + x}{x^2 + xy + x + y}\right)\right], \tag{162}$$

where $x = R/(a_1 + a_2)$, $y = a_1/a_2$, and A_{12} is an interaction constant (Hamaker constant) [146] between the two bodies 1 and 2. The Hamaker constant is related to a molecular description term in the following way:

$$A_{12} = \pi^2 N_1 N_2 \lambda_{12} = \frac{3}{2}\pi^2 \frac{I_1 I_2}{I_1 + I_2} \alpha_1 \alpha_2 \iint \rho_1 \rho_2 \, dV_1 \, dV_2. \quad (163)$$

where N_1 and N_2 are the numbers of molecules containing in bodies 1 and 2, respectively.

Some of the Hamaker constants are listed in Table 9. For equal spheres consisting of the same substance, Eqn. (162) becomes

$$U = -\frac{A}{12}\left\{ \frac{1}{x(x+2)} + \frac{1}{(x+1)^2} + 2\ln\frac{x(x+2)}{(x+1)^2} \right\}. \quad (164)$$

If a small interparticle separation is assumed ($R \ll a$, or $x \ll 1$), Eqn. (164) is reduced to,

$$U = -\frac{Aa}{12R}. \quad (165)$$

For the case of two identical parallel infinite plates at a separation distance R, the interaction energy is

$$U = -\frac{A}{12\pi R^2}. \quad (166)$$

The resultant forces ($-\partial U/\partial x$) of the London dispersion interaction are long-range forces which decay rather slowly with the interseparation distance R. Lifshitz (1955) [148] and co-workers developed a theory based on a macroscopic field equation, which avoids the calculation of individual intermolecular interactions. Their analysis indicates that the new theory accounts for many-body interactions. Other van der Waals terms are shorter range forces and their contributions to the total van der Waals interaction energy is usually minor compared with the London dispersion term.

When two interacting bodies are in a solution of a certain dielectric constant $\varepsilon > 1$, which is usual for colloid suspensions, the magnitude of the dispersion interaction is smaller than is the case in vacuum. In such a case, the dispersion energy W is expressed by:

$$W_{102} = W_{12} - W_{00} - [(W_{10} - W_{00}) + (W_{20} - W_{00})]$$
$$= W_{12} + W_{00} - W_{10} - W_{10} \quad (167)$$

Table 9. Values of Hamaker constants

Substance	A_{11}(micro)[a] ($\times 10^{13}$ ergs)	A_{11}(macro)[a]	A_{101}*(0 water)
Water	3.3–6.4	3.0–6.1	
Hydrocarbon	4.6–10	6.3	0.002–0.6
Polystyrene	6.2–16.8	5.6–6.4	0.04–0.4
Metals	7.6–15.9	27.1	5–9

*Calculated from the values for macroscopic A_{11}.
[a]Ref. [147].

where W_{12} refers to the interaction between the two bodies, 1 and 2, W_{10} is the interaction between body 1 and the medium 0, which is replaced by an equal volume of the other body 2. W_{00} is the interaction between the two media replaced by the two bodies.

Then, the Hamaker constant corresponding to the expression of Eqn. (167) is

$$A = A_{102} = A_{12} + A_{00} - A_{10} - A_{20}. \tag{168}$$

If the attraction between unlike phases is taken to be the geometrical mean of the attraction of each phase to itself, the following approximation may be used:

$$A_{ij} = (A_{ii} \times A_{jj})^{1/2}. \tag{169}$$

Then, Eqn. (168) becomes

$$A = A_{102} = (A_{11}^{1/2} - A_{00}^{1/2})(A_{22}^{1/2} - A_{00}^{1/2}). \tag{170}$$

For example, the Hamaker constant for the decane|vacuum|decane system is $A = 50 \times 10^{-14}$ ergs and for the decane|H_2O|decane system $A = 2.8 \times 10^{-14}$ ergs [150].

5.1.3. DLVO theory: The free energy of interaction of two interacting bodies may be expressed as a sum of the electrical double layer interaction energy (in general, repulsive) V, and the van der Waals interaction energy (in general attractive) U:

$$G^{tot} = V + U. \tag{171}$$

By using the above equations, Derjaguin and Landau [139], and Verwey and Overbeek [134] (DLVO), proposed a system for the process of adhesion. According to the work of Derjaguin and Landau, the interaction of two diffuse electrical double layers would cause the formation of an energy barrier between the two interacting particles. This energy barrier would lie between two minima (primary and secondary minima shown in Fig. 18).

For the two particles to come into close contact they would have to cross this energy barrier. Figure 18 shows a schematic drawing of the system. Curves A and B represent cases where the minima are separated by a maximum (energy barrier). Curve C represents a special case where the disappearance of the force barrier has taken place, and no energy barrier exists between the two minima. Derjaguin and Landau [139] derived the following criterion for the adhesion reaction of two identical spheres to occur

$$\frac{1}{\gamma n_1} = c f(\beta) \frac{A^2 e^6 z_1^6}{\varepsilon^3 (kT)^6} \tag{172}$$

where A is the Hamaker constant, e is the charge of an electron,

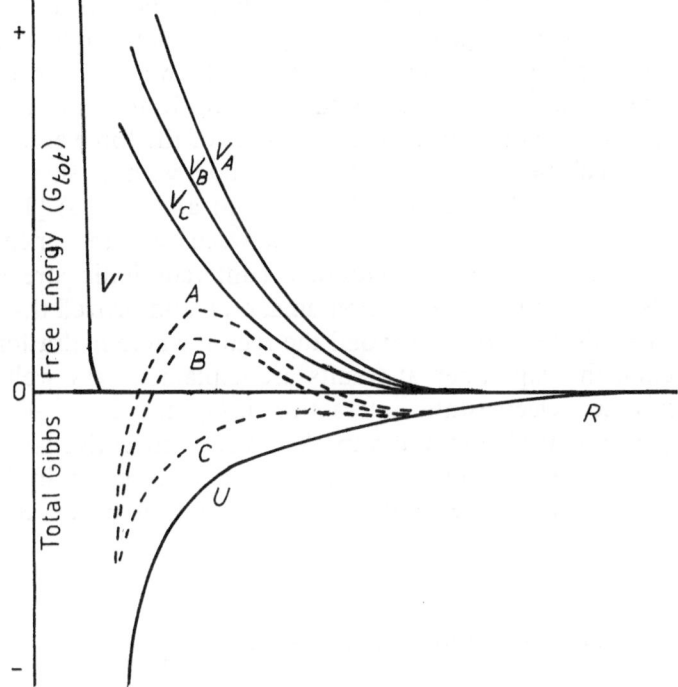

Fig. 18. A schematic diagram of the total interaction energy of two bodies. U refers to the energy due to the van der Waals attractive force between two interacting bodies, V refers to the energy due to the electrostatic interaction (repulsive) force, V' is the short-range repulsive interaction energy. R is the distance between the two bodies. The total interaction energy $G^{tot} = V' + U + V$ (dotted curves). Curves A and B are the cases where there is a maximum (primary maximum) between two areas of minimum (primary minimum and secondary minimum). Curve C is the case where there is no maximum (no barrier) so that the two bodies could come in close contact with each other at the primary minimum region.

$c = ((8\pi)^3/k^3 A)(\gamma n, kT)^2 \cong$ a constant, γ is the concentration of the electrolyte in moles per cm^3, n_1 is the number of dominating ions in one molecule, ε the dielectric constant, β the ratio of the valency of the auxiliary ion Z_2 to that of the dominating ion Z_1, and $f(\beta)$ is a function of the "asymmetry of the electrolyte". The function $f(1) = 1$, retaining the same order of magnitude for other values of β. While theoretical analysis in the field of colloid science had not by any means been completed, these early papers that presented the basic ideas were necessary for the theoretical biologists and the biophysicists to proceed further.

For the last decade, experimental techniques have been developed to measure the interaction energy between two interacting membranes, and the theory presented above has come to be tested. Cowley et al. have measured the repulsive force between acidic lipid membranes using an osmotic stress technique and an x-ray diffraction method [151]. Apart from the expected double layer force, they have reported the existence of an additional short-range hydration repulsion below a bilayer separation of 30 Å. The latter repulsive force exerted between two membranes will be discussed later, but it had been found by Rand and Parsegian's group earlier [152]. More direct measurements of the force and distance between charged membranes have been made with a surface force measuring apparatus (Marra and Israelachvili and others) [117, 118]. They have measured the forces exerted between two lipid membranes under various environmental conditions (different ionic strengths or ionic species, with and without ion binding, etc), and proved that, in the range of separation distances, beyond the region where hydration interaction occurs, the experimental results are explained very well by the above mentioned electrostatic interaction using the Gouy–Chapman double layer potential and the van der Waals attractive interaction energy terms [118]. One example of the results is given in Fig. 19. DLVO theory is applicable to systems interacting at distances greater than 20 Å.

5.2. DLVO theory applied to biological systems

Although the DLVO theory has many shortcomings, many attempts have been made to apply it to interaction phenomena in biological systems [14]. Since all mammalian cells examined thus far carry a net negative charge at their surfaces, contact between these cells can be considered as the interaction of two electrical double layers. The DLVO theory tells us that electrostatic forces of double layers tend to keep cells apart, while attractive forces, principally van der Waals forces, which

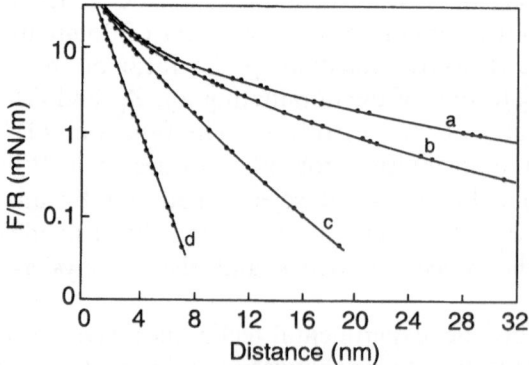

Fig. 19. Measured forces between two DSPG bilayers (22°C, pH 6.9) at various NaCl concentrations. The solid lines are the theoretically predicted forces assuming fully charged bilayers and a Hamaker constant $A = 6 \times 10^{-21}$ J. (a) 0.3 mM NaCl. (b) 1.1 mM NaCl. (c) 9.2 mM NaCl. (d) 100 mM NaCl. (From Marra [118].)

tend to favor the close approach of two cells, would be hindered by a considerable energy barrier between the cells due to electrostatic repulsive forces. For close contact to take place, this energy barrier of repulsion must be overcome. In a state where the membranes are separated by a few angströms, they are held together by attractive forces with an energy equal to that of any chemical bond which could be formed between them. In a primary energy minimum, close contact of the cells can occur. It should be noted, however, that if the interacting bodies are large enough, it is also possible that cell adhesion occurs at the secondary minimum because the depth of the secondary minimum could be large compared to the thermal energy kT [153, 154].

While it has been stated that there are inherent limitations in the DLVO theory, it does appear to be useful in explaining adhesion and contact in some ʻbiological cell systems. An early paper by van den Tempel in 1958 [155] showed that the theory could be used to analyze systems that were similar to colloidal ones. His work on emulsified oil globules, in relation to contact phenomena, enabled him to set up equations of repulsion and attraction resulting from the double layer. These equations, which are a direct result of the DLVO theory, have been applied with great success to biological systems. Van den Tempel was able to measure the thickness of the double layer and he confirmed that the secondary minimum predicted in the DLVO theory does exist.

Puck and others [156] used the theories of colloid stability to explain the adsorption of bacteriophage by bacteria. Their findings appeared to be compatible with those of the basic colloid theory. This work and that of Valentine and Allison [157] led to the novel result that the initial contact between virus particles and bacterial surfaces is regulated by electrostatic repulsion and the forces that govern the double layer. While these early studies were not as successful as one might have wished, they did show a correlation between the DLVO theory and the experimental data. Though the evidence is not conclusive, the work by Curtis [158] and Pethica [159] also tends to show that the DLVO theory is relevant to some biological systems. Other studies done at the time tended to show little or no correlation between the DLVO theory and the experimental data [160]. Using other types of cellular systems, Wilkins, Ottewill, and Bangham [161] were able to attain partial agreement between electrophoretic studies and flocculation data on sheep leucocytes.

Recent work has suggested that probes with low radius of curvature are important in crossing the primary maximum [62, 63]. From Eqn. [155] it is clear that the repulsive interaction energy is proportional to the radius of curvature of the interacting bodies. It is also believed that approach and contact can be made by microvilli with a low radius of curvature ($<0.1\ \mu$m). Microvilli of these dimensions encounter significantly less electrostatic repulsion (Eqn. 155) than larger pseudopodial type projections, and are therefore better able to overcome the potential energy barrier that opposes close cell contact. The formation of microvilli appears to be related to the thickness of the cell coat, and work with virus-induced fusion has played a large role in this work [162].

However, most biological cell membranes are not perfect two-dimensional molecular arrays. Most surfaces are composed of heterogeneous molecular assemblies and have three-dimensional molecular roughness. In some cases, some molecular components of membranes extend their molecular residues into the extracellular phase at a considerable distance $(50 \sim 100 \text{ Å})$ from the membrane surface (e.g. glycoprotein residues) [163–166] (Fig. 20). Such extensions of polymer molecules (proteoglycans) may exert a repulsive force between two closely opposed membranes because of electrostatic interaction due to the fixed charge on the molecule or because of steric effects resulting from the exclusion volume of such molecular extensions (based on the principle of steric exclusion in colloid chemistry) [167]. In some cells, the plasma membranes are covered by a so-called non-structured cell-coating material (mucopolysaccharides). In such cases, the straight application of the DLVO theory obviously does not hold. The adhesion between cell surfaces may be mediated by a quite different process. There may be specific interactions between specific molecular components [160].

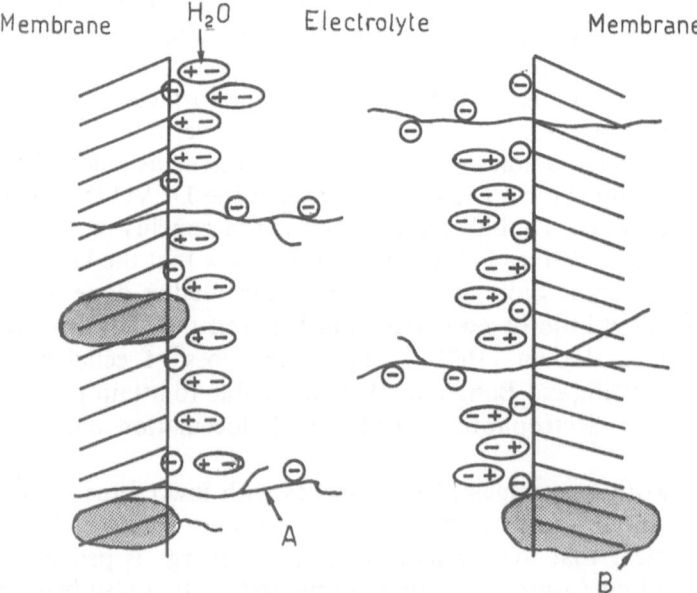

Fig. 20. A diagram of two apposed cell membrane surfaces, showing heterogeneous surfaces (oriented adsorbed water H_2O, molecular extensions of membrane protein moieties (A), clustered proteins imbedded in the membranes (B), etc.).

5.3. Hydration interaction energy

In addition to the above considerations, the polar surfaces of membranes are believed to be surrounded by oriented water molecules whose physical state is quite different from that of free water molecules in the bulk aqueous phase. The degree of hydration and the degree of orientation (water structure) may vary from one membrane surface to another. Recently, it was found that the structural water on the membrane surface exerts a strong repulsive interaction force between the two interacting surfaces [152, 167–170]. This force is a rather short range one. As the two surfaces approach within less than 20 Å distance it develops sharply and rapidly. The force, F, which is exerted between two apposed hydrated membranes, such as phosphatidylcholine bilayers, decays exponentially with respect to the interbilayer distance:

$$F = F_o \exp(-R/l) \tag{173}$$

where R is the interseparation distance and l the decay length. The decay length for various phospholipid membrane systems has been obtained by several workers [169, 170]. It varies for different types of lipid membrane, but it is in the range of $2.0 \sim 3.0$ Å. Since the force is repulsive due to the bound water or structured water at the membrane surface, it is called the hydration (repulsive) force. This energy is usually much greater than the van der Waals attractive and electrical repulsive energies, which have been mentioned earlier. Therefore, this energy term becomes important when the two surfaces are in close approach. A schematic graph of the hydration energy with respect to other van der Waals forces is shown in Fig. 21. It is seen that the electrostatic repulsive forces, including the double layer interaction force and the van der Waals attractive force, may play an important role in the stability of biocolloids at some large distance >20 Å even when the hydration force exists for the interaction of the two colloidal surfaces. At a short separation distance, the above two forces (electrostatic repulsive and van der Waals attractive) become important only with respect to the magnitude of hydration repulsive forces (i.e., when the magnitude of hydration repulsive force is weak compared to the former forces).

It has recently been shown that when the surfaces of the interacting membranes become more hydrophobic, the hydration repulsive interaction energy is reduced, and when the surface hydrophobicity is greater than a certain value, the hydration energy even becomes negative [171]. In the theory, it is assumed that the membrane surface is composed of a mixture of a hydrocarbon-like phase and a water-like phase in a fractional ratio p. The fractional ratio p is called the hydrophobic index. For example, when the membrane surface is a hydrocarbon phase, the

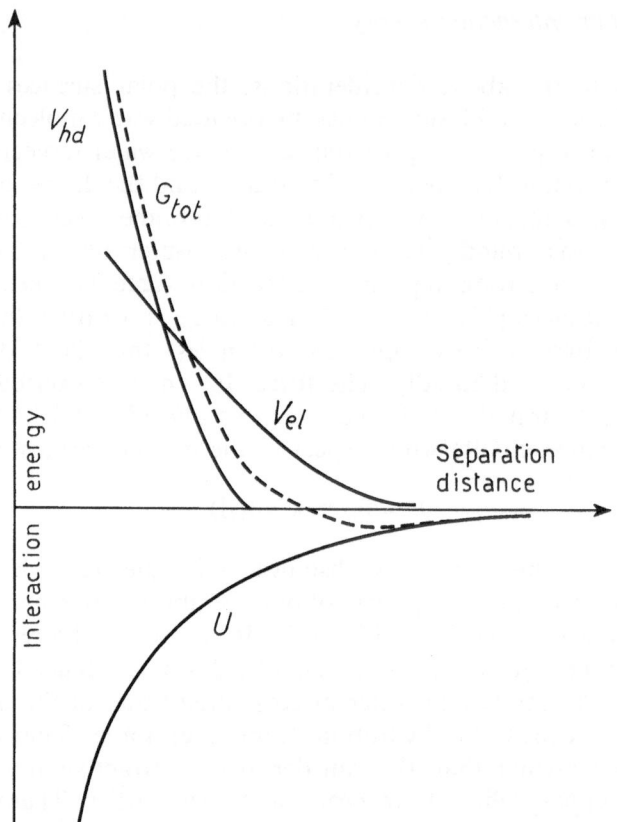

Fig. 21. A schematic diagram of the total interaction energy of two bodies. U refers to the energy due to the van der Waals attractive force, V_{el} refers to the energy due to electrical double layer interaction, and V_{hd} refers to the hydration interaction energy which increases sharply and rapidly at an interseparation distance smaller than about 20 Å. The magnitudes of V_{el} and V_{hd} vary greatly depending upon the surface density and the nature of the membrane surface.

hydrophobic index is 1.0, and when it is a water phase, the hydrophobic index is zero. Usually the hydrophobic index of the membrane surface is between 1.0 and zero. The total interaction energy of two interacting membranes which consists of the van der Waals, electrostatic repulsive, and hydration interaction energies, can then be calculated with respect to the separation distance R for different values of surface hydrophobicity (Fig. 22) [172, 173]. In the calculation, an exponential decay function with a short range decay constant ($1 = 2$ Å) is used for the repulsive hydration interaction, but an exponential decay function with a longer decay constant ($1 = 10$ Å) is used for the attractive hydration energy interaction. The longer decay constant for the attractive hydration energy has been reported by other authors [174]. A similar behavior of

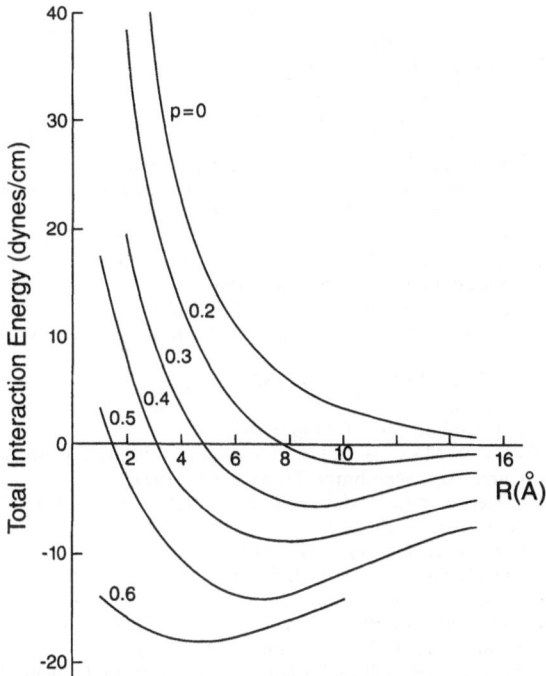

Fig. 22. The total interaction energy of two phosphatidylserine bilayers at close distance as a function of separation distance R, for various values of the hydrophobic index, p. The hydration interaction energy is the major contributor to the interaction energy within a separation distance of 15 Å. When a flat phosphatidylserine membrane is placed in 0.1 M NaCl/1 mM Ca^{2+}/pH 7.0 solution, the hydrophobic index of the membrane is calculated as approximately p = 2.0. With phosphatidylserine vesicles of small radius (200 Å) in the same salt solution as above, the hydrophobic index of the vesicle membrane is approximately p = 2.5. That of the rim area (the boundary of close contact becomes as high as p = 5.0. In such a case, the two vesicles would initiate membrane fusion through such boundary areas.

hydration energy due to the nature of the surface has been proposed by introducing the concept of a polar-apolar surface based on acid-base theory [175].

References

1. PC Hiemenz, *Principles of Colloid and Surface Chemistry*. Marcel Dekker, Inc. New York, 1977.
2. J Edsall, and J Wyman, *Biophysical Chemistry* Vol. 1, Academic Press, New York, 1956.
3. MS Fernandes and P Fromherz, J. Phys. Chem. 81 (1977) 1755.
4. S Ohki and R Kurland, Biochim. Biophys. Acta 645 (1981) 170.
5. JThG Overbeek, Electrochemistry of the Double Layer, in *Colloid Science*, HR Kruyte (ed), Elsevier, Amsterdam, Vol. 1, p. 115.
6. DJ Shaw, *Introduction to Colloid and Surface Chemistry*, Butterworths, London, 1980.
7. G Gouy, J. Phys. Rad. 9 (1910) 457.
8. DLL Chapman, Phil. Mag. 25 (1913) 475.
9. DC Grahame, Chemical Rev. 41 (1947) 441.

10. B Abraham-Shrauner, J. Math. Biol. 2 (1975) 333.
11. H Müller, Kolloidchem. Beihefte 26 (1928) 257.
12. O Stern, Z. Electrochem. 30 (1924) 508.
13. AN Frumkin, Phys. Chem. 109 (1924) 34.
14. JT Davies and EK Rideal, *Interfacial Phenomena*, Academic Press, New York, 1963.
15. DA Haydon, Recent Prog. Surface Sci. Vol. 1, JF Danielli, KGA Pankhurst, and AC Riddiford (eds), Academic Press, New York, 1964, p. 94.
16. JO'M Bockris, BE Conway and E Yeager (eds), *Comprehensive Treatise of Electrochemistry*, Vol. 1. Plenum Publ. Co., New York, 1980.
17. I Prigogine, P Mazur and R Defay, J. Chem. Phys. 50 (1953) 146.
18. DA Haydon and FH Taylor, Phil. Trans. A253 (1953) 146.
19. S Carnie and S McLaughlin, Biophys. J 44 (1983) 325.
20. H Ohshima and S Ohki, J. Colloid Interface Sci. 142 (1991) 596.
21. S Ohki, E Marcus, DK Sukumaran and K Arnold, Biochim. Biophys Acta 1194 (1994) 223.
22. F Booth, J. Chem. Phys. 19 (1951) 391, 1327.
23. JJ Bikerman, Phil. Mag. 33 (1942) 384.
24. VS Vaidhyanathan, Colloid Surf. 6 (1983) 291.
25. M Langer, D Cafiso, S Marcelja and S McLaughlin, Biophys. J. 57 (1990) 335.
26. S McLaughlin, Curr. Top. Membrane Transport 9 (1977) 71.
27. CM Drain, B Christenson and D Mauzerall, Proc. Nat'l. Acad. Sci. USA 86 (1989) 6959.
28. AL Loeb, J. Colloid. Sci. 6 (1953) 75.
29. WE Williams, Proc. Phys. Soc. 66 (1953) 372.
30. H Ohshima, TW Healy and LR White, J. Chem. Soc. Faraday Trans. 2, 79 (1983) 1613.
31. H Ohshima, TW Healy and LR White, J. Colloid Interface Sci. 90 (1982) 17.
32. S Levine, S Mingins and GM Bell, J. Phys. Chem. 67 (1963) 2095.
33. FP Buff and FH Stillinger, Jr, J. Chem. Phys. 39 (1963) 1911.
34. S Levine, K Robinson, GM Bell and J Mingins, J. Electronal Chem. 38 (1972) 253.
35. RH Brown Jr, Prog. Biophys. Mol. Biol. 28 (1974) 341.
36. AP Nelson and DA McQuarrie, J. Theor. Biol. 55 (1975) 13.
37. R Sauve and S Ohki, J. Theor. Biol. 81 (1979) 157.
38. AP Winiski, AC McLaughlin, RV McDaniel, M Eisenbergand and S McLaughlin, Biochemistry 259 (1986) 8206.
39. S Levine, Proc. Phys. Soc. 39 (1953) 1897.
40. MV Smoluchowski, Bull. Inter. Acad. Sci. (Cracovie), 184 (1903).
41. E Hückel, Physik. Z. 25 (1924) 204.
42. DC Henry, Trans. Faraday Soc. 44 (1948) 1021.
43. AD Bangham, BA Pethica, and GVF Seaman, Biochem. J 69 (1958) 12.
44. RJ Hunter, Arch. Biochm. Biophys. 88 (1960) 308.
45. GVF Seaman and VF Cook, in *Cell Electrophoresis*, EJ Ambrose (ed), Little Brown, Boston, 1965, p. 78.
46. AD Bangham and D Papahadjopoulos, Biochim. Biphys. Acta 126 (1966) 181.
47. D Papahadjopoulos, Biochim. Biophys. Acta 163 (1968) 240.
48. DO Shah and JH Schulman, J. Lipid Research 8 (1967) 227.
49. G Colacicco, Chem. Phys. Lipid 10 (1973) 66.
50. S Ohki and R Sauve, Biochim. Biophys. Acta 511 (1978) 377.
51. MC Phillips, Progr. Surface & Membrane Sci. 5 (1972) 139.
52. T Guilmin, E Goormaghtigh, R Brasseur, J Caspers and JM Ruysschaert, Biochim. Biophys. Acta 685 (1982) 169.
53. S McLaughlin, G Szabo and G Eisenman, J. Gen. Physiol. 58 (1971) 667.
54. G Eisenman and S Krasne, in *MTP International Review of Science, Biochemistry Series, Vol. 2.* CF Fox (ed), Butterworths, London, 1975, p. 17.
55. JD Castle and WL Hubbell, Biochemistry 15 (1975) 4818.
56. SC Hartsel and DS Cafiso, Biochemistry 25 (1986) 8214.
57. GV Sherbet, *The Biophysical Characterization of the Cell Surface*, Adacemic Press, New York, 1978.
58. H Alpes and WG Phol, Naturwissenschaften 65 (1978) 652.
59. R Pal, WA Petri, Y Barenholz and R Wagner, Biochim. Biophys. Acta 729 (1983) 185.
60. T Birshtein and O. Ptitsyn, *Conformations of Macromolecules*, Wiley, New York, 1966.
61. C Tanford, in *Adv. in Protein Chemistry*, CB Anfinsen, Jr. et al. (eds), Academic Press, New York, 1962, p. 69.

62. S Ohki, D. Doyle, T Flanagan, SW Hui and R Mayhew (eds), *Molecular Mechanisms of Membrane Fusion*, Plenum Publ. Corp., New York, 1988.
63. T Stegmann, RW Doms and A Helenius, Ann. Rev. Biophys. Biophys. Chem. 3 (1989) 187.
64. DE Goldman, J. gen. Physiol. 27 (1943) 37.
65. AL Hodgkin and B Katz, J. Physiol. (London) 108 (1949) 37.
66. T Teorell, Progr. Biophys. Chem. 3 (1953) 305.
67. MJ Polissar, in *Kinetic Basis of Molecular Biology*, FH Johnson, H Eyring and MJ Polissar (eds), Wiley, New York, 1954, p. 515.
68. RC MacDonald and AD Bangham, J Membrane Biol. 7 (1972) 29.
69. G Colacicco, Nature (London) 207 (1965) 936.
70. DL Gilbert, in *Biophysics and Physiology of Excitable Membranes*, WJ Adelman Jr. (ed), Van Nostrand Reinhold, New York, 1971, p. 264.
71. O Aono and S Ohki, J. Theor. Biol. 37 (1972) 273.
72. S Ohki, Physiol. Chem. and Phys. 13 (1981) 195.
73. S Ohki, in *Electrical Double Layers in Biology*, M Blank (ed), Plenum Press, New York, 1985, p. 103.
74. I Tasaki, A Watanabe and T Takenaka, Proc. Natl. Acad. Sci. U.S.A. 48 (1962) 117.
75. I Tasaki, and T Takenaka, Proc. Natl. Acad. Sci. U.S.A. 50 (1963) 619.
76. S Ohki and O Aono, Jap. J. Physiol. 29 (1979) 373.
77. PF Baker, AL Hodgkin and TL Shaw, J. Physiol. 164 (1962) 1177.
78. N Lakshminarayanaiah, Bull. Math. Biol. 39 (1977) 643.
79. S Ohki, in *Surface and Membrane Science* Vol. 10, DA Cadenhead and JF Danielli (eds), Academic Press, New York, 1976, p. 117.
80. RF Zwaal, B Roelofsen and CM Colley, Biochim. Biophys. Acta 300 (1973) 159.
81. LD Bergelson and LI Barsukov, Science 197 (1970) 224.
82. B Hille, *Ionic Channels of Excitable Membranes*, Sinauer, Sunderland, Mass. 1984,
83. S. Nir, C. Newton and D. Papahadjopoulos, Bioelectrochem. Bioener. 5 (1978) 116.
84. J Bentz, J. Colloid Interface Sci. 80 (1981) 179.
85. M Eisenberg, T Gresalfi, T Riccio and S McLaughlin, Biochemistry 18 (1979) 5213.
86. A Lau, A McLaughlin and S McLaughlin, Biochim. Biophys. Acta 645 (1981) 279.
87. MM Hammoudah, S Nir, J Bentz, E. Mayhew, TP Stewart, SW Hui and R Kurland, Biochim. Biophys. Acta 645 (1981) 102.
88. S McLaughlin, N Mulrine, T Gresalfi, G Vaoi and A McLaughlin, J. Gen. Physiol. 77 (1981) 445.
89. S Ohki, N Duzgunes and K Leonards, Biochemistry 21 (1982) 206.
90. S Ohki, Biochim. Biophys. Acta 777 (1984) 56.
91. JA Cohen and M Cohen, Biophys. J. 36 (1981) 623.
92. S Ohki and H Ohshima, Biochim. Biophys. Acta 812 (1985) 147.
93. F Oosawa, in *Polyelectrolytes*, Marcel Dekker, New York, 1971, p 160.
94. GS Manning, Ann. Rev. Phys. Chem. 23 (1972) 117.
95. H. Wennerstrom, B Lindman, G Lindblon and GJ Tiddy, J. Chem. Soc. Faraday Trans. 75 (1979) 663.
96. S Engstrom and H Wennerstrom, J. Phys. Chem. 82 (1978) 2711.
97. R Kurland, S Ohki and S Nir, in *Solution Behavior of Surfactants*, Vol. 2, KL Mittal and EJ Fendler (eds), Plenum Press, New York (1982).
98. M Mosior and S McLaughlin, in *Protein Kinase C: Current Concepts and Future Perspective*, DS Lester and RM Epand (eds), Ellis Wood, New York, 1992, p. 157.
99. PJ Goetz and JG Penniman, Jr., *A New Technique for Micro-electrophoretic Measurements*, American Lab. Oct. (1976).
100. P McFadyen, American Lab. April (1987) p. 64.
101. M Hook, L Kjelle'n, S Johanssen and J Robinson, Ann. Rev. Biochem. 53 (1984) 847.
102. JThG Overbeek, Quantitative Interpretation of the Electrophoretic Velocity of Colloids, in *Advances in Colloid Science*, Vol. 3, H Mark and EJW Verwey (eds), Wiley-Interscience, New York, 1950.
103. PH Wiersman, AL Loeb and JThG Overbeek, J. Colloid Interface Sci. 22 (1966) 78.
104. AL Loeb, JThG Overbeek and PH Wiersman, *The Electric Double Layer around a Spherical Colloid Particle*, MIT Press, Cambridge, Mass., 1960.
105. F Booth, Nature (London) 161 (1948) 83.

106. DC Grahame, Z. Electrochem. 62 (1958) 264.
107. CA Barlow and JR MacDonald, Adva. Electrochem. Eng. 6 (1971) 1.
108. KS Cole, Biophys. J. 9 (1969) 465.
109. A Voigt and E Donath, in *Biophysics of the Cell Surface*, R Glaser and D Gingell (eds), Springer-Verlag, London, 1990, p. 75.
110. FH Stillinger, J. Chem. Phys. 35 (1961) 1584.
111. VG Levich and Y Yalamov, Z. Fiz. Chim. 36 (1962) 1096.
112. S Aytyan, SS Dukhin and Y Chizmadshev, Elektrochimi 13 (1977) 779.
113. DT Enos and DA McQuarrie, J. Theor. Biol. 93 (1981) 499.
114. E Matijevic (ed), *Surface and Colloid Science, Vol. 7*, Wiley, New York, 1974, p. 81.
115. S McLaughlin, C Grathwohl and S McLaughlin, Biochim. Biophys. Acta 513 (1978) 338.
116. S Oka, Proc. Phys. Math. Soc. Japan 14 (1932) 649.
117. J Marra and J Israelachvili, Biochemistry 24 (1985) 4908.
118. J Marra, Biophys. J. 50 (1986) 815.
119. ME Loosley–Miillman, RP Rand and VA Parsegian, Biophys. J. 40 (1982) 221.
120. EA Evans and VA Parsegian, Pro :. Natl. Acad. Sci. USA 83 (1986) 7132.
121. MP Hentschel, M Mischel, RC Oberthur and G Buldt, FEBS Lett. 193 (1985) 236.
122. W van Megen and J Snook, J. Chem. Phys. 73 (1980) 4656.
123. GM Torrie and JP Valleau, J. Chem. Phys. 73 (1980) 5807.
124. SL Carnie and GM Torrie, Adv. Chem. Phys. 56 (1984) 141.
125. VA Parsegian and D Gingell, J Adhesion 4 (1972) 283.
126. EH Eylar, MA Madoff, OV Brady and JL Oncley, J. Biol. Chem. 237 (1962) 1992.
127. DA Haydon, Biochim. Biophys. Acta 50 (1961) 450.
128. E Donath and V Pastushenko, Bioelectrochem. Bioenergy. 7 (1979) 543.
129. RW Wunderlich, J. Colloid Interface Sci. 88 (1982) 385.
130. S Levine, M Levine, KA Sharp and DE Brooks Biophys. J. 42 (1983) 127.
131. V Pastushenko and E Donath, Studia Biophys. 56 (1976) 9.
132. H Ohshima and S Ohki, Bioelectrochem. Bioenerg. 15 (1986) 173.
133. S. Ohki and H. Ohshima, in *Electrical Double Layers in Biology*, M Blank (ed), Plenum Press, New York, 1985, p. 1.
134. EJA Verwey and JThG Overbeek, in *Theory of the Stability of Lyophobic Colloids*, Elsevier, Amsterdam, 1948, p. 25.
135. S Ohki, J. Phys. Soc. Japan 20 (1965) 1674.
136. A Mauro, Biophys. J. 2 (1962) 179.
137. VA Parsegian, Ann. Rev. Biophys. Bioengen. 2 (1973) 221.
138. H Ohshima and S Ohki, Biophys. J. 47 (1985) 673.
139. BV Derjaguin and L Landau, Acta Phys. Chim. URSS 14 (1941) 633.
140. Y Ikeda, J. Phys. Soc. Japan 8 (1953) 49.
141. OF Devereux and PL de Bruyn, *Interaction of Parallel Double Layers*, MIT Press, Boston, Mass., 1963.
142. S Usui, J. Colloid Interface Sci. 44 (1973) 107.
143. R. Hogg, TW Healy and D Fuerstenan, Trans. Faraday Soc. 62 (1966) 1638.
144. GR Wiese and TW Healy, Trans. Faraday Soc. 66 (1970) 490.
145. F London, Trans. Faraday Soc. 33 (1937) 8.
146. HC Hamaker, Physica 4 (1937) 1058.
147. J Visser, in *Surface and Colloid Science Vol. 8*, E. Matijevic (ed), Wiley-Interscience, 1976, p. 3.
148. EM Lifshitz, J. Exp. Theor. Phys. 29 (1955) 94.
149. IE Dzyaloshiuskii, EM Lifshitz and LP Pitaevskii, Adv. Phys. 10 (1961) 165.
150. S Nir, Progr. Surface Sci. 8 (1977) 1.
151. AC Cowley, NL Fuller, RP Rand and VA Parsegian, Biochemistry 17 (1978) 3163.
152. DM LeNeveu, RP Rand and VA Parsegian, Nature (London) 259 (1976) 601.
153. S Nir and J Bentz, J. Colloid Interface Sci. 65 (1978) 399.
154. S Ohki, S Roy, H. Ohshima and K Leonards, Biochemistry 23 (1984) 6126.
155. M van den Tempel, J. Colloid Sci. 13 (1958) 125.
156. TT Puck and LT Tolmach, Arch. Biochem. Biophys. 51 (1954) 229.
157. RC Valentine and AC Allison, Biochim. Biophys. Acta 34 (1959) 10.
158. ASG Curtis, Amer. Nat. 94 (1960) 37.
159. BA Pethica, Exp. Cell Res. Suppl. 8 (1961) 123.

160. L Weiss, *The Cell Periphery Metastasis and Other Contact Phenomena*, North-Holland Press, Amsterdam, 1967.
161. DJ Wilkins, RH Ottewill and AD Bangham, J. Theor. Biol. 2 (1962) 165.
162. G Poste, Int. Rev. Cytol. 33 (1972) 157.
163. L Weiss and JP Harlos, Progr. Surface Sci. 1 (1972) 335.
164. U Aebi and ZJ Engel (eds), *Cytoskeletal and Extracellular Proteins*, Springer-Verlag, Berlin, 1989.
165. R Harrison and GG Lunt, *Biological Membranes*, Blackie, Glasgow, 1980.
166. PM Kraemer, in *Surface of Normal & Malignant Cells*, RO Hynes (ed), Wiley, New York, 1979.
167. NG Maroudas, J. Theor. Biol. 49 (1975) 417.
168. S Marcelja and N Radic, Chem. Phys. Lett. 42 (1976) 129.
169. RP Rand, Annu. Rev. Biophys. Bioeng. 10 (1981) 277.
170. TJ McIntosh, AD Magid and SA Simon, in *Cell and Model Membrane Interactions*, S. Ohki (ed), Plenum Publ. Co., New York, 1991, p. 249.
171. S Ohki, Studia Biophysica 110 (1985) 95.
172. S Ohki, in *Cell and Model Membrane Interactions*, S Ohki (ed), Plenum Publ. Co. New York, 1991, p. 267.
173. S Ohki, Membrane 19 (1994) 48.
174. J Israelachvili, *Intermolecular and Surface Forces*, Academic Press, San Diego, 1991.
175. CJ van Oss, *Interfacial Forces in Aqueous Media*, Marcel Dekker, Inc., New York, 1994.
176. JR Lakowicz, *Principles of Fluorescence Spectroscopy*. Plenum Publ. Co., New York, 1983, p. 496.
177. DM Crothers and H Metzger, Immunochem. 9 (1972) 341.
178. JA Reynolds, Biochemistry 18 (1979) 204.
179. AV Hill, J. Physiol (London) 40 (1910) 4.

Bioelectrochemistry: General Introduction
ed. by S. R. Caplan, I. R. Miller and G. Milazzot
© 1995 Birkhäuser Verlag Basel/Switzerland

CHAPTER 5
Adsorption and surface reactions

Israel R. Miller

Department of Membrane Research and Biophysics, The Weizmann Institute of Science, Rehovot, Israel

1. Introduction

A significant proportion of biological processes occur at interfaces. The predominant interface in biological systems is that between an aqueous solution and a cell membrane, though very important processes also occur at surface boundaries between cells and air, e.g. animal skin surfaces, lung alveoli, plant leaf surfaces, etc. Adsorption/desorption processes and surface reactions take place at these interfaces, accompanied by diffusion of reagents normal to the surface; lateral diffusion and the passage of substances across the interface also sometimes occur. Interfacial reactions at heterogeneous phase boundaries are vectorial in nature, whose rate constants have the dimensions of velocity. The vectorial nature of interfacial reactions is expressed not only by the direction of the reagents diffusing to and possibly penetrating the interface, but also by orientation of the adsorbed reagents and catalytic sites on the surface. In the case of electrochemical processes the electrons exchanged at an interface can be considered equivalent to any other reagents taking part in a surface reaction. Interfacial electrochemical processes take place on both electrode surfaces and between redox systems on the two sides of a phase boundary. An electron will generally cross an interface, while the other reagents may be reflected after their electron gain or loss. At the interface of a biological membrane electron exchange may occur between water-soluble components like cytochrome c, NADH, and ascorbic acid, and between membrane-bound components such as ubiquinones, cytochrome oxidase, etc.

Given the fraction of biological processes which proceed at interfaces, it is desirable for biologists to acquire an adequate knowledge of basic interfacial science, including adsorption phenomena and interfacial electrochemical processes.

2. Adsorption at interfaces

2.1. The definition of interfaces

2.1.1. Interfaces: general description.
An interface is a boundary plane between two homogeneous phases. The three states of condensation of matter, solid, liquid, and gas, give rise to five conventional types of interfaces: 1. solid-solid, 2. solid-liquid, 3. solid-gas, 4. liquid-liquid, and 5. liquid-gas. The existence of crystalline and amorphous solids and of liquid crystals yields some subclasses of interfaces besides the conventional ones. The boundary between two homogeneous phases should not be regarded as a simple geometrical plane; the forces acting on the layers of molecules at the interface differ from the forces acting on molecules in the bulk phases, so the surface plane of the interface has

different properties from the two bulk phases. There is uncertainty with respect to the thickness of the surface layer, though the anisotropic directional cohesion forces are chiefly exerted on the outer-most molecular layer of each bulk phase (Fig. 1), with the second and possibly third layers probably also being affected, albeit to a lesser extent. In the case of liquid-gas interfaces (Fig. 1a) the cohesion forces between molecules on the surface of the liquid and between the molecules of the gaseous phase and liquid phase approach zero, and a force tending to minimize the surface area arises. Thus, the surface energy is always positive and decreases with decreasing surface area. Negative surface energies may arise transiently, in which case the interface expands spontaneously either by emulsification or by decreasing the average size of the dispersed units, until the interfacial tension becomes positive or zero. Any liquid in the absence of gravitational forces will assume a shape with a minimal area to volume ratio, namely a sphere. The cohesion forces between molecules of the two phases constituting the interface between two immiscible liquids are also non-isotropic (Fig. 1b).

Fowkes [1] suggested that the force of attraction between the two boundary monolayers in phase 1 and phase 2 is the geometric mean of the dispersion contributions to the surface tensions of the two phases, $\sqrt{(\gamma_1^d \gamma_2^d)}$. Thus, the interfacial tension would be the sum of the surface tensions of the two phases less twice the attractive forces between the two monolayers:

$$\gamma_{1,2} = \gamma_1 + \gamma_2 - 2\sqrt{\gamma_1^d \cdot \gamma_2^d} \qquad (1)$$

In hydrocarbons the dispersion contribution is nearly equal to the total surface tension, whereas for water the dispersion contribution is

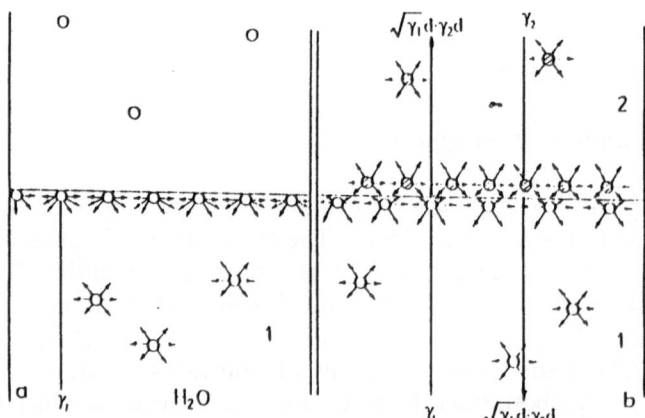

Fig. 1. Molecules at liquid/gas (a) and at liquid/liquid (b) interfaces. The forces acting on the molecules at the interior of the liquid phases are isotropic; they are directional at the interfaces with the resultant force normal to the surface.

about 30% (21.8 $\pm$ 0.7 dyne/cm) of the total surface tension ($\sim$ 72 dyne/cm). This shows a very pronounced contribution of hydrogen bonds to the cohesion forces. The interfacial tension between water and hydrocarbons is always around 51 dyne/cm, even though the surface tension for hydrocarbons varies from 18 for pentane and hexane to > 26 for hexadecane. Here also the system will seek shapes with a minimal interfacial area, viz. spherical drops of one liquid in the other. The preference of liquids to assume a spherical shape depends on the molecular structures, which determine how the molecules will best fit into the concave and convex planes to assure minimal surface energies.

The cohesion forces between the identical molecules, at least in one of the phases, are always larger than those between the non-identical molecules on the two sides of the interface; otherwise the boundary would vanish and the two liquids would mix. Table 1 lists interfacial tensions at 20° between water and organic liquids of increasing polarity and the solubility of those organic liquids in water.

Ethyl ether is a pronounced exception in this series, being much more soluble in water than n-hexanol while maintaining a higher interfacial tension at its aqueous boundary than the latter. This is because the polar residue of the ether is in the middle of the molecule and cannot be exposed exclusively to the water surface. The normal aliphatic alcohols are amphipatic, with a polar water-soluble hydroxyl group which at the interface is oriented towards the aqueous phase. When using eq. 1 to calculate the interfacial tension $\gamma_{1,2}$ the surface tensions γ_1 and γ_2 are of liquid phases 1 and 2 *saturated by the liquid on the other side of the interface*, and the geometric mean is *not* merely that of the dispersion forces.

The interfacial tension is related to the excess Helmholtz free energy F_σ. The total energy of the interface U_σ is larger than F_σ from which it may be calculated using the relation

$$F_\sigma = U_\sigma - TS_\sigma \tag{2}$$

where the surface entropy S_σ is given by the partial derivatives at

Table 1. Interfacial tension and solubility of different organic molecules in water

	$\gamma 20°$ (dynes cm^{-1})	Solubility in water g/100 ml
hexane	51	0.014
carbon tetrachloride	45.1	0.08
benzene	35	0.082
nitrobenzene	26	0.19
ethyl ether	10.7	7.5
n-hexanol	6.8	0.59
n-butanol	1.6	7.9

constant volume V and constant number of molecules n

$$S_\sigma = -\left(\frac{\partial F_\sigma}{\partial T}\right)_{V,n} = -\left(\frac{\partial \gamma_0}{\partial T}\right)_{V,n} \tag{3}$$

and

$$U_\sigma = \gamma_0 - T\left(\frac{\partial \gamma_0}{\partial T}\right)_{V,n} \tag{4}$$

For the air/water interface at $20°$ the internal energy

$$U_\sigma = 73 - 295(-0.154) = 118 \text{ ergs cm}^{-2}$$

Frenkel [2], Langmuir [3] and others suggested simple methods for establishing the relation between the total surface energy U_σ and the intermolecular cohesion forces. According to Frenkel

$$U_\sigma = u(Z - Z')v_\sigma \tag{5}$$

where u is the intermolecular cohesion energy, Z is the number of interacting neighbors in the bulk, Z' is their number in the surface and v_σ is the number of molecules per cm^2. The value of u is inferred from the heat of evaporation

$$L = \tfrac{1}{2}Zuv_b \tag{6}$$

where v_b and L are the number of molecules and the heat of evaporation per cm^3. The factor 1/2 arises from the fact that each cohesion energy is between two molecules. From here:

$$U_\sigma = 2L\left(\frac{Z - Z'}{Z}\right)v_\sigma^{-1/3} \tag{7}$$

This expression gives good results for non-polar molecules but over-estimates values for water or alcohol. For water, assuming $Z = 6$ and $Z' = 5$, U_σ comes out to be 240 m joules/m^2, twice the measured value, probably because the molecules on the surface are oriented and interact with a larger number of neighboring molecules.

According to Langmuir, each individual molecule possesses a spherical surface with a specific internal surface energy U_σ. The area of a water molecule is $47 \times 10^{-16} \text{ cm}^2$. Thus the internal energy of 1 cm^3 water is U_σ $47 \times 10^{-16}N/18$, where N is Avogadro's number. This energy is equal to the latent heat of evaporation, 2.4×10^{10} erg/cm^3. The value of U_σ calculated by this approach is about 150 ergs/cm^2. There are more sophisticated formulations, including quantum mechanical ones, which give better agreement with experimental values, but most of these include adjustable parameters.

Molecules at the interface exchange very rapidly with molecules in the bulk phase. The rate of exchange is usually determined by diffusion, and the numbers of molecules exchanged per cm^2 per second may vary in ordinary liquids between 10^{15} and 10^{25}; in other words, the dwelling time of a molecule in the surface may vary from a fraction of a nanosecond to a second. In spite of this vigorous molecular movement the surface remains microscopically quite planar. Any microscopic deviation from planarity is energetically costly. Formation of a hemispherical bulge or indentation with a radius of 1 nm would require about 0.8 kT per dyne/cm of interfacial tension. Even at a relatively low-energy hydrocarbon/air interface (~ 20 mN/m) such bulges or indentations would be relatively rare, occupying only $\sim 10^{-7}$ of the surface. The interfacial tension of many biological surfaces is below 1 mN/m, and fluctuating deviations from planarity may be much more frequent; however, the surface energy of these interfaces increases as they expand and the hydrocarbon interior becomes exposed to the aqueous medium, converting part of the surface into a hydrocarbon/water interface. Expansion of the surface area of the fluctuating domain by 5% may increase the interfacial tension by up to 1 dyne/cm.

2.1.2. Biological interfaces: Biological surfaces are sites of most of the biological processes and they comprise a large fraction of biological tissues. Unlike the interfaces between two homogeneous phases, biological interfaces may have a distinct architecture determined by some very specific interactions. One can speak about a surface of a biological macromolecule and about interfaces between interacting macromolecules, but the composition and properties of these interfaces change from site to site, and a great deal of abstraction is required to deal with these interfaces in general terms. The same is true about cell membrane surfaces and cytosol-membrane interfaces. Membranes contain different lipid head groups, polypeptide chains, and carbohydrate chains, some of which serve as receptors specific for particular substances. The interfaces between cell organelles like nuclei, mitochondria, chloroplasts and endoplasmic reticula, and the cytoplasm are of a similar nature. There are also tissue-gas biological interfaces, e.g. skin and lung alveoli, and tissue-solid biological interfaces, e.g. muscle or tendon-bone junctions. The tissue surface is a very complex one with different degrees of hydration. The study of biological interfaces requires not only a physico-chemical approach but also some biochemical and biological knowledge and techniques. As most of the biological reactions take place on interfaces which serve as catalytic sites, studies of their kinetics require the use of tools of heterogeneous catalysis irrespective of the detailed biochemical reaction involved.

2.2. Physics and thermodynamics of interfaces

2.2.1. Interfacial tension and its measurement: The anisotropy of cohesion forces at the interface confers upon the surface an energy proportional to its area. The surface tension counteracts any attempt to increase the area and thus the energy of the system. Any surface tension measurement is based upon measuring the force counteracting an area increase. The Wilhelmy plate and capillary rise methods are the best ways to quantify surface tension. In the Wilhelmy plate method a plate is immersed into the solution and the force acting on it is measured. After subtracting the weight and the buoyancy force, one obtains the force exerted by the surface tension only, F_r, from the equation

$$F_r = 2(1 + d)\gamma \cos \alpha \tag{8}$$

where γ is the surface tension, l and d are the width and thickness, respectively, of the plate, and α is the contact angle (Fig. 2). It's best to choose plates which produce contact angles approaching zero, such as roughened platinum or even filter paper. The capillary rise method is predicated on establishing a balance between forces exerted by surface

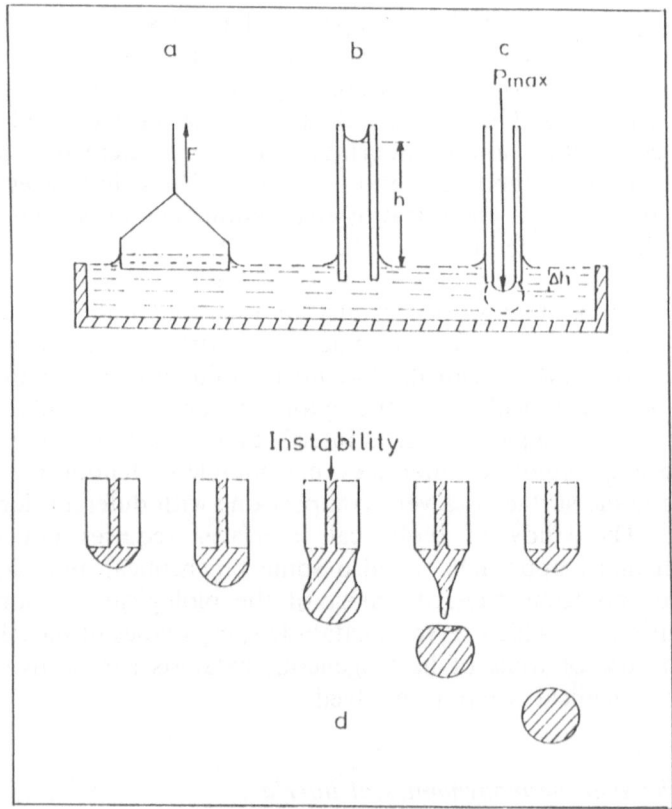

Fig. 2. Methods for surface tension measurements. a. Wilhemy plate; b. capillary size; c. maximal bubble pressure; d. drop weight method.

tension and by hydrostatic pressure (Fig. 2b)

$$2\pi r\gamma \cos \alpha = \pi r^2 h\rho g$$

or

$$\gamma = \frac{rh\rho g}{2 \cos \alpha} \qquad (9)$$

where h is the rise of the liquid in the capillary, r the capillary's internal radius, ρ the density of the liquid, and g the gravitational acceleration constant.

Other methods for measuring surface tension, like the maximal bubble pressure and drop weight methods, are based on the same principle as the capillary rise technique. When the pressure within a capillary immersed into a liquid is increased, the hydrostatic head within the capillary is lowered until the meniscus reaches the bottom of the capillary and the pressure reaches its maximum value. Then a bubble increasing in radius starts to form with a concomitant decrease in pressure. The maximal pressure p_{max} relates to the surface tension if Δh is the depth of the immersion of the capillary and the radius of the meniscus as follows:

$$\frac{2\gamma}{r} = P_{max} - \Delta h\rho g \qquad (10)$$

Equations (9) and (10) are correct for sufficiently narrow capillaries where the capillary rise h $\gg$ r or $(r/h)^{1/2} < 0.12$. For large capillary radii the hemispherical shape of the meniscus becomes distorted and correction factors need to be used [6].

The maximal bubble pressure method can be used to determine interfacial tensions of lipid bilayers [4] and biological membranes. A lipid bilayer is formed from a hydrocarbon solution across a hole (0.1–2 mm in diameter) punched in a Teflon septum separating two aqueous compartments. When the pressure P is increased in one compartment, the bilayer bulges until a hemispherical shape is obtained, at which point $P = P_{max}$. The interfacial tension on each side of the bilayer is obtained from the difference in pressure between the two compartments,

$$\Delta P_{max} = 2(2\gamma/r) = 4\gamma/r \qquad (11)$$

where r is the radius of the hole. Similar information on cell membrane tensions can be obtained by measuring the pressure difference required to suck a cell into a thin capillary (Fig. 3d).

In the drop weight or drop volume method, the effective weight of the detaching drop from a capillary with an outer radius r is considered to be equal to the surface force

$$HV\Delta\rho \cdot g = 2\pi r\gamma \qquad (12)$$

where V is the drop volume and $\Delta\rho$ is the density difference between the two liquid phases if a drop of one liquid is formed in the other one and H is a correction factor arising from the fact that the radius of the neck of the hanging drop before its detachment (Fig. 2d) is smaller than the radius r of its cylindrical support. The correction factor H was originally determined empirically by Harkins and Brown [5]; since then, different ways to calculate H have been reported. Details and tabulated values of H as a function of $(r/V^{1/3})$ can be found in Adamson's *Physical Chemistry of Surfaces* [6].

Surface tension changes during polymer adsorption are generally slow, and biopolymer adsorption accompanied by conformational changes are particularly slow. Suitable methods are required to measure slow changes in surface tension and to eliminate artifacts which may arise from solvent evaporation and "drying" of the surface layer or from changes in the supporting surface (as in the capillary rise method). Measuring shape changes of sessile and pendant drops kept in a constant atmosphere are suitable for this purpose. Figs. 3a–c are drawings of sessile and pendant drops from a gaseous environment and of a pendant drop of a less dense liquid formed in a denser one. For very large drops where the radius of curvature in only one plane is to be

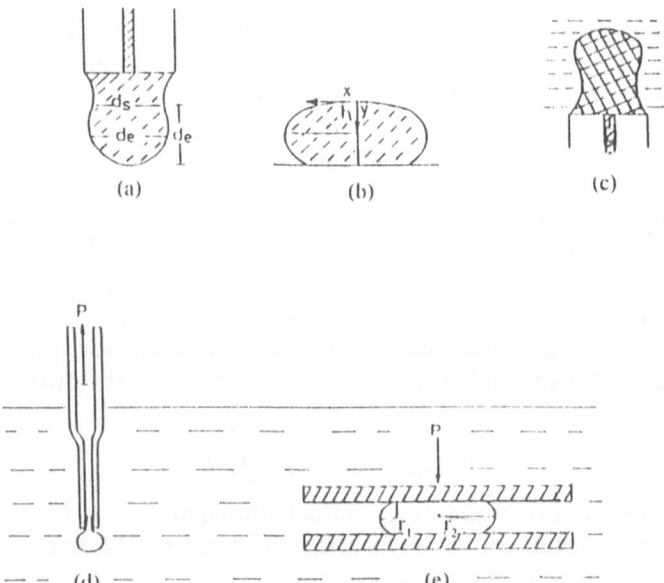

Fig. 3. Stationary methods for interfacial tension measurements. a) pendant drop; b) sessile drop; c) sessile drop of a low-density liquid in a high-density liquid; d) interfacial tension of a cell membrane sucked into a capillary of radius r by a pressure P; e) interfacial tension of a membrane of a cell compressed between two plates at a pressure P, inferred from the radii of curvature r_1 and r_2.

considered, the following differential equation describes the contours:

$$\Delta\rho\, gy = \gamma(y''/(1 + y'^2)^{3/2}) \tag{13}$$

where $y' = dy/dx$ and $y'' = d^2y/dx^2$. After integration we obtain

$$y^2 a^2 = -1/(1 + y'^2)^{1/2} + \text{constant} \tag{14}$$

where a is the capillary constant and $a^2 = 2\gamma/\Delta\rho g$.

If, for a sessile drop, h denotes the distance from the apex $y = 0$ to the equatorial plane, (see Fig. 3b), then at the equatorial plane $h = y$, $y' = \infty$ and Eqn. (14) becomes

$$y^2 a^2 - h^2 a^2 = -1/(1 + y'^2)^{1/2} \tag{15}$$

At the apex of the sessile drop $y = 0$, $y' = 0$ and hence

$$h = a = (2\gamma/\Delta\rho g)^{1/2} \tag{16}$$

Thus, measured values of h, the distance between the equatorial plane and the apex, suffice for the determination of the surface tension and its variation over time. The sessile drop method can also be applied to the determination of surface tensions of spherical cells.

It is also possible to determine the surface tension of a pendant drop and its variation with time from its shape. A conveniently-measurable shape parameter is $S = d_3/d_e$. As shown in Fig. 3a, d_e is the equatorial diameter and d_s the diameter measured at the distance d_e from the bottom of the drop. The surface tension is given by

$$\gamma = \Delta\rho\, gd_e^2/H \tag{17}$$

and the relationship between the shape-dependent quantity H and the shape parameter S has been tabulated [6].

The cell membrane tension is the sum of an interfacial tension and elastic tension. The two components of the membrane tension increase with the extension of the surface area; therefore, all membrane tension measurements have to be accomplished under conditions of minimal area extension to avoid artifically high values. This can be done if one considers a cell on a planar support to be a sessile drop, or if one uses Cole's compression method [7] at low pressures. In the latter method the cell is compressed between two plates and the two radii of curvature r_1 and r_2 as shown in Fig. 3e are determined. Then, the pressure applied P is related to r_1 and r_2 by

$$P = \gamma\left(\frac{1}{r_1} + \frac{1}{r_2}\right) \tag{18}$$

2.2.2. Thermodynamics of adsorption: J. W. Gibbs derived an equation relating the change in interfacial tension to the change in chemical or electrochemical potentials of the components [6, 11]. Figure 4 shows an

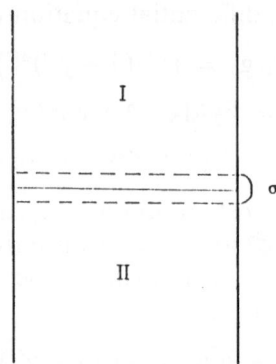

Fig. 4. Interfacial region between phase I and II of an arbitrary thickness σ.

interface of arbitrary thickness σ at the phase boundary between two phases, denoted I and II. The total excess energy of the interface U^σ, less the energy of the original bulk phases within the same interfacial region, is given by

$$U^\sigma = TS^\sigma - PV^\sigma + \gamma A + \sum_i \tilde{\mu}_i n_i^\sigma \qquad (19)$$

Differentiation of Eqn. (19) yields

$$dU^\sigma = TdS^\sigma + S^\sigma dT - PdV^\sigma - V^\sigma dP$$

$$+ \gamma dA + Ad\gamma + \sum \tilde{\mu}_i dn_i^\sigma + \sum n_i^\sigma d\tilde{\mu}_i \qquad (20)$$

However, dU^σ is also obtained by differential change of all the extensive properties of the surface layers

$$dU^\sigma = TdS^\sigma - PdV^\sigma + \gamma dA + \sum \tilde{\mu}_i dn_i^\sigma \qquad (21)$$

Subtracting Eqn. (21) from Eqn. (20), one obtains (for constant temperature and pressure)

$$-d\gamma = \sum \Gamma_i d\tilde{\mu}_i \qquad (22)$$

where Γ_i is the surface excess concentration of component i. Γ_i can be either positive or negative. Assuming incompressibility, the volumes of the positive and negative excess concentrations in the surface have to be balanced:

$$\sum \Gamma_i v_i = 0 \qquad (23)$$

The electrochemical potential can be factored into a chemical and an electrical component:

$$\sum \Gamma_i d\tilde{\mu}_i = RT \sum \Gamma_i d \ln a_i - \sum z_i e \Gamma_i d\psi \qquad (24)$$

where z_i is the sign and valency of the ion and e is the electronic charge. If the surface is electroneutral, then $\sum z_i e_i \Gamma i = 0$. The interface may be charged either by applying an external potential to a polarizable electrode (e.g. mercury) or by spreading an insoluble charged monolayer at the interface, or by changing the concentration of the ionic component of a reversible electrode, e.g. Ag^+ or Cl^- in contact with a Ag/AgCl electrode. For charged interfaces the surface charge density q is

$$q = -\sum z_i e_i \Gamma_i \qquad (25)$$

Differentiating Eqn. 22 with respect to the chemical potential of a solution of a single salt, keeping the potential constant with respect to an electrode reversible to one ionic component, one obtains the surface excess of the other ionic component

$$-\left(\frac{\partial \gamma}{\partial \mu}\right)_{\psi(+)} = \Gamma_{(-)}/z_{(-)}$$

$$-\left(\frac{\partial \gamma}{\partial \mu}\right)_{\psi(-)} = \Gamma_{(+)}/z_{(+)} \qquad (26)$$

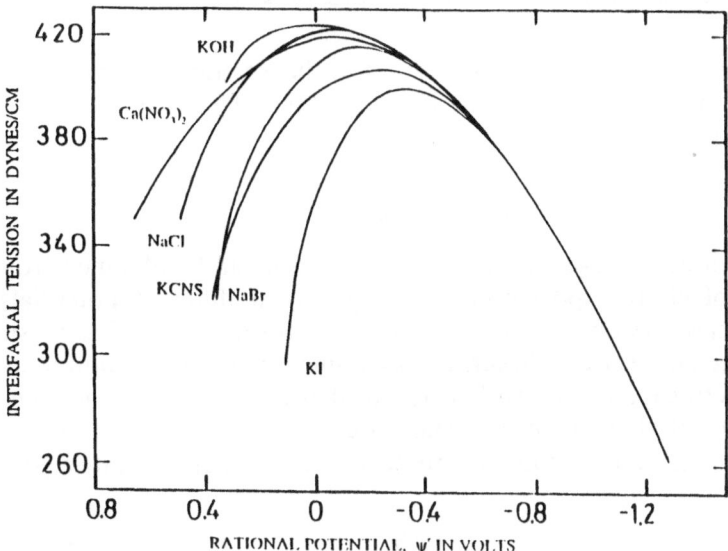

Fig. 5a. Electrocapillary curves [8], with permission. Interfacial tension of mercury in contact with aqueous solutions of the salts named. $T = 18°C$. Abscissas are measured relative to a "rational" scale in which the potential difference between the mercury and a capillary-inactive electrolyte is arbitrarily set equal to zero at the electrocapillary maximum.

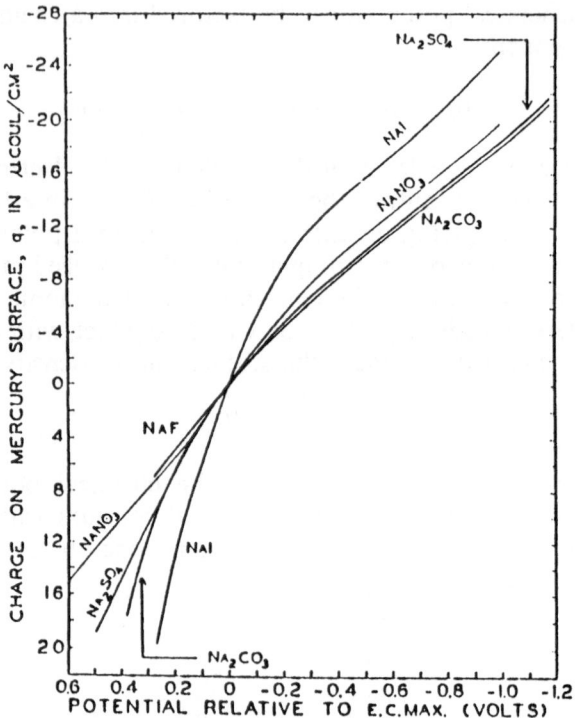

Fig. 5b. Surface change density curves [8], with permission. Electronic charge on mercury surface in contact with uninormal aqueous solutions of the salts named. T = 25°C. Curves would coincide at the right if "rational" potential scale had been chosen.

Differentiating Eqn. (22) with respect to the electrical potential, keeping the chemical potentials constant, one obtains the Lipman equation for surface charge density [8]

$$-\left(\frac{d\gamma}{\partial\psi}\right) = q \tag{27}$$

Thus, surface charge density curves (Fig. 5b) can be obtained from the slopes of electrocapillary curves (Fig. 5a) at different potentials [5]. Electrocapillary curves are obtained by measuring the surface tension of polarized liquid electrode surfaces as a function of polarization potential. The electrocapillary maximum is called the zero charge point because $(d\gamma/d\psi) = 0$ when the surface charge density q = 0. The second derivative of the interfacial tension with respect to the electric potential gives the (differential) capacitance.

$$\left(\frac{\partial^2\gamma}{\partial\psi^2}\right) = C \tag{28}$$

If the surface tension dependence on potential is measured, then the surface charge density and the capacitance can be obtained by differentiation; conversely, if the measured quantity is capacitance, then the surface charge density and interfacial tension differences can be obtained by integration, provided that the integration constant, the zero charge potential, can be determined independently. Only liquid metal electrode/water interfaces are amenable to surface tension as well as capacitance measurements. Capacitance values of solid metal electrodes are easily obtainable but to compute surface charge densities, the zero charge potential must be known. The latter can be estimated from the minimum in the diffuse double layer capacitance at low salt concentrations (see below).

The differential capacitance is not constant over the whole polarization potential region, as it depends on the structure of the electrical double layer, which in turn depends on the potential. The structure of the ionic layer on an electrode surface is complex. First there are the planes of closest approach for the different ions; for a simple salt solution containing one kind of anion and one kind of cation Grahame [8] defined the inner (anion) and the outer (cation) Helmholtz planes with their respective capacitances C_i and C_o. Beyond these two planes is the diffuse part of the double layer [Fig. 6, reconstructed from the calculated surface excesses (Eqn. 26) and the diffuse double layer components]. The overall capacitance of the electrical double layer is determined by these three double layer components in series,

$$1/C = 1/C_i + 1/C_o + 1/C_d \qquad (29)$$

where C_i, C_o and C_d are the capacitances of the inner Helmholz plane, the outer Helmholtz plane, and the diffuse double layer component. Except at very low salt concentrations and polarization potentials, $C_d \gg C_i > C_o$, and the contribution of the diffuse double layer to the total capacitance is small. However, at salt concentrations of around 1 mM and near the zero charge point, the contribution to the total capacitance of the diffuse double layer, which has a minimum at the zero charge point [4], becomes predominant (Fig. 7). Hence the zero charge potential can be identified with the potential of the capacitance minimum at low salt concentrations. This potential may vary with salt concentration, especially when one of the ionic components is surface-active.

The surface potential and the surface charge of colloidal AgI particles, which can be considered as reversible electrode surfaces, can be monitored by the concentration of Ag^+ or of I^- in the solution. The potential is obtained directly from the Nernst equation or the measured electrode potential, and the surface charge is inferred from the electrophoretic mobility and from Ag^+ or I^- adsorption [9].

In the case of a spread charged monolayer the potential can be related to the surface charge density through the Gouy-Chapman diffuse

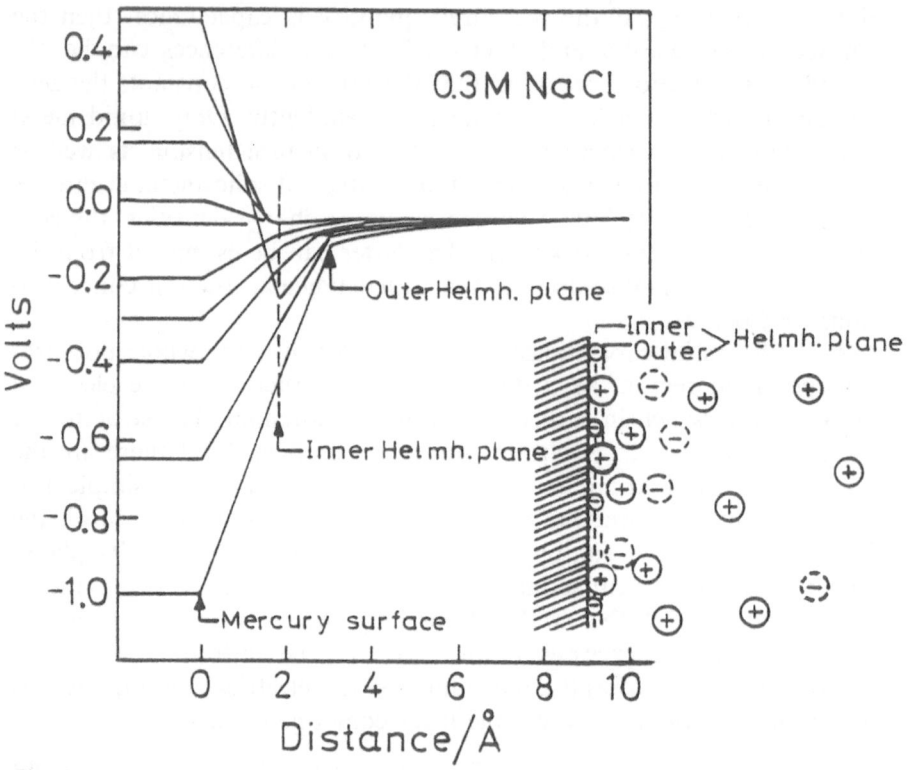

Fig. 6. Potential profiles across the mercury surface, the inner Helmholtz plane, the outer Helmholtz plane, and the diffuse double layer. As shown in the inset, the inner Helmholtz plane is at the distance of nearest access of anionic charge from the mercury surface, and the outer Helmholtz plane is at the distance of nearest access of cationic charges from the mercury surface [8], with permission.

double layer theory, based on the solution of the one dimensional Poisson-Boltzmann equation

$$\frac{d^2\psi}{dx^2} = -\frac{4\pi e}{D} \sum_i n_i Z_i \exp(-Z_i e\psi/kT) \qquad (30)$$

where e is the charge of an electron, D the dielectric constant, n_i the number of the ions per cm³, and Z_i their valence and sign (+ for cations, − for anions). For univalent ions the solution of Eqn. (30) yields [8]

$$q = 11.72\sqrt{c_i} \sinh 19.46\psi \qquad (31)$$

where c_i is the molar salt concentration, ψ is the potential in volts, and q, the surface concentration of the spread monolayer multiplied by its charge, is in μC/cm². As the measured quantity in this case is $d\gamma/dq$,

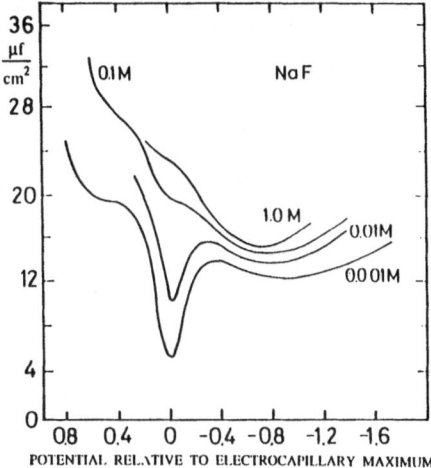

Fig. 7. Differential capacitance curves of the mercury electrode in the presence of different concentrations of NaF. At very low NaF concentrations and at low surface charge densities the capacitance values approach those of the diffuse double layer [8], with permission.

we can write

$$\frac{d\gamma}{d\psi} = \frac{\partial\gamma}{\partial q} \cdot \frac{\partial q}{\partial\psi} = \frac{\partial\gamma}{\partial q}(19.46)(11.72)\sqrt{c_i}\cosh(19.46\psi) = q$$

$$= \frac{\partial\gamma}{\partial q}(19.46)(11.72)\sqrt{c_i}\cosh\left[\sinh^{-1}\left(\frac{q}{11.72\sqrt{c_i}}\right)\right] \qquad (32)$$

from which

$$\frac{\partial\gamma}{\partial q} = \frac{q}{228\sqrt{c_i}\cosh\left[\sinh^{-1}\left(\frac{q}{\beta}\right)\right]} = \frac{q}{228\sqrt{c_i}\left(1 + \left(\frac{q}{\beta}\right)^2\right)^{1/2}}$$

$$(33)$$

where $\beta = 11.72\sqrt{c_i}$ (in $\mu C/cm^2$)

From the calculated values of $d\gamma/dq$ for different surface charge densities q, the dependence of γ and of ψ on q can be established for different ionic strengths (Fig. 8). The total change in surface tension up to very high surface charge densities, corresponding to a fully compressed charged monolayer, is very modest; this is due to the very high values of the diffuse double layer capacitances.

$$C_d = \left(\frac{\partial q}{\partial\psi}\right)_d = 228\sqrt{c_i}\cosh 19.46\psi(\mu f/cm^2) \qquad (34)$$

At a charge density of 30 $\mu C/cm^2$, $C = 398$ $\mu f/cm^2$ at a salt concentration of 10^{-3} M and $C = 568$ $\mu f/cm^2$ at a salt concentration of 0.1 M, which is over an order of magnitude higher than the double layer capacitance

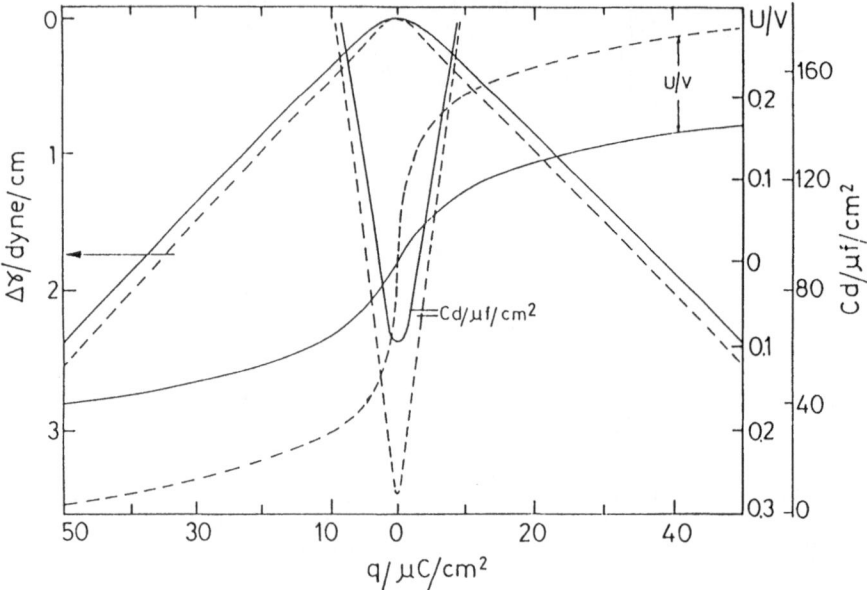

Fig. 8. Diffuse double layer potential and its contribution to the lowering of surface tension $\Delta\gamma$ and to the differential capacitance C_d as a function of the surface charge density. Concentration of uni-univalent salt: full lines, 10^{-1} M; dashed lines, 10^{-3} M.

on a mercury electrode at corresponding salt concentrations and potentials (123 mV and 26 mV, respectively). High capacitance means that large changes in surface charge density correspond to only very small changes in potential. Figure 8 shows that this is indeed the case, except for the narrow region of very low charge densities where the differential capacitance is very low, particularly at low ionic strengths. C is at its minimum at q = 0. At high surface charge densities, γ varies nearly linearly with q, reaching values of ca. 2 dyne/cm at 40 μC/cm², corresponding to ~ 42 Å² per charge.

2.2.3. Adsorption at the liquid/liquid or liquid/gas interfaces: The Gibbs adsorption isotherm (Eqn. 14) defines the surface excesses, whether they be large or small, positive or negative. Assuming incompressibility of the two phases across the interface, $\Sigma\, v_i \Gamma i = 0$, and the volume of adsorbed components equals the volume of desorbed ones. The components that tend to adsorb have a special affinity for the interface determined by the nature of the interface and of the adsorbed substance. In general, the electrochemical potentials of component i in the bulk and in the surface are equal. Replacing concentrations by molar fractions (Γ_i/Γ_{max}) and X_i, adsorption as a surface exchange reaction of an adsorbent a and a solvent molecule of equal volumes and areas can be expressed as

$$a^b + s^\sigma = a^\sigma + s^b$$

where superscripts b and σ indicate the bulk and the surface phase, respectively. The subscripts s and a in the standard chemical potential μ_0 are for the solvent and adsorbent, respectively.

$$\mu_{s,0}^b + RT \ln X_s + \mu_{s,0}^\sigma + RT \ln\left(\frac{\Gamma_a}{\Gamma_{max}}\right)$$

$$= \mu_{a,0}^b + F\Delta\psi + RT \ln X_a + \mu_{s,0}^\sigma + RT \ln \frac{\Gamma_s}{\Gamma_{max}} \tag{35}$$

Denoting Γ_a/Γ_{max} by θ; (Γ_s/Γ_{max}) by $1 - \theta$, and X_2 by $1 - X_a$

$$\frac{\theta}{1-\theta} = \frac{X_a}{1-X_a} \exp \frac{\Delta\mu_0 + F\Delta\psi}{RT} \tag{36}$$

where $\Delta\psi = \psi_b - \psi_\sigma$ is the potential difference between the adsorption layer and the bulk, $F\Delta\psi$ is positive when the sign of the charge of one Faraday (F) is the same as the sign of $\Delta\psi$, and $\mu_0 = \mu_{a,0}^b - \mu_{a,0}^\sigma + \mu_{s,0}^\sigma - \mu_{s,0}^b$

As the term in the exponent is a partial free energy, one often sees the term $a\gamma$ added to it, a being the molar area. This is wrong, because adsorption is accompanied by desorption from the same area and the consequent change in surface energy $a\Delta\gamma$ becomes an integral part of the standard free energy of adsorption $\Delta\mu_0$.

In most of the cases of interest $X_a \ll 1$, thus $(1 - X_a) \sim 1$ and Eqn. 36 assumes the form of the Langmuir adsorption isotherm, as follows:

$$\frac{\Gamma_i}{X_a} = \frac{\Gamma_{max} \exp \dfrac{\Delta E^\sigma}{RT}}{1 + X_a \exp \dfrac{\Delta E^\sigma}{RT}} \tag{37}$$

where $\Delta E_\sigma = \Delta\mu_0 + F\Delta\psi$.

The standard free energy of adsorption (ΔE^σ) is considered to be independent of surface concentration, and any lateral interaction between the adsorbed molecules is neglected. Frumkin (10) took into account the lateral interactions in the surface by adding a term $\varepsilon\Gamma$, proportional to the surface concentration, to the adsorption energy. If the surface potential is determined by the charge of the adsorbed molecules the electrostatic term of the standard free energy of adsorption also becomes a function of Γ_i. Adding the terms accounting for these effects to ΔE^σ in Eqn. (37),

$$\Delta E^\sigma = \Delta\mu_0 + \varepsilon\Gamma_i - \frac{F}{19.46} \sinh^{-1} \frac{\Gamma_i \cdot 10^{11}}{11.72\sqrt{c_i}} \tag{38}$$

The last term in Eqn. (38) was taken from Eqn. (31) after multiplication of Γ_i (moles/cm^2) by the charge of a Faraday which is 10^{11} μC/cm^2. After insertion of Eqn. (38), Eqn. (37) can be solved numerically. It is

evident from eqn (38) that the interaction energy ε is attractive and
increases the standard free energy of adsorption, while the electrostatic
term is repulsive and decreases the tendency of a charged substance to
adsorb into a monolayer of those same charge.

The main driving force for adsorption is the difference in standard
chemical potential between the adsorbing molecule in the bulk and in
the surface, $\Delta\mu_0$. The main contribution to $\Delta\mu_0$ in aqueous solution is
the hydrophobic force. Amphipathic molecules contain a polar segment
which tends to keep them in a aqueous solution and a non-polar
segment which tends to be adsorbed to hydrophobic surfaces. The
hydrophobic energy can be identified with the surface energy between
the non-polar part of the molecule and water.

Let us consider the two extreme cases depicted in Fig. 9b for the
water oil interface when the isolated adsorbed molecule lies either flat in
the surface or is oriented normally to the surface when its hydrocarbon
chain is fully immersed in oil. For simplification let us assume that the
oil/water, the oil/polar group, and the hydrocarbon chain/water interfacial energies per unit area are the same and polar group/water interfacial energy is negligible. Let us also assume that the surface domain
where the molecule is adsorbed retains its planarity. The adsorption
energy of the molecule lying in the surface is then $A_h\gamma_{\omega/\eta}$ where A_h is the
area of the hydrocarbon chain, as only the hydrocarbon chain/water
interface was annihilated in this process and the oil/water interface did
not change. When the adsorbed molecule is immersed perpendicularly
into the oil phase the hydrocarbon/water interface decreases by

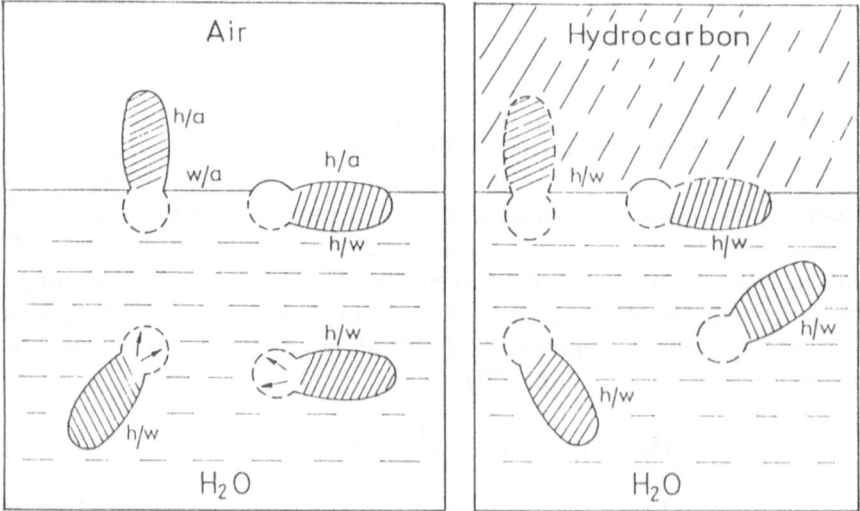

Fig. 9. Schematic illustration of adsorption of amphipathic molecules having a polar head
group and a non-polar tail (shaded) at the air/water (a) and oil/water (b) interfaces.

$(A_h + A_c)$ and the adsorption energy becomes $(A_c + A_h)\gamma_{\omega/\eta}$ where A_c is the cross-sectional area of the adsorbed molecule. Considering the two extreme cases for adsorption at the air/water interface, the hydrophobic adsorption energy for the hydrocarbon chain sticking out from the surface will be $A_h (\gamma_{\eta/\omega} - \gamma_{h/a}) + A_c\gamma_{\alpha/\omega}$ and $1/2 A_h (g_{h/w} - \gamma_{h/a} + \gamma_{a/w})$ for the molecule lying flat in the surface. How does this relate to Traube's rule that the increase in adsorption energy per $- CH_2 -$ in short chain alcohols or acids (up to $8 - CH_2 -$ units) is about 1.1 kT? The increase in the molecular hydrophobic interfacial area per $- CH_2 -$ is about $10 \, \text{Å}^2 = 10^{-15} \, \text{cm}^2$. For adsorption at the oil/water interface the increase in the hydrophobic adsorption energy per CH_2 residue is $\sim 50 \times 10^{-15}$ ergs or ~ 1.25 kT; for adsorption at the air interface the hydrophobic adsorption energy per CH_2 residue is only slightly more than half this value. However, Traube's measurements were done under conditions of relatively high surface concentrations, when a large fraction of the hydrocarbon chains are in contact and the conditions of an oil/water interface are approached.

3. Adsorption of polymers

The adsorption of polymeric molecules is a function of the standard free energy of adsorption of their monomeric units as well as of their configuration. Hydrophobic residues which may have a very high standard free energy of adsorption, will not contribute to adsorption if they are in the middle of a spherical macromolecule with no possibility of reaching the interfacial plane. From a configurational perspective polymers or macromolecules can be subdivided into three groups: Random coils, rodlike molecules, and globular molecules with some fixed configuration (e.g. globular proteins).

The entropy contribution to the increase in free energy associated with the transfer of a macromolecule from a dilute solution is, for all practical purposes, independent of the macromolecule's size or molecular weight. Every fraction of the molecular surface that reaches the interface and adheres to it contributes to the free energy of adsorption. The adsorbed molecular segments bring about the desorption of solvent molecules, the combined volume of which is equal to that of the adsorbed polymer segments, and a proportional decrease in free energy. Macromolecules are thus in a very favorable position in the adsorbed state; they are extremely surface active, and adsorb very strongly even if the individual adsorbed groups or segments each have only a very small standard free energy of adsorption. It is the *large number* of these individual small energy contributions that combine to favor adherence of the macromolecule to the interface. As a result, the surface can become saturated even at extremely low bulk concentrations, irrespective of the conformation of the macromolecular particle. Although only a fraction of the adsorbed molecule adheres to the surface and contributes to the standard free energy of adsorption, the whole molecule is considered to be adsorbed. Hence the surface phase or region is generally thicker than the extent of the surface force field and may reach macromolecular dimensions. In every mathematical treatment the thickness of the interfacial region and the segment density profile have to be considered together with the surface concentration Γ of the macromolecule.

Macromolecules may be either homopolymers composed of the same repeating monomeric unit, or heteropolymers containing different monomeric units. Biopolymers such as proteins, polypeptides, and nucleic acids are heteropolymers, while many types of polysacharides (for example, starch and cellulose) are homopolymers. Many biological macromolecules, like native DNA, collagen, and cellulose, are rodlike. Native DNA, though a heteropolymer has a homopolymeric hydrophilic surface because the various purine and pyrimidine bases are directed towards the interior of the DNA cylinder.

3.1. Adsorption of cylindrical rodlike macromolecules

A cylindrical rodlike molecule will adsorb parallel to the interface, except when it has an extremely surface-active residue at its end. In the latter case adsorption does not differ fundamentally from the adsorption of an ordinary amphipatic molecule.

Let us first consider a stiff extended chain, of a thickness comparable to the diameter d of a solvent molecule and of length Pxd which adsorbs parallel to an interfacial surface. The adsorption process can be represented by an exchange reaction where one polymeric chain displaces P solvent molecules from the surface [11, 12].

$$p^b + Ps^\sigma \Leftrightarrow p^\sigma + Ps^b \tag{39}$$

At equilibrium, the relation between the chemical potential of the solvent s and the chemical potential of the polymer p (composed of P monomeric units equal to size to the solvent molecule) in the bulk (b) and surface (σ) phases is:

$$P\mu_s^\sigma - \mu_p^\sigma = \mu_p^b - P\mu_s^b \tag{40}$$

The chemical potentials of the polymer and of the solvent in the bulk and in the surface can be written as follows using the Flory approximation:

$$\mu_s^b = \mu_s^{0,b} + RT \ln \phi_s^b + RT\left(1 - \frac{1}{P}\right)\phi_p^b + \alpha(\phi_p^b)^2 \tag{41}$$

$$\mu_p^b = P\mu_p^{0,b} + RT \ln \phi_p^b + RT(P - 1)\phi_s^b + P\alpha(\phi_s^b)^2 \tag{42}$$

$$\mu_s^\sigma = \mu_s^{0,\sigma} + RT \ln \phi_s^\sigma + RT\left(1 - \frac{1}{P}\right)\phi_p^\sigma + \alpha e(\phi_p^\sigma)^2 + \alpha m(\phi_p^b)^2 \qquad (43)$$

$$\mu_p^\sigma = P\mu_p^{0,\sigma} + RT \ln \phi_p^\sigma + RT(P-1)\phi_s^\sigma + P\alpha e(\phi_s^\sigma)^2 + P\alpha m(\phi_s^b)^2 + P\gamma a \qquad (44)$$

where $\mu_s^{0,b}$ and $\mu_p^{0,b}$ are the standard chemical potentials of the solvent and polymer in the bulk phase; $\mu_s^{0,\sigma}$ and $\mu_p^{0,\sigma}$ are the standard chemical potentials of the solvent and polymer in the surface phase; and ϕ_s^b, ϕ_p^b, ϕ_s^σ, and ϕ_p^σ are the volume fractions of the solvent and polymer in the bulk and surface phases, respectively. α is the configurational solvent/solute interaction energy

$$\alpha = NZ[\varepsilon_{s,p} - \tfrac{1}{2}(\varepsilon_{s,s} + \varepsilon_{p,p})] \qquad (45)$$

where N is Avogadro's number and $\varepsilon_{i,j}$ is the interaction energy between pairs of lattice sites, for which i and j may be either a solvent molecule or a polymer segment p. Z is the total number of nearest neighbors on the lattice. e is the fraction of neighbor sites in the plane parallel to the surface, and m is the fraction of neighbor sites in one of the adjacent planes. It follows, then, that

$$e + 2m = 1 \qquad (46)$$

The surface adsorption equation (40) can now be rewritten as

$$\frac{\left(\dfrac{\mu_p^{0,\sigma} - \mu_p^{0,b}}{p}\right) - \mu_s^{0,\sigma} + \mu_s^{0,b}}{RT} = \frac{\Delta\mu_0}{RT} = \ln K = \ln\left(\frac{\phi_p^b}{\phi_p^\sigma}\right)^{1/P} + \ln\left(\frac{\phi_s^\sigma}{\phi_s^b}\right)$$

$$+ \frac{\alpha}{RT}\{e[(\phi_p^\sigma)^2 - (\phi_s^\sigma)^2]$$

$$- (e+m)[(\phi_p^b)^2 - (\phi_s^b)^2]\} \qquad (47)$$

The last term in Eqn. (47) is the difference in interaction energy with neighbor molecules upon adsorption. In the case of adsorption of an organic molecule or polymeric segment from an aqueous solution, this interaction energy is essentially a hydrophobic force considered to be part of the standard free energy of adsorption, and is a function of concentration. Since $\phi_p + \phi_s = 1$ and $\phi_p^b \ll 1$, the change in interaction energy between solutes and solvents upon adsorption can be reduced to

$$\alpha\{1[(\phi_p^\sigma)^2 - (\phi_s^\sigma)^2] - (1+m)[(\phi_p^b) - (\phi_s^b)^2]\} \approx \alpha(m + 2e\phi_p^\sigma) \qquad (48)$$

and Eqn. (47) can be written as:

$$\left(\frac{\phi_s^\sigma}{\phi_s^b}\right)\left(\frac{\phi_p^b}{\phi_p^a}\right)^{1/p} = K \exp - \frac{\alpha}{RT}(m + 2e\phi_p^\sigma) = \exp\frac{\Delta\mu_0 - \alpha(m + 2e\phi_p^\sigma)}{RT}$$

Because $\phi_s^b \approx 1$,

$$K' = K^P \exp \frac{-\alpha(m + 2e\theta)P}{RT} = \phi_p^b \frac{(1 - \theta)^P}{\theta} \qquad (49)$$

where $\theta = \phi_p^a$

Equation (49) is a Frumkin-type adsorption isotherm for a thin rodlike polymer molecule when the degree of surface coverage θ corresponds to the volume fraction in the surface. For the adsorption of a hydrophobic molecule from an aqueous solution, there is a hydrophobic contribution to both the standard free energy of adsorption $\Delta\mu_0$ and the proportionality factor α of the energy term which is linearly dependent on surface concentration.

3.2. Adsorption of rigid bulky molecules

Rigid bulky molecules include any macromolecules with a fixed shape such that only a fraction of the monomeric segments can adhere to the interfacial surface. Examples of rigid bulky molecules from among biological macromolecules are globular proteins. As only a fraction of the adsorbed molecule is in the interface proper, most of the molecular volume of the adsorbed molecule is accomodated in the relatively thick layer adjacent to the surface (Fig. 10). We shall call this thick surface layer the *surface phase*, Σ. Only those segments of the molecule accommodated in the interfacial layer and in direct contact with the adsorbing phase contribute to the standard free energy of adsorption. The contribution of the surface phase is entropic and through solvent/polymer interactions at the high concentrations prevailing in this domain. If the number of polymer segments per molecule in the interfacial layer is S and $(P - S)$ segments are in the surface phase, then the chemical

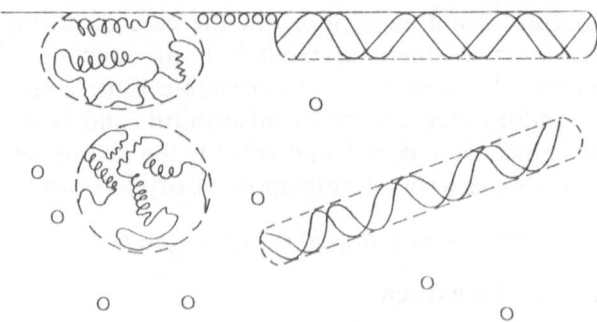

Fig. 10. Relatively rigid rodlike and globular polymer molecules in solution and at a surface. The relative size of solvent molecules is represented by small circles.

potential of polymer molecules accommodated in the surface is

$$\mu_p^{\sigma,\Sigma} = \sum_s \mu_p^{0,\sigma} + \sum_{p-s} \mu_p^{0,b} + RTn\phi_p^\sigma\phi_p^\Sigma + RT[(S-1)(1-\phi_p^\sigma)$$
$$+ (P-S-1)(1-\phi_p^\Sigma)]$$
$$+ Se\alpha(\phi_s^\sigma)^2 + Sma(\phi_s^\Sigma)^2 + (P-S)\alpha(\phi_s^\Sigma)^2 \qquad (50)$$

$$\mu_s^\sigma = \mu_s^{0,\sigma} + RT\ln(1-\phi_p^\sigma) + RT\left(1-\frac{1}{S}\right)\phi_p^\sigma + \alpha e(\phi_p^\sigma)^2 + \alpha m(\phi_p^\Sigma)^2$$
$$\qquad (51)$$

$$\mu_s^\Sigma = \mu_s^{0,b} + RT\ln(1-\phi_p^\Sigma) + RT\left(1-\frac{1}{P-S}\right)\phi_p^\Sigma + \alpha(\phi_p^\Sigma)^2 \quad (52)$$

The chemical potentials in the bulk phase are given by Eqns. (41) and (42). Applying these together with Eqns. (52), (50), and (51) to the exchange reaction

$$\mu_p^b + S\mu_s^\sigma + (P-S)\mu_s^\Sigma = \mu_p^{\sigma,\Sigma} + P\mu_s^b \qquad (53)$$

and assuming $\phi_p^b \ll 1$, we obtain the adsorption isotherm

$$K = \frac{(1-\theta)^s(1-\phi_p^\Sigma)^{(p-s)}}{\theta \cdot \phi_p^\Sigma} \phi_p^b$$
$$= \exp\frac{S(\Delta\mu_0 + \alpha m) - 2\alpha(Se\phi_\rho^\sigma + (P-S)\phi_\rho^\Sigma)}{RT} \qquad (54)$$

where $\theta = \phi_p^a$. In this case, the Frumkin surface interaction term also includes interactions in the concentrated surface phase. The fixed shape of the adsorbed polymer may be not completely rigid, and it may be distorted to expose a larger number of polymer segments in the surface, thereby increasing the free energy of adsorption. This introduces an elastic strain on the molecule which increases with S. For instance, small deformations in a globular protein resulting in a moderate increase in S may be easily produced without energetically costly deformations in the tertiary structure. On the other hand, a complete break down of the tertiary structure may involve the rupture of relatively high-energy disulphide bonds. Moreover, the energy gain in transferring some hydrophobic residues to the surface would not always suffice to compensate for the energy required to extricate them from the shielded hydrophobic interior of the molecule. The latter energy would have to be invested in breaking hydrophobic and hydrogen bonds which maintain the protein's secondary structure. Thus, no general theoretical treatment of adsorption of biological heteropolymers is possible. Each biological heteropolymer shows a large degree of specificity when adsorbing on surfaces in general and when binding to active sites in particular. In the latter case binding requires complete conformational compatibility between the binding sites of the receptor or ligator and of the ligand.

3.3. Conformation of randomly coiled macromolecules in the surface

Most synthetic polymers and certain biopolymers such as RNA and some natural polysaccharides and polypeptides, are random coils in solution. Even rodlike molecules like double-stranded DNA have some bends, at least some fraction of the secondary structure of most globular proteins is random coil. The random coil structure becomes more abundant in denatured bipolymers or in some transition states during biological synthesis and translation across membranes before completion of final folding. Randomly coiled macromolecules are pliable and can accommodate different spatial restrictions; this may be why the random coil model at interfaces attracted considerable attention, and a great deal of effort has been devoted to its theoretical treatment.

Four parameters are of interest when studying adsorption of coiled macromolecular chains: (1) the total number of molecular segments adsorbed per unit area; (2) the fraction of these segments (p) in direct contact with the surface and their degree of coverage θ of the surface; (3) the thickness of the surface region produced by out loops and loose tails attached to the adsorbed segments; (4) the concentration profile in this surface phase (Fig. 11). Most theoretical and experimental studies of adsorption of flexible macromolecules are concerned with determining these parameters.

The shape of a randomly-coiled macromolecule changes quite drastically upon adsorption. The center of gravity which in solution is in the middle of a random coil, shifts towards the surface when the molecule flattens along the interface as it attaches to it. Even at a very moderate

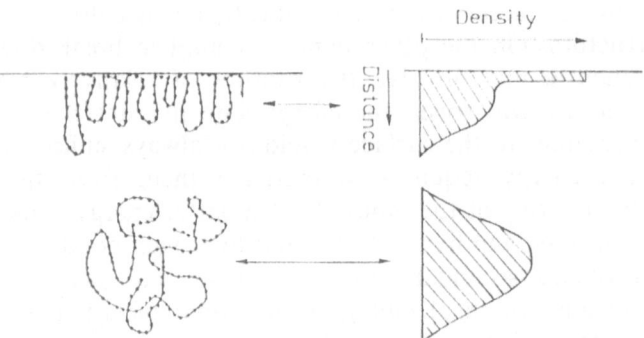

Fig. 11. A randomly-coiled polymer molecule in solution and adsorbed onto a surface. The right side of the figure shows the segment density distribution of these molecules.

standard free energy per segment, the interfacial area becomes saturated at very low bulk concentrations of macromolecule, resulting in surface concentrations of $0.1-0.2\ \mu g/cm^2$. Under these conditions the coverage of a smooth liquid/liquid interface by polymers adsorbed from aqueous solutions may reach the values of $0.85-0.9$, with the fraction of polymer *segments* actually adhering to the surface (p) varying between 0.1 and 0.4 [13–16]. Values of p up to 0.9 have been reported for different solid surfaces [17, 18].

Theoretical treatments of polymer adsorption use different approaches and sometimes address different aspects of the phenomenon. In general all possible configurations of the flexible chain near the surface are counted, registering all segments which are adsorbed onto the surface and which contribute to the adsorption energy. It is possible to assign a weighted probability to each configuration and to calculate the average properties of the adsorbed polymer chains as well as the free energy of the system. Statistical theories based on these considerations were developed using either analytical methods or Monte Carlo calculations [19].

The first analytical model discussed by Frish, Simha, and Eirich [20] was that of a reflecting wall with a polymer attached to it by at least one segment. According to this model the fraction attached to the surface, p, is very small:

$$p = \frac{1}{\sqrt{P}} \qquad (55)$$

$$S = a \cdot \sqrt{P} \qquad (56)$$

where P is the number of units per molecular chain, and S is the extent of protrusion of the adsorbed molecule when a is the length of the unit. This model was subsequently improved by allowing the reflecting wall to exert forces [21], and by accounting for the energy-rich configurations of the adsorbed sequences [22, 23]. A series of papers was published, treating the conformation of a macromolecule near a fully- or partially-adsorbing wall. Possible configurations on a lattice were counted, assuming alterations of adsorbed trains and protruding loops of segments, and deriving the most probable configuration as a function of the adsorption energy per segment. Roe [24] employed the "divergent generalized partition function" method to derive the average conformation of an isolated polymer molecule on the surface. The generating function $\Gamma(x)$ for the partition function Z_n is represented by the sequence-generating functions of adsorbed segments $V(x)$ and loops $U(x)$

$$\Gamma(x) = \sum_{n=1}^{\infty} Z_n X^{-n} \qquad (57)$$

The sequence generating function was then constructed from the respective functions $v(a)$ and $\mu(d)$:

$$U(x) = \sum_{d=1}^{\infty} \mu(d)x^{-d} \tag{58}$$

$$V(x) = \sum_{a=1}^{\infty} v(a)x^{-a} \tag{59}$$

where a and d are the numbers of segments in the adsorbed trains and in the loops protruding into the solution, respectively. The dependence of the configuration of the adsorbed polymer on the adsorption energy, as derived by the different methods, is very similar.

The detailed configuration of an isolated polymer molecule on a surface, taking into account the attraction to the surface, the excluded volume, and transition probabilities from loops to surface trains and vice versa, was treated by Silberberg [25, 26]. According to this treatment, a macromolecule composed of P identical segments is assumed to divide upon adsorption into m adsorbed trains and in loops (considering one tail sequence to be a train and the other a loop):

$$\sum m_{\sigma,i} = \sum m_{Bj} \tag{60}$$

Thus,

$$\sum_i im_{\sigma i} + \sum_j jm_{Bj} = P = P_\sigma + P_B \tag{61}$$

where P_σ is the average length of the adsorbed trains and P_B is the average length of the loops.

A train of i segments in the surface can be arranged in $\omega_\sigma(i)$ configurations and a loop of j segments in $\omega_B(j)$ configurations. The partition function of the isolated macromolecule on the surface ξ_σ is then:

$$\xi_s = \xi^* \left(\frac{1}{\Omega_p^*}\right) \sum \Omega_p^\sigma(m_{\sigma,i} m_{Bj}) \exp\left(\sum_i im_{\sigma i} \chi_\sigma\right) \tag{62}$$

where ξ^* is the partition function of the isolated macromolecule in the bulk solvent, χ_σ is the energy gain (in units of kT) when a polymer segment replaces a solvent molecule in the surface

$$\chi_\sigma = -\left(\frac{\Delta F_s}{kT}\right) + \ln \frac{U_B}{U_\sigma} \tag{63}$$

where Ω_p^σ and Ω_p^* are the number of arrangement possibilities in the surface and in the bulk, respectively, and U_B and U_σ are the effective numbers of ways of adding a segment to a chain in the bulk and in the surface. It follows, then, that

$$\Omega_p^* = \zeta P^{-a,B}(U_B)^P \tag{64}$$

$$\Omega_p^\sigma(m_{\sigma i}, m_{Bj}) = \left[\frac{\Pi_i(\omega_{\sigma i})^{m_{\sigma i}}}{m_{\sigma i}!}\right] \cdot \left[\frac{\Pi_j(\omega_{Bj})^{m_{Bj}}}{m_{Bj}!}\right]$$

$$\cdot \left[\left(\sum_i m_{\sigma i}\right) \cdot \left[\left(\sum_j m_{Bj}\right)!\right]\right]$$

$$\omega_{\sigma i} = \zeta_s i^{-a_s}(U_\sigma)^i$$

$$\omega_{Bj} = \zeta_B J^{-a_B}(U_B)^j \tag{65}$$

The parameters a_B and a_σ allow for excluded volumes (precluding adsorption to an already-occupied site); they can be compensated or even overcompensated for by intermolecular attraction, and they are very different in the surface and in the bulk. ζ makes allowance for end effects in the bulk and ζ_B and ζ_s consider the transition from adsorbed sequences to loops and vice versa. From the maximal term of the partition function and using Langrangian multipliers, the values of P_B and P_σ are obtained for different values of ζ_B, ζ_σ, a_s and a_B. In Fig. (12) the quantity $p = P_\sigma/(P_\sigma + P_B)$ is plotted against $-\Delta F_s/kT$ for different values of $\zeta_B \cdot \zeta_\sigma$, where $\zeta_B \cdot \zeta_\sigma$ is the probability of transition between the surface and the bulk state. For low probabilities of transition, the transition from full adsorption to full desorption is abrupt at $\Delta F_s/kT = 0$. If the transition probability is equal to the probability of staying in the respective state, $\zeta_B \cdot \zeta_\sigma = 1$, the adsorbed fraction p varies gradually with $(\Delta F_s/kT)$. It is not only p that varies with the transition probability, but also P_σ and P_B. As intuitively expected, both P_σ and P_B increase as the transfer probability $\zeta_B \cdot \zeta_\sigma$ decreases.

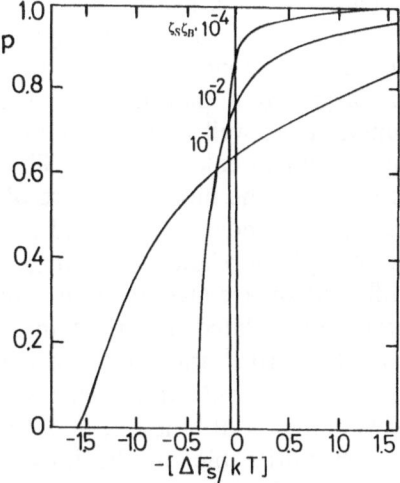

Fig. 12. Fraction of number of segments p for the isolated polymer ($P = \infty$) in contact with surface as a function of (ΔFs/kT); $(a_B, \bar{a}_s) = (1/3, -1/3)$ [26], with permission.

3.4. Adsorption from solutions of macromolecules

How does the segment concentration profile near the surface change when a saturated monolayer composed of isolated polymer molecules is in contact with a dilute polymer solution? What is the driving force for further adsorption? The segment concentration profile of an isolated polymer molecule is influenced by the lack of competition for surface sites from other adsorbed molecules. If the whole surface is covered by such "isolated" molecules, up to above 80% of the surface sites are occupied. For further adsorption either the subsequent adsorbing molecule will stick to some of the free sites on the surface by a few "tentacles" protruding from the second layer, or else there must be a total rearrangement of the molecules in the surface layer to make place for the additional adsorbing molecule. Equilibration in the surface layer should favor the second alternative, which would change the profile of the surface layer. What is the energy balance favoring further adsorption with consecutive increase of the surface layer thickness?

Transfer of a molecular particle from a dilute solution into a condensed surface layer requires entropic energy. In poor solvents this may be counteracted by the energy gain from the polymer segment contacts. As will be shown in paragraph 4 there is only a small decrease of segments in the adsorbed layer contributing to the adsorption energy (θ does not increase above 0.9), and this source adds only little to the increase in surface layer thickness. The main driving force for further adsorption and for the increase in monolayer thickness is derived from the entropy increase when the flattened adsorbed molecules expand in the direction normal to the surface. As is evident from Fig. 13, the fraction "p" of the segments adhering to the surface decreases with concentration, for any value of adsorption energy χ_s, corresponding to an increase in the thickness and concentration of the surface phase. The effect is as expected: larger in a bad than in a good solvent. The complete partition function, as well as the equations representing the detailed results of the analytical calculation of Silberberg [27] are too involved to be discussed here. The most salient results of these calculations are presented in Figs. 13 and 14. In Fig. 13 p is given as function of the adsorption energy (χ_s) for a good solvent ($\chi = 0$) and for a bad solvent ($\chi = 0.5$) at different bulk concentrations. Note the large difference between the curves for isolated adsorbed molecules and for any very low concentrations ($\phi = 10^{-3}$ and 10^{-4}), in which region there is hardly any dependence on concentration. This shows that a nearly saturated surface layer is already attained at $\phi \ll 10^{-4}$. The concentration dependence becomes noticeable on the linear scale at $\phi > 10^{-2}$ when the bulk concentration becomes comparable to that in the loop region. In every case p reaches an asymptotic value at high values of χ_s. In Fig. 14 the dependence of the loop length (which is proportional to

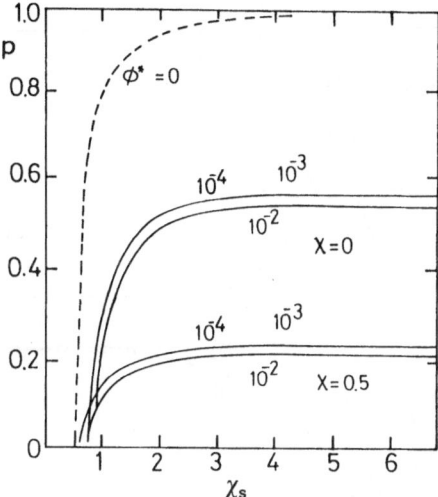

Fig. 13. Fraction of the number of segments of adsorbed macromolecules p in contact with surface as a function of χ_s in the case $P = 10^6$, ζ_B; $\zeta_s = 0.1$ for $\chi = 0.5$ and 0, and $\phi^* = 10^{-2}$, 10^{-3}, 10^{-4}, and 0. The adsorption energy χ_s per segment is in kT units [27], with permission.

the thickness of the surface layer) on the degree of polymerization P is given for different values of $\zeta_B \cdot \zeta_\sigma$ and for different bulk concentrations. The thickness of the surface layer depends strongly on $\zeta_B \cdot \zeta_\sigma$ and reaches an asymptotic value at high molecular weights.

Similar results were obtained by de Gennes, who estimated the segment concentration profile of a polymer layer adsorbed from a solution using a scaling argument based on local conformational freedom [28, 29]. The interfacial tension γ is split into two parts:

$$\gamma = \gamma_d(\theta_\sigma) + I(\phi_0, \phi^*) \tag{66}$$

where θ_σ is the area fraction in immediate contact with the wall that is occupied by polymer segments. The surface occupancy θ_σ in the adsorption layer is

$$\theta_\sigma = \frac{\bar{m}_\sigma}{2} \phi_0 \tag{67}$$

and the segment concentration per unit area is $(\bar{m}_\sigma/2)\phi_0 a^{-2} \times \bar{m}_\sigma$ is the average length of the adsorbed segment trains according to Silberberg (18), ϕ_0 is the extrapolated value of ϕ_z at $z = 0$, and γ_d describes interactions in the surface.

$$\delta_d = \gamma_0 + (\gamma_p - \gamma_0)\left(\frac{\bar{m}_\sigma}{2}\phi_o\right) = \gamma_0 + \Delta\gamma\left(\frac{\bar{m}_\sigma}{2}\right)\phi_o \tag{68}$$

where γ_0 is the interfacial energy of the solvent and γ_p the interfacial

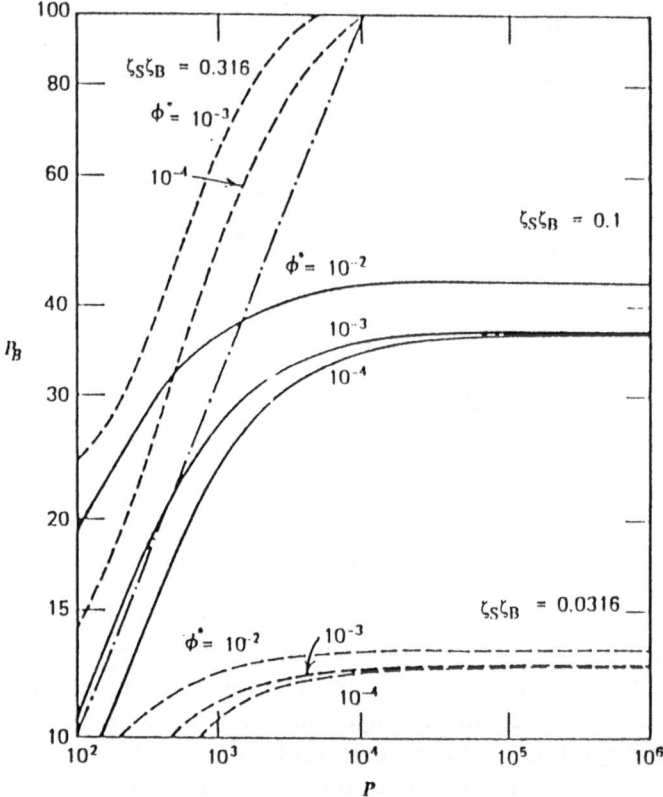

Fig. 14. Dependence of average loop length P_B (proportional to film thickness) on the degree of polymerization P in the case $c = 0.5$, $\chi_s = \infty$ ($\theta = 1$) for $\xi_s^{\cdot}\xi_B = 0.0316$, 0.1, 0.316, and $\phi^* = 10^{-4}$, 10^{-3}, and 10^{-2}. The straight line ($-\cdot-\cdot-$) corresponds to the relation $P_B \sim p^{1/2}$ [27], with permission.

energy of the adsorbed polymer (adsorption occurs if $\Delta\gamma$ is negative). The term $I(\phi_o, \phi^*)$ describes the energy contribution of the concentration profile $\phi(z)$ at larger distances, particularly the contribution of its deviation from the volume fraction ϕ^* in the bulk. An adsorbed chain of P segments has a finite fraction p of its monomeric segments in direct contact with the surface. The reversible adsorption energy per polymer chain is $pP\Delta\gamma a^2$, where a is the lattice unit and a^2 is the area occupied by a monomeric segment or a solvent molecule. The partition coefficient f of the polymer chains between the bulk and the surface phase is proportional to $\exp(-k_0 pP(\Delta\gamma a^2/kT)$, where k_0 is a numerical constant accounting for adsorption entropy. Even when $\Delta\gamma a^2/RT$ is very small, large values of P assure huge values of f and consequently complete adsorption.

It is assumed that the adsorbed polymeric chain has loops extending to the thickness δ from the adsorbing plane. In quantitative terms it means that the concentration profile has the scaling structure

$$\phi(z) = \phi_o g(z/\delta) \quad (z > a) \tag{69}$$

where $g(z/\delta) \to 0$ when $z/\delta \gg 1$. There are many functions that fulfill this requirement; $g(z/\delta)$ was chosen to correspond with the local correlation length associated with the local concentration $\phi(z)$.

The thickness δ is defined by the boundary condition

$$\left| \frac{1}{\phi} \frac{d\phi}{dz} \right|_{z \to 0} = \frac{1}{\delta} \tag{70}$$

With this boundary condition, Eqn. (69) becomes

$$\phi(z) = [a/(z + (4/3)\delta)]^{4/3}$$

and

$$\phi_o = \phi_{(z=0)} = \left(\frac{3}{4} \frac{a}{\delta} \right)^{4/3} \tag{71}$$

The total number of segments per unit area Γ is then

$$\Gamma = \frac{\bar{m}_\sigma}{2} \left(\frac{3}{4} \frac{a}{\delta} \right)^{4/3} a^{-2} \int_0^\infty \phi(z) a^{-3} \, dz = \phi_o a^{-2} \left(\frac{\bar{m}_\sigma}{2} + 4 \frac{a}{\delta} \right) \tag{72}$$

and

$$p = \frac{\dfrac{\bar{m}_\sigma}{2} \dfrac{a}{\delta}}{\left(\dfrac{\bar{m}_\sigma}{2} \dfrac{a}{\delta} + 4 \right)} \tag{73}$$

The free energy F_c per polymer chain can now be written as

$$\frac{F_c}{kT} = \left(\frac{R_F}{\delta} \right)^{5/3} - P \frac{\Delta \gamma a^2}{RT} \frac{\bar{m}_\sigma}{2} \bigg/ \left(4 \frac{a}{\delta} + \frac{\bar{m}_\sigma}{2} \right) \tag{74}$$

The first term is the confinement energy, where R_F ($= P^{3/5}a$) is the Flory radius of the chain in the bulk solvent. The second term is the free energy associated with the surface contact points.

The fraction of the surface $\theta = \bar{m}_\sigma/2\phi_o$ covered by polymer segments is given by

$$\frac{\theta}{1-\theta} = \phi_o \exp - \frac{\Delta \gamma a^2}{kT} \tag{75}$$

from which it follows that

$$\theta = \frac{\bar{m}_\sigma}{2} \phi_o = \frac{\phi_o \exp - \dfrac{\Delta \gamma a^2}{kT}}{1 + \phi_o \exp - \left(\dfrac{\Delta \gamma a^2}{kT} \right)} \tag{76}$$

and

$$\frac{2}{\bar{m}_\sigma} = \phi_o + \exp + \frac{\Delta\gamma a^2}{kT} \tag{77}$$

For large adsorption energies $-\Delta\gamma a^2 \gg kT$, in which case $\bar{m}_\sigma = 2/\phi_o$ and $\theta \to 1$.

Experimental values of θ never exceed 0.9, indicating that steric hindrances in the polymeric chain do not allow complete filling of the surface.

The osmotic energy term $I(\phi_o, \phi^*)$ in Eqn. (66) for low bulk concentrations of polymer (ϕ^*) depends only on $\phi(z)$ near the surface. The local osmotic energy density is kT/ξ^3, where ξ is the correlation length $\xi(\phi) = a\phi^{3/4}$. Integrating over a thickness δ we obtain Eqn. (70) and

$$I(\phi_o) = \frac{kT}{a^2} \phi_o^{3/2} = \left(\frac{3}{4}\right)^2 \frac{kT}{\delta^2} \tag{78}$$

Inserting Eqns. (71) and (72), the osmotic energy per polymer segment $I(\phi_o)/\Gamma$ becomes

$$\left(\frac{3}{4}\right)^2 \frac{kT}{\delta^2} \bigg/ \phi_o a^{-2}\left(\frac{\bar{m}_\sigma}{2} + 4\frac{\delta}{a}\right) = \left(\frac{3}{4}\frac{a}{\delta}\right)^{2/3} kT \bigg/ \left(\frac{\bar{m}_\sigma}{2} + 4\frac{\delta}{a}\right) \tag{79}$$

Thus, the adsorption equilibrium in a dilute solution is given by combining Eqn. (79) with Eqn. (74) to obtain the energy of the polymer in the surface. Assuming the free energy of the polymer in the solution to be $RT \ln \phi^*$ and assuming equilibrium with the polymer in the surface layer, we obtain:

$$RT \ln \phi^* = 1(\phi_o) + F_c$$

and

$$\ln \phi^* = P\left[\left(\frac{a}{\delta}\right)^{5/3} + \left\{\left(\frac{3}{4}\frac{a}{\delta}\right)^{2/3} - \frac{\Delta\gamma a^2 m_\sigma}{2RT}\right\} \bigg/ \left\{\frac{4\delta}{a} + \frac{\bar{m}_\sigma}{2}\right\}\right] \tag{80}$$

3.5. Adsorption of polyelectrolytes

The condition of electroneutrality results in the movement of a cloud of counterions together with a charged molecule. If such a charged macromolecule adsorbs at an uncharged surface, the particles removed from the solution will include its counterions as well. The effectiveness of the linkage between counterions and charged macromolecules or charged surfaces depends on the amount of added salt. In solutions of low salt concentration the counterions constitute the principal source of ions and are strongly held to the fixed charges of the macromolecules and the charged surface. At low ionic strengths, therefore, adsorption of polyelectrolytes to neutral and to identically-charged surfaces is poor, whereas polyelectrolyte adsorption to highly- but oppositely-charged surfaces is very pronounced. At high salt concentrations the counterions are overwhelmed by the salt ions, which shield the potential and weaken the attachment of counterions to the polyelectrolyte. If the adsorbing surface has a charge opposite to that of the polyelectrolyte and has counterions of identical charge to that of the adsorbing polyelectrolyte molecule, charge neutralization occurs at the surface with concomitant release of neutral salt molecules. The detailed entropy balance will depend on the charge density profiles near the surface and around the polyelectrolyte. In this case the charge interaction and the release of ions from their containment is the main driving force of adsorption. For a p+ valent positively charged polyelectrolyte carrying along with it pA− univalent counterions adsorbing on an area domain with S negative surface charges and SC+ small univalent surface counterions mutually neutralizing part of their charges and releasing μ neutral univalent salt molecules into the bulk solution one can write.

$$[(s \cdot C^+) + S^{s-}]_s + [PE^{p+} + pA^-)]_B$$
$$= [(S \cdot PE)^{[(s-\mu)^- + (p-\mu)^+]} + (s-\mu)C^+$$
$$+ (p-\mu)A^-]_s + [\mu(C^+ + A^-)]_B \qquad (81)$$

The entropic adsorption driving force will be positive if there is a net release of small ions from the concentrated surface phase into the dilute bulk phase or $2\mu >$ fp, where f is the osmotic factor [30]. The value of f for polyelectrolytes with an identical charge on each monomeric segment is ~ 0.2. The entropic adsorption driving force decreases and even may invert its sign with decreasing surface charge density. If $s \ll p$ and all the surface charges are neutralized the following situation may arise: $2\mu = 2s <$ fp. In this case the polyelectrolyte carries with it more counter ions into the concentrated surface phase than are released by neutralization of the surface charge. Increase in ionic strength decreases adsorption when $2\mu >$ fp but it enhances it when fp $> 2\mu$. The second case seems paradoxical, but it occurs when highly charged polyelectrolytes adsorb on polarized mercury surfaces with relatively low surface charge density with average distances of 20 Å to 30 Å between surface charges as compared to ~ 5 Å on the polyelectrolyte [56, 57]. In this case an excess of polyelectrolyte charges, compared to what is required for neutralization of the surface charge, is adsorbed, which inverts the sign of the surface charge and has an additional impeding effect on further polyelectrolyte adsorption. Increase of ionic strength facilitates adsorption against the inverted surface potential induced by polyelectrolytes which have already been adsorbed. In either case the final result depends on the superposition of the electrostatic and of the other kinds (hydration, hydrophobic, structure-specific) of adsorption forces. The different adsorption forces may be interrelated; for example, increasing the ionic strength may lower hydration-induced repulsion forces and increase hydrophobic ("salting out") forces.

The energy of adsorption of a polyelectrolyte is the sum of the adsorption energy of the same polymer without charges, and of all the contributions to the standard free energy of adsorption and the change in electrostatic free energies ΔFe of the surface and the polyelectrolyte as a result of adsorption. (The change in the entropic contribution of the counterions is included in ΔFe). The change in the electrostatic free energy of the surface can be calculated from the change in surface charge density upon adsorption according to the electric double layer theory (see eqns 31 and 38). The electrostatic free energies of polyelectrolytes are calculated and presented in some classical publications [30, 31].

3.6. The problems of equilibrium and reversibility in polymer adsorption

All the theoretical treatments of polymer adsorption assume that adsorption is a reversible process and that equilibrium conditions are attained. This may be the case if sufficient time is allowed for equilibration; however, most of the methods used for surface tension or surface energy measurements perturb the equilibrium conditions, including static methods like the pendant vesicle drop method. As the drops strive towards their equilibrium shape, the drop surface is perturbed, and consequently so is its approach towards equilibrium. It turns out that surface tension measurements are a poor way of obtaining surface excess concentration and surface profiles of adsorbed polymers. Moreover, if the standard free energy of adsorption per monomeric segment is small, traces of highly surface-active impurities may contribute to the equilibrium surface tension value attained after a long time. Therefore, all the published surface excess values of polymers inferred from measured "equilibrium" surface tension values must be considered with caution. Direct methods for determining surface concentrations, surface thickness, and concentration profiles across the surface phase are to be preferred.

4. Experimental methods for determining surface concentration and surface structure

The following chapters review different methods for obtaining adsorption parameters. We shall concentrate on adsorption from aqueous solutions on the following interfaces: (1) air/water; (2) liquid electrode/water; (3) amorphous solid/water; (4) smooth crystal (cleaved mica)/water; and (5) water with lipid monolayers, lipid bilayers, and biological membranes.

4.1. Surface concentration determination

4.1.1. Use of radiotracer methods:
This method has been discussed very thoroughly in a review article by Muramatsu [32]. The emphasis in this chapter will be on adsorption at pure- and lipid monolayer-covered air/water interfaces as well as at biological surfaces. The special problems encountered during adsorption of polymer substances at these surfaces will be also discussed.

Since the radiotracer method is a way to determine directly the amount of a specific substance at a certain location, it is essential that the labelled compound be chemically identical with the main compound, or at least have identical adsorption properties. Only under these conditions can the proportionality between radioactivity and quantity at every location be assured (33).

The specific activity S is given by:

$$S = N'/(N + N') \qquad (82)$$

where N' is the number of labelled molecules and N is the number of non-labelled molecules.

The sensitivity of the determination of the labelled compound N' is given by the decomposition rate $- (dN'/dt)$ measurable above background radioactivity:

$$N' = \tau \frac{(-dN'/dt)}{\ln 2} \qquad (83)$$

where τ is the decay half-time. For ^{35}S, for which $t = 87$ days $(1.25 \times 10^4 \text{ min})$, 100 dpm above background corresponds to $N' \sim 1.8 \times 10^6$ atoms or 3×10^{-18} moles. The mean deviation of the measured radioactivity R is $\pm\sqrt{R}$

$$R = \eta\left(-\frac{dN'}{dt}\right) \qquad (84)$$

where η is the detection coefficient; $\eta < 1$ and depends on the source and the counting method. The detection coefficient is a product of different factors, e.g. geometric factors, self-absorption, attenuation, and counter efficiency factors, all of which are < 1, as well as scattering factors which may be > 1.

One type of counting device is the Geiger-Müller counter, based on discharge measurements when a disintegration particle (β or α) enters a gas-filled tube. The tube may be equipped with an end window or it may be windowless. The tube may be permanently filled with a specific gas, or gas of a desired composition can flow through it. Another type of counting device is based on scintillation induced by discharged elementary particles; the scintillant may be either a liquid or a solid. Semi-conductors can also serve as solid state counters. The choice of radioactivity counter will depend on the experimental requirements.

Adsorption at solid/liquid interfaces and on the large surfaces of dispersed particles. The process here is very simple. Adsorption is completed under predetermined conditions, e.g. adsorbent concentration, pH, or even imposed potential if a metallic or semiconductor surface is being used. The surface is then removed from the solution, blotted, and the surface radioactivity measured with a windowless counter or in a scintillation counter; if necessary, correction for the radioactivity of the solution retained on the surface can be made by determining its quantity by weight difference after solvent evaporation. In the case of large areas of dispersed particles the amount adsorbed can be determined both by measuring the concentration (radioactivity) depletion in the solution and by separating the particles by filtration and measuring their radioactivity. Independent determination of the specific surface areas of these particles is required; this is done by saturation of some well-characterized adsorbates, such as strongly surface-active dyes whose molecular areas are known. This method is also used to investigate adsorption on biological surfaces and binding of agonists to receptors on biological membranes. In these cases, it might be possible to distinguish between different types of binding sites with different affinities for various ligands. Ligand binding to different types of binding sites is discerned by studying the concentration dependence of ligand binding.

Adsorption at the air/water interface and on lipid monolayers spread over a water surface. An advantage in using radiotracers for adsorption studies is that the measurements can be made in situ, then not only can the equilibrium adsorption be determined, but the adsorption kinetics can also be resolved. For reliable measurements of radioactivity from adsorbed monolayers, the radioactivity in the bulk of the solution has to be either negligible or fully accountable. The second condition can be fulfilled only if the radioactivity in the bulk is comparable or lower than the radioactivity in the monolayer. This can be achieved by using very low bulk concentrations and by using emitters whose radiation is absorbed in a very thin layer of solution (α-particles or very low energy β-particles). The maximal range in water of β-particles for nuclides which may be inherent constituents of adsorbing biopolymers, or which may be bound to them, is given in Table 2. The maximal penetration range increases with the energy of the disintegration particle.

The maximal range of radiation determines the maximal allowed bulk concentration of a nuclide for surface concentration determination to be feasible. For convenient surface concentration determination, the total amount of nuclide in a layer whose thickness is equal to the maximal radiation range should not exceed the amount in the adsorbed

Table 2. Energies and half-times of β rays from different surveys

Nuclide	Disintegration particle	Half-time	E_{max} (MeV)	Maximum range in water (μm)
^{45}Ca	β-	164 days	0.26	650
^{35}S	β-	87.1 days	0.169	340
^{14}C	β-	5.570 years	0.156	300
^{63}Ni	β-	85 years	0.065	67
^{3}H	β-	12.46 years	0.0179	6

monolayer of an equal area. For a monolayer concentration of 0.1 μg/cm^2, the bulk concentration of a compound labelled by ^{3}H can be up to 150 μg/ml, but the bulk concentration of a compound labelled by ^{14}C can range only up to 3 μg/ml, and for ^{45}Ca^{2+} the bulk concentration should not exceed 1.5 μg/ml.

There is of course a great advantage in using ^{3}H as the radiolabel, as one can measure adsorption from relatively concentrated solutions. The disadvantage is the low counting efficiency of the tritium low-energy β-particles. The window of a gas flow counter has to be thinner than 1 μm to pass >20% of the radiation. Moreover, an air gap of 1 mm also allows less than 20% of the radiation to pass. To work with ^{3}H one has to use very thin end windows (100–200 nm), and the air replaced by He, whose density is about one-fourth that of air [33–35]. He, being the main component of counter gas, also diminishes the chances of air and water vapor leakage into the counter, which might impair its function. In this way the combined barrier of the gap and of the window permits up to 30% of the radiation to pass. However, the attenuation factor and the geometric factor do not allow the total counting efficiency from the surface to exceed 7%, and 5% is considered to be a satisfactory efficiency [34, 35].

The surface concentration Γ is related to the surface radioactivity R_t according to the equation

$$\Gamma = \frac{S_m}{S} \frac{\Gamma_m}{R_m} (R_t - R_i) \qquad (85)$$

where Γ_m is the surface concentration of a spread monolayer (e.g. stearic or oleic acid) used for calibration, R_m is its surface radioactivity, S_m is the specific radioactivity of the substance in the spread monolayer, S is the specific radioactivity of the investigated substance, and R_i is the radioactivity of the solution interior. R_i can either be calculated from the bulk concentration of the substance or determined using the same nuclide in a system where there is no adsorption, e.g. ^{3}H-dodecanol in dodecanol. The best way to assure constant counting efficiency is to determine R_i in the same system in the absence of adsorption. This can be done by spreading a monolayer which does not adsorb the investigated substance and which cannot be displaced by it.

The labelling of biopolymers can be carried out by chemical modification. It is essential that chemical modification does not alter the chemical properties or adsorbability of the biopolymer. In the case of the polypeptide antibiotic polymyxin B, for example, which bears five positive charges and is active against Gram-negative bacteria, labelling was accomplished by acetylation of one of the lysyl residues using ^{14}C- or ^{3}H-acetyl chloride. However, it turned out that the native polymyxin B has a five-fold greater affinity to negatively-charged lipid monolayers than the monoacetylated one. Therefore, radioactive monoacetylated polymyxin can be used as a tracer for adsorption studies with cold monoacetylated polymyxin B but not for native polymyxin B. The time course of surface radioactivity studied using radiolabelled acetyl polymyxin B (Fig. 15) showed that at the beginning, when the adsorption layer was far from saturated labelled monoacetylated polymyxin adsorbed concurrently with native polymyxin B. At longer times, when the surface became saturated, native polymyxin B, which binds with higher affinity to the negatively-charged monolayer, displaced bound radioactive monoacetylated polymyxin B, and the surface radioactivity started to decrease, leveling out at a surface radioactivity corresponding to an acetyl-polymyxin B concentration in the surface ~ 5.5 times

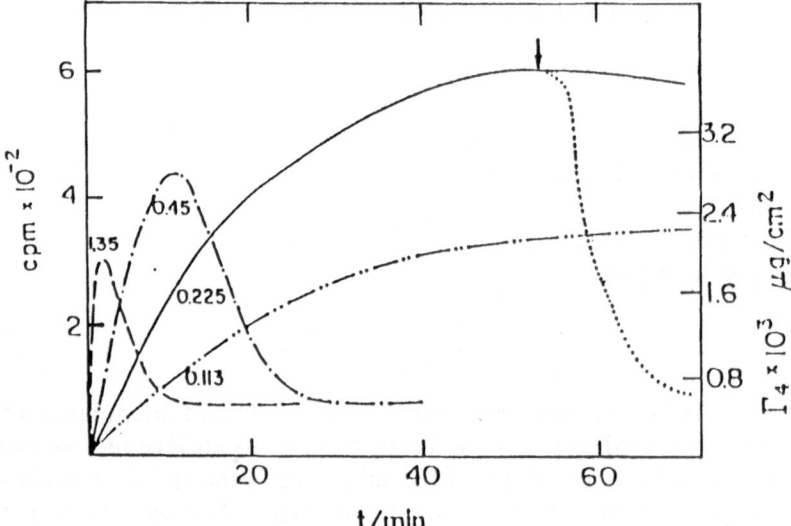

Fig. 15. Time dependence of surface radioactivity and concentration of ^{14}C-acetyl PX (Γ_4) adsorbed from a 10^{-2} M phosphate buffer solution at pH 7.2 in the presence of a 19-fold excess of native PX on the condensed monolayer of phosphatidylinositol. Total PX concentrations: $-$, 0.225 μg/ml; $-\cdot-$, 0.45 μg/ml; $--$, 1.35 μg/ml. At a line indicated by the arrow, the concentration was raised by another 0.225 μg/ml, and then Γ_4 corresponding to the surface c.p.m. is given by the dotted line [33]. PX = polymyxin B.

lower than in the bulk. Polymyxin B has a terminal octyl group and is surface-active, adsorbing at the air/water interface. However, mono-acetylated polymyxin B is more surface-active than the native molecule, and the relative concentration of acetyl-polymyxin B on the aqueous surface is ca. 50% greater than in the bulk.

A change in the charge or hydrophobicity of a single amino acid can affect the pH-dependent surface activity of a whole protein molecule. Thus, even minute changes in chemical composition, which do not affect the biological activity of the molecules, may affect their surface activity at different pHs, leading to a lower or higher content of the labelled component in the surface than in the bulk solution. Re-equilibration rates may be considerably slower than that observed for polymyxin. To avoid this type of problem, labelling has to be performed in a way which does not affect the chemical composition of the biopolymer. This can be achieved, for example, by growing an organism in an appropri-ately-labelled nutrient medium. Alternatively, an exchange reaction can be used to replace a non-radioactive terminal amino acid by a radioac-tive one. There are proteins containing attached non-peptide moieties, e.g. prothrombin contains sialic acid. The sialic acid moiety can be oxidized to an aldehyde and re-reduced with ^{3}H-borohydride, thus producing ^{3}H-prothrombin without changing its physico-chemical prop-erties [35].

4.1.2. Optical methods

Multiple attenuated total reflection (ATR). Absorbance and fluores-cence, which are generally used for concentration determination, can also be used to assess surface concentrations. However, even moderate surface concentrations will yield quenched fluorescence, which impairs quantitative determination. Nevertheless, surface fluorescence measure-ments even from air/water interfaces have been reported. Light ab-sorbance of a single monolayer is very weak even for adsorbed molecules with very high extinction coefficients. A fully condensed monolayer of 10^{-9} mole/cm^2 of a dye with a molar extinction coefficient of 10^4 will have an optical density of 10^{-2} Au. To increase the sensitivity of the determination multiple reflection techniques are used. This is done using attenuated total reflection conditions known as the ATR method (for details see Ref. 36). In this procedure a thin, optically dense, transparent parallel prism is used to multiply the total reflection of light impinging on an interface with a less optically dense medium from which the chromophores are adsorbed (Fig. 16). Radiation from the more dense medium of refractive index n_1, at angles of incidence $\alpha_1 > \alpha_c$, is totally reflected from the interface with the less dense medium of refractive index n_2:

$$\sin \alpha_c = n_2/n_1$$

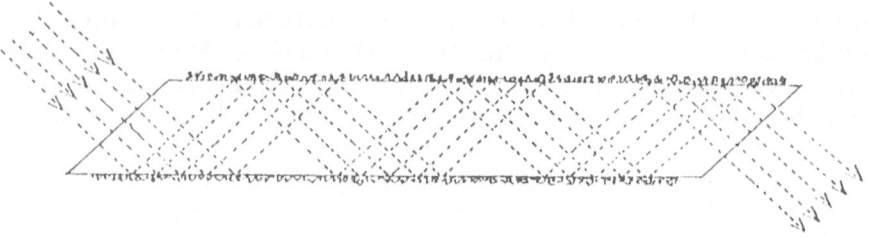

Fig. 16. A multiple reflection prism at 45° for attenuated total reflection measurements. Adsorbed polymer layer, attenuating the reflection, is shown.

However, when the less dense medium (or a layer in it near the surface) absorbs, part of the radiation is absorbed and the energy of the reflected light decreases. Absorption, and thus the attenuation of reflection, is largest near the critical angle α_c and larger for parallel polarization with respect to the plane of incidence, as seen in the Fresnel equations (86) and (87) for the reflectances of beam components polarized perpendicular ($R_\perp$) and parallel to the plane of incidence ($R_\parallel$)

$$R_\perp = \left[\frac{n_i \cos \alpha_1 - \lfloor (n_2 - i\chi_2)^2 - n_1^2 \sin^2 \alpha_1 \rfloor^{1/2}}{n_1 \cos \alpha_1 + [(n_2 - i\chi_2)^2 - n_1^2 \sin^2 \alpha_1]^{1/2}} \right]^2 \tag{86}$$

$$R_\parallel = \left[\frac{(n_2 - i\chi_2)^2 \cos \alpha_1 - n_1[(n_2 - i\chi_2)^2 - n_1^2 \sin^2 \alpha_1]^{1/2}}{(n_2 - i\chi_2)^2 \cos \alpha_1 + n_1[(n_2 - i\chi_2)^2 - n_1^2 \sin^2 \alpha_1]^{1/2}} \right]^2 \tag{87}$$

These relations were derived considering the complex nature of the refractive index n of an adsorbing medium with an extinction index $\chi : n = n(1 - i\chi)$. The attenuation or absorption parameter a is defined by the relation $a \equiv (100 - R)\%$. The attenuation shown in Fig. 17 is a function of the angle of incidence and the absorption modulus m_n. The absorption calculated from Eqns. (86) and (87) first increases with the absorption modulus m_n ($m_n = 4\pi n\chi/\lambda = 2.303 \times \Sigma c$ where Σ is the molar extinction coefficient and c is the concentration), then passes through a maximum, and finally falls to zero for extremely strong absorption ($\chi > 10$) (Fig. 18).

According to Maxwell's theory standing waves that penetrate into the low refractive non-adsorbing medium are formed perpendicular to the reflecting surface. The amplitude of the electric field falls off exponentially with distance z from the phase boundary as follows:

$$\vec{E} = \vec{E}_0 \exp\left[\frac{-z}{d_p} \right] \tag{88}$$

The quantity d_p, the distance within which $\vec{E}$ decreases to E/e, is given for $\alpha > \alpha_c$ by

$$d_p = \frac{\lambda}{2\pi[\sin^2 \Phi_1 - (n_2/n_1)^2]^{1/2}} \tag{89}$$

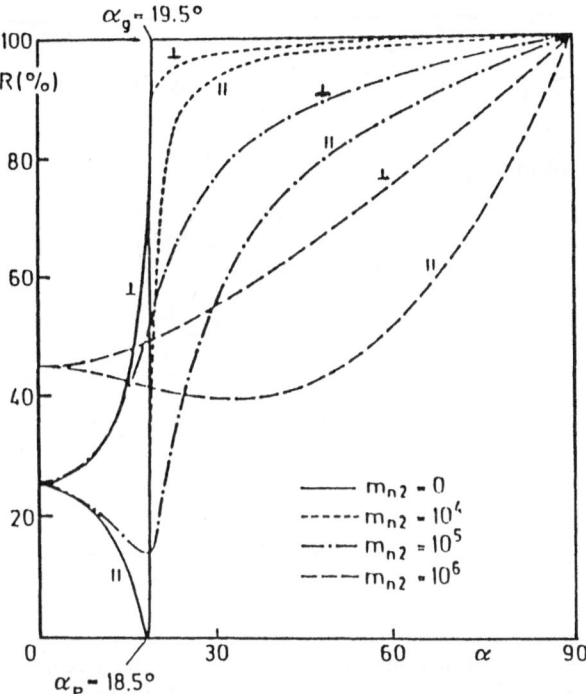

Fig. 17. Internal reflectance of a phase boundary with $n_2/n_1 = 0.333$ as a function of the angle of incidence α at $\lambda = 400$ nm and various values of the extinction modulus m_{n2} as parameter [36], with permission.

If the optically-thinner medium is also absorbing, the penetrating wave becomes attenuated and the light intensity I becomes

$$I = I_0 e^{-m_n d} \approx I_0(1 - m_n d) \qquad (90)$$

The reflectance decreases in a similar manner, according to the form

$$R \cong 1 - m_n d_e \qquad (91)$$

where d_e, the effective layer thickness, is related to the absorption parameter a by $d_e = a/m_n$. The effective thickness d_e represents the thickness of a layer giving the same degree of extinction at transmittance as obtained at total reflectance. It can be obtained with the aid of Eqn. (92):

$$d_e = \frac{n_2/n_1}{\cos \Phi_1} E_0^2 \int_0^\infty \exp\left(\frac{-\chi}{d_p}\right)^2 dx = \frac{n_2/n_1}{2 \cos \Phi_1} E_0^2 d_p \qquad (92)$$

where E_0 is the amplitude of the electric field in the optically thinner medium at the phase boundary. It is larger for parallel-polarization than for perpendicular-polarization. In Fig. 19 the relative depth of penetration, d_p, and d_e expressed in wave length λ units are shown as a

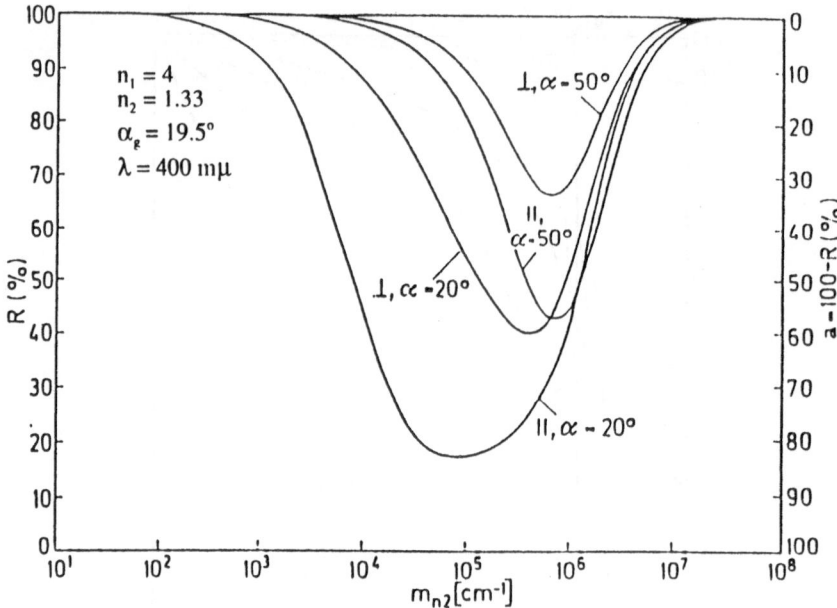

Fig. 18. Calculated dependence of the internal reflectance R or adsorption parameter a ($=(100 - \text{R})$) from the extinction modulus of the optically-rarer medium at two different angles of incidence. $\lambda = 400$ nm; $n_2/n_1 = 0.333$ [36], with permission.

function of the angle of incidence α. To assure a longer depth of penetration than the thickness of the polymeric layer one should work at angles of incidence only moderately higher than α_c. In the case of adsorption from a dilute medium we are interested in a thin film of thickness $d < d_e$. The electric field can be considered constant over the film thickness, and therefore thin films are ideal for absorbance measurements of surface concentration by ATR. The attenuation after N reflections is given by

$$R^N = (1 - m_n d)^N$$

Different plates and planar prisms have been devised for fixed and varying angles. The method is most widely used in the infrared region, but can also be used in the visible and UV regions with plates or prisms made of materials with different refractive indices n such as quartz ($n = 1.5$), Al_2O_3 ($n = 1.8$), AgCl ($n = 2$), Si ($n = 3.4$), and Ge ($n = 4$). In the sample compartments of most spectrometers attachments are available for ATR measurements.

Ellipsometry. Ellipsometry as a reflection method can be used for determination of optical spectra of highly-absorbing solids, including metals and semiconductors, but its most important use is for the optical

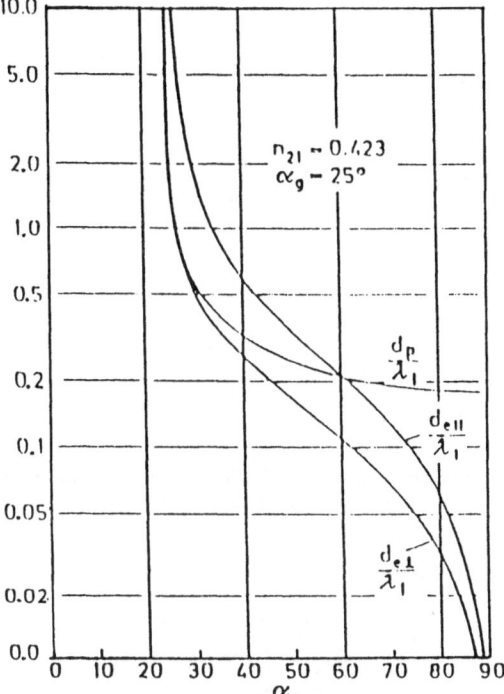

Fig. 19. Relative depth of penetration and effective layer thickness of the waves penetrating into the optically-rarer medium as a function of the angle of incidence α at the phase boundary [36], with permission.

characterization of thin films. It provides information on the thickness of the film as well as data on the real and imaginary components of the complex refractive index. With the recent development of automatic ellipsometry very fast changes in film thicknesses during adsorption, electrosorption, or desorption processes can be followed. Only a very brief account of the principles of the method and of the scope of its application will be presented here. A detailed discussion of the method has been given by Gottesfeld [37].

We are interested in defining the optical parameters of a thin film adsorbed onto an optically dense absorbing support (Fig. 20). Incident light is reflected into the solution; the refractive indices of the solution, the adsorbed layer, and the solid support are n_1, n_3, and n_2, respectively. If the support and the film are adsorbing their refractive indices become complex

$$\hat{n}_2 = n_2 - i\chi_2$$

$$\hat{n}_3 = n_3 - i\chi_3,$$

The respective Fresnel reflection coefficients $r_\parallel$ and $r_\perp$ parallel and

METAL-AMBIENT   METAL-FILM-AMBIENT

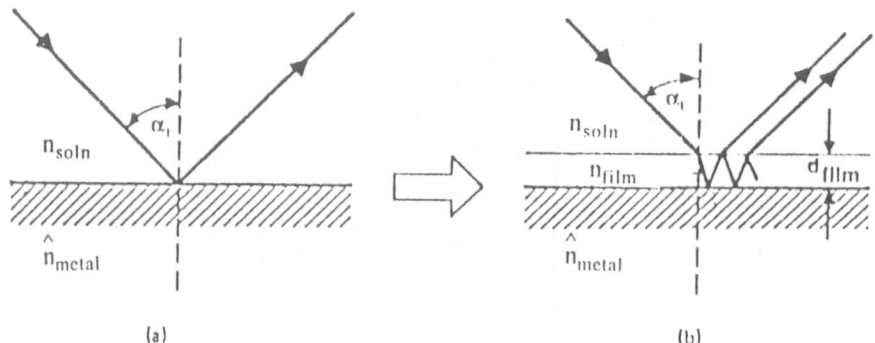

(a)   (b)

Fig. 20. Schematic presentation of specular reflection at the electrode-electrolyte interface (a) in the absence of any surface film, (b) in the presence of a surface film.

perpendicular to the plane of incidence for the base surface are then

$$r_\parallel^{1,2} = \frac{n_1 \cos \alpha_2 - \hat{n}_2 \cos \alpha_1}{n_1 \cos \alpha_2 + \hat{n}_2 \cos \alpha_1}$$

$$r_\perp^{1,2} = \frac{\hat{n}_2 \cos \alpha_2 - n_1 \cos \alpha_1}{\hat{n}_2 \cos \alpha_2 + n_1 \cos \alpha_1}. \tag{93}$$

The Fresnel reflection coefficients of the film-covered surface are obtained by summing up all the reflected rays from the film/metal interface. Thus, the Drude equations are obtained

$$r_\parallel^{1,2,3} = \frac{r_\parallel^{1,3} + r_\parallel^{3,2} \exp(-i\delta)}{1 + r_\parallel^{1,3} r_\parallel^{3,2} \exp(-i\delta)}$$

$$r_\perp^{1,2,3} = \frac{r_\perp^{1,3} + r_\perp^{3,2} \exp(-i\delta)}{1 + r_\perp^{1,3} r_\perp^{3,2} \exp(-i\delta)} \tag{94}$$

where superscripts 1, 2, and 3 refer to the solution, film, and support, respectively. They contain the reflection coefficients of the solution interface film and the film/support interface, as well as the phase delay δ by the optical thickness of the film. $\delta = (4\pi/\lambda)\,\hat{n}_3 d \cos \alpha_3$, where λ is the wavelength and d is the thickness of the film.

An ellipsometer measures the ratio between the Fresnel reflection coefficients $r_\parallel/r_\perp$. The ellipsometric parameters ψ and Δ are defined by

$$(\hat{r}_\parallel/\hat{r}_\perp) = \tan \psi \, \exp(i\Delta) \tag{95}$$

For a two phase system without a film

$$r_\parallel^{1,2}/r_\perp^{1,2} = \tan \psi_0 \exp(-i\Delta_0)$$

$$= \frac{(n_1 \cos \alpha_2 - n_2 \cos \alpha_1)(n_1 \cos \alpha_1 + n_2 \cos \alpha_2)}{(n_1 \cos \alpha_2 + n_2 \cos \alpha_1)(n_2 \cos \alpha_2 - n_1 \cos \alpha_1)}$$

and for a three phase system containing the surface film

$$(r_\parallel^{1,3,2}/r_\perp^{1,3,2}) = \tan \psi \, \exp(-i\Delta)$$

$$= \frac{[r_\parallel^{1,3} + r_\parallel^{3,2} \exp(-i\delta)][1 + r_\parallel^{1,3} r_\parallel^{3,2} \exp(-i\delta)]}{[r_\perp^{1,3} + r_\perp^{3,2} \exp(-i\delta)][1 + r_\parallel^{1,3} r_\parallel^{3,2} \exp(-i\delta)]} \quad (97)$$

To characterize the film properties, the shifts $\delta\psi$ and $\delta\Delta$ and their dependence on the optical characteristics of the whole system $n_1(\lambda)$, $\hat{n}_2(\lambda)$, α, and λ must be determined, from which n_3 and d can be found. The ellipsometric measurements taken at a single λ and α_1 supply only two parameters $\delta\psi$ and $\delta\Delta$. For the complex refractive index $\hat{n}_3 = n_3 - i\chi_3$ when the surface film is absorbing, we have at least three unknown parameters if the film is uniform. There are several possible ways to use experimental data to define more precisely the optical properties of the films. However, in the case of an adsorbed polymer where the effective value of $\hat{n}_3$ may vary along the film thickness, the simultaneous determination of film thickness and its density profile becomes a formidable problem.

To obtain experimental data for determining the relevant optical parameters one can measure reflectance using the same instrument; in fact, one really measures the difference in reflected light intensity δI in the presence and absence of film at a constant incident beam intensity.

$$\frac{\delta I_{ref}}{I_{ref}} = \frac{|r_{\parallel,\perp}|^2 - |r_{\parallel,\perp}|_0^2}{|r_{\parallel,\perp}|_0^2} = \frac{\delta R}{R_0} \quad (98)$$

The reliability of the reflectance measurement depends on the stabilities of the incident beam and the system. For analysis, the results have to be fitted by computer to the data calculated from Eqns. (96) and (97).

Instrumentation in ellipsometry

i. Null technique. Classical ellipsometry is based on this technique. Ellipticity is introduced in the incident beam until the difference in phase shift in the parallel and perpendicular polarizations yields a linearly polarized reflected beam which can be completely extinguished with an adjusted polarizer. To accomplish this, the linearly polarized incident beam passes through a compensator of birefringent material which introduces a phase difference between the two perpendicular components of the electric field. After reflection from the sample the beam passes through another linear polarizer ("analyzer") to a linear photodetector. The orientations of the optical axes of the polarizer, the compensator, and the analyzer are adjustable manually. For the case of a constant azimuth of 45° of the compensator with respect to the phase of incidence, the angles of the polarizer and the analyzer are adjusted to P_{null} and A_{null}, at which the intensity of light reaching the photodetector is zero (null point). Under such conditions the values of ψ and Δ for the reflecting surface are related to P_{null} and to A_{null} with respect to the plane of incidence by:

$$\Delta = 270° - 2P_{null} \qquad (99)$$

and

$$\psi = A_{null}$$

The manual null apparatus is very tedious to operate, but automatic versions have recently been developed (38). Automated nulling is achieved by adding electronically controlled optical rotation induced by Faraday cells added to both the polarizer and the analyzer.

ii. Automatic ellipsometer with a rotating analyzer. This type of automatic ellipsometer (39) is now available commercially (Rudolph Research Gaernter Scientific). The incident collimated monochromatic beam is linearly polarized with the plane of polarization, usually at 45° to the plane of incidence. Under these conditions the ratio of parallel to perpendicular electric field components of the incidence beam, $E_\parallel / E_\perp$, is 1 and the phase difference $\Delta\delta$ is zero. The ratio of the complex electric field vectors of the reflected beam in the parallel and perpendicular directions then becomes

$$\left(\frac{\hat{E}_\parallel}{\hat{E}_\perp}\right)_{refl} = \left(\frac{E_\parallel}{E_\perp}\right)\frac{r_\parallel}{r_\perp} = \tan\psi \, \exp(-i\Delta)$$

Due to the phase difference introduced by reflection from the interface the beam is elliptically polarized, i.e., the tip of the electric field vector rotates in an elliptical orbit in a plane perpendicular to the direction of propagation of the reflected beam (Fig. 21a). The values of ψ and of Δ can be inferred from the characteristics of the ellipse. Characterization of the elliptically-polarized reflected beam is done by passing the beam through a linear analysing polarizer which rotates at a constant angular velocity (~ 100 rpm) and samples the reflected intensity along the orientation θ of the analyzer's optical axis. The intensity of light emerging from the rotating analyzer is a sinusoidal function of θ (Fig. 21b). The ratio of minimum to maximum in the sinusoidal intensity, designated by $\tan\gamma$ is equal to the ratio of the minor and major axes of the ellipse described by the parallel and perpendicularly polarized components of E_{refl}. The phase of the intensity waveform, α, is the angle between the major axis of the ellipse and the plane of incidence. The ellipse described by the electric field vector of the reflected beam is thus defined by α and $\tan\gamma$, from which Δ and ψ may be derived:

$$\tan\Delta = -\frac{\tan 2\gamma}{\sin 2\alpha}$$

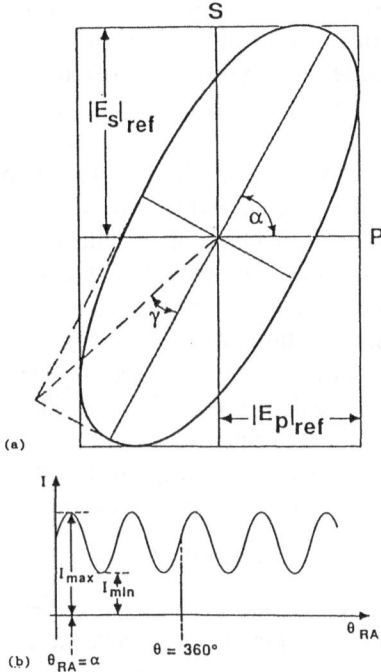

Fig. 21(a). Ellipse described by the tip of the E_{ref} vector in an elliptically-polarized light beam, shown with the characteristic parameters α and $\tan \gamma$, which define it in the Cartesian coordinate system. Also shown are the amplitudes of the components of E_{ref} along the p and s orientations, the combination of which, with a phase difference Δ, results in the same elliptical rotation. $|E_p|/|E_s| = \tan \psi$. (b) Light intensity vs. analyzer orientation for the elliptically-polarized reflected beam, as detected after passing through a rotating analyzer.

and

$$\cos 2\psi = \cos 2\gamma \cos 2\alpha \qquad (100)$$

Evaluation of the ellipsometric parameters from the sinusoidal intensity waveform is done nowadays with digital phase-sensitive detectors employing microprocessors. Digital processing of the sinusoidal intensity waveform can be performed by Fourier transform techniques.

There are also automatic ellipsometers based on periodic modulation of polarization; those interested in this technique and in ellipsometry generally should consult Ref. (37). Ellipsometry is an excellent technique for characterizing surface films so long as the films are uniform in their thickness and refractive index.

4.1.3. Gravimetric determination of adsorption: Gravimetric determination of adsorption is straightforward if one deals with large specific surface areas, i.e. very small particles in suspension or a porous material which can be separated from the solution and the amount of material adsorbed determined by weighing. However, mass adsorption can also be determined on small pure or covered surfaces of piezoelectric wafers such as those sliced from a single crystal of quartz. Since quartz is most suitable for this purpose, the method is often called "quartz crystal microbalance" (QCM) [40]. A quartz wafer is sandwiched between two electrodes bonded to its surface. These electrodes induce an oscillating electric field across the wafer, which in turn produces a mechanical standing wave within the quartz wafer. For a quartz wafer of a given thickness the amplitude of mechanical oscillation varies with the frequency. A resonant oscillation is achieved when the frequency of the electrical and mechanical oscillations centers at a characteristic fundamental frequency. This fundamental frequency depends on the wafer thickness and varies usually between 2 and 20 MHz. A wafer of thickness 320 μm oscillates near 5 mHz. The oscillation frequency is a function of, amongst other things, the mass of adsorbed or deposited material on the wafer surface. The change in frequency Δf due to added rigid mass Δm is given by the Sauerbrey equation [41].

$$\Delta f = -2\Delta m n f_0^2 / A \sqrt{\mu_q \rho_q} \qquad (101)$$

where f_0 is the fundamental frequency of QCM, n is the overtone number (the sensitivity is larger at overtones than at the fundamental frequency), and μ_q and ρ_q are the shear modulus ($\mu_q = 2.947 \cdot 10^{11}$ g cm^{-1} s^{-2}) and density ($\rho_q = 2.648$ cm^{-3}) of quartz, respectively. With high resolution counters such as the Philips PM6654 series, the frequency Δf can be measured to within 0.1 Hz. For a QCM oscillating at a fundamental frequency of 5 MHz, the sensitivity is 18 ng cm^{-2} Hz^{-1}. The Sauerbrey equation is correct for rigid films. For plastic polymer surface layers a more complex theory is required in which the density and shear modulus must be considered. The frequency of the QCM is also dependent on the contact with solution and on changes in solution viscosity; therefore, the environment has to be kept constant when a surface layer is adsorbed or deposited.

Practically speaking, to follow adsorption or electrosorption on a metal electrode or on any other surface, the metal or other surface material is first deposited onto a quartz wafer, and then the frequency shift during adsorption at a fixed potential or electrodeposition at changing potential is determined.

4.2. The profile of adsorbed macromolecules on solid adsorbents

The radiotracer, optical, and QCM methods described above provide a means of determining the overall surface concentration or the overall thickness of the surface layer. To obtain information on the more detailed structure of surface layers in general and of adsorbed layers of polymers or biopolymers in particular, other techniques are required.

Some optical methods which are sensitive to the orientation or immobilization of groups adhering to a surface can be used. Double layer capacitance measurements of metal electrode/water interfaces are also very sensitive to the composition of the adsorption layer adjacent to the electrode surface. In contrast, hydrodynamic methods and force measurements between two polymer-coated surfaces are sensitive to the extent of protrusion of polymer chains from the surface into the bulk of the solution.

4.2.1. Determination of the polymer fraction adsorbed and surface fraction occupied: The polymer fraction adhering to the surface, p, and the fraction of the surface occupied by polymer segments, θ, are the most important parameters describing the profile of an adsorbed polymer layer. Calorimetric measurements in combination with polymer adsorption determinations are the most general means of determining p and θ, and are predicated on knowing the enthalpy per adsorbed segment. The latter can be estimated by assuming that it has to be nearly equal to that of a chemically identical monomeric equivalent; in that case, θ is simply the ratio of the adsorption enthalpies of the polymer and of the equivalent monomer at saturation concentrations, and p is $\theta \Gamma_m^{max}/P\Gamma_p$ where Γ_p is the surface concentration of the polymer expressed as monomeric units.

The major problems with equating the enthalpy of adsorption of a monomer and the enthalpy of adsorption of a monomeric segment in a polymer are (i) the orientational restrictions imposed on adsorbing segments in polymeric chains, and (ii) the different number in the monomer solvent contacts in the two systems, as a monomer in a polymer chain is in contact with other monomers even in dilute solutions. Spectroscopic techniques are useful for assessing p and θ if the surface areas and concentrations are known independently; the most promising spectroscopic methods are those sensitive to the orientation and mobilities of adsorbed constituents, such as vibrational (IR or Raman) spectra. A shift in the frequency of some characteristic IR adsorption peak for an adsorbed polymer can serve as an indicator of which moieties are adhering to the surface. This shift can be observed in ATR measurements, but in order to achieve the required sensitivity dispersed systems are usually used. The pioneering work in this area was by Fontana and Thomas [42], who observed a shift of ~ 25 cm^{-1} in the carbonyl vibration frequency (~ 1740 cm^{-1}) induced by hydrogen bonding between polyalkylmethacrylate segments and the hydroxyl groups on the surface of a particulate silica adsorbent. While these early investigations were restricted to non-aqueous systems, with the advent of Fourier transform IR (FTIR) and with improvements in facilities for background subtraction, adsorption of biopolymers from aqueous media can likewise be studied by vibrational spectroscopy. Moreover, IR techniques can be used for the estimation of θ. For example, from the decrease in the free Si-OH band at 3695 cm^{-1} and the concurrent increase in the hydrogen bonding band at 3300 cm^{-1}, the values of θ for polyethylene oxide adsorbed on a silica surface was estimated under the assumption of hydrogen bonding between adsorbed -CH$_2$CH$_2$O- moieties and SiOH moieties on the silica surface [43].

Electron paramagnetic resonance (EPR) and nuclear magnetic resonance (NMR) have also been used to determine p. These methods rely on the assumption that segments in the adsorbed trains have a lower mobility than those in loops. This assumption is not perfect, as the motion of loop segments near the surface is probably also restricted, albeit to a lesser extent. Consequently the fraction of the polymer immobilized by the surface inferred from methods sensitive to the mobility of the *whole* segment is always higher than that obtained from IR or Raman spectroscopy. For instance, IR determination of p for polyvinylpyrrolidone adsorbed from D$_2$O onto silica yielded values of between 0.12 and 0.2 [44], while p for the same polymer adsorbed from aqueous solutions determined by EPR had values of 0.67–0.97 when adsorbed onto carbon and 0.6–0.9 when adsorbed onto silica [45]. Although it is generally less sensitive, NMR has two major advantages over EPR: (1) it does not require the attachment of a chemical spin label which can change the chemical nature and adsorptive propensity of the polymer; and (2) the NMR method is absolute and does not depend on comparisons with the behavior of other materials. Most adsorption studies using resonance techniques were performed in non-aqueous media; consequently, little data on the anatomy of surface layers of biopolymers on solid surfaces have been reported by these methods. Another problem with p, θ, and the structure of the layer adjacent to the solid surface lies in the inhomogeneity of the solid interface. It is not ideally planar and not all the surface sites are equivalent. The density and the distribution of active sites, which are usually distinct chemical groups, may depend on the preparation of the surface samples. Consequently the results reflect the chemical architecture of the adsorbing system as well as the properties of the adsorbing polymer molecule.

4.2.2. The thickness of adsorbed monolayers of polymeric substances

i. Hydrodynamic thickness. "Thickness", which is determined by the maximal protrusion of loops or tails from adsorbed polymers, may refer to either *average thickness* or *hydrodynamic thickness*. Ellipsometry provides information on the *average* thickness. The term *hydrodynamic thickness* derives from the hydrodynamic methods which were first employed for its determination. Hydrodynamic thickness was originally inferred from the increase in flow time of a pure solvent or a very dilute solution when the viscometer capillary was covered by a polymer monolayer [46]. Compared with flow through a glass capillary without an adsorbed surface phase, the time of flow in the presence of a polymeric surface layer is altered by a factor

$$t/t_0 = \left(1 - \frac{d}{R}\right)^{-4} \approx 1 + \frac{4d}{R} \qquad (102)$$

where R is the inner radius of the glass tube and d is the hydrodynamic thickness of the surface layer. The accuracy of flow time determination is around 0.1 s, and if the flow time is on the order of 10^3 s, then $4d/R \approx 10^{-4}$. If the capillary radius $R = 2.10^{-2}$ cm, d can be determined within 50 Å, or a layer of 50 nm can be determined with an accuracy of $\pm 10\%$. There can be very little improvement in this accuracy, since the use of narrower capillaries with polymer solutions is not practical. Improvement of flow time determination is limited by the reproducibility of the fluid volume retained on viscometer bulb surfaces which is affected by the presence of the adsorbed polymer film. Moreover, the relationship between the experimentally-determined values of d and the actual thickness of the protruding loops is not well established. The hydrodynamic thickness is generally larger than the thickness derived from the segment distribution normal to the surface, which is related to the radius of gyration of the molecule in the solution [47].

There are other methods for evaluating hydrodynamic thickness. One can determine the frictional coefficient of small ($r \approx 10 \, \mu m$ radius) beads by visually measuring their rate of precipitation in solution [48]. The frictional coefficient is proportional to the radius, and adsorption causes an effective increase in bead radius. If one starts with "bare" beads, the growth of the adsorbed layer can be followed by measuring the change in precipitation rate with time. One can use smaller and smaller beads to obtain higher ratios of film thickness to bead radius. With very small beads determination of frictional coefficients has to be done by quasi-elastic laser light scattering techniques. Problems inherent in these sorts of determinations are the usual radius distribution in a population of beads and the fact that when the radius of curvature of adsorbing beads becomes too small both adsorption and the thickness of adsorbed layers becomes a significant function of the radius.

Macromolecular adsorption also alters the original volume fraction ϕ of very small spherical beads of original radius r in a dilute suspension. The relative increase in volume of suspended beads as a result of polymer adsorption, $(1 + d/r)^3 \approx 1 + 3d/r$, can be computed from the increase in viscosity of the suspension by using the Einstein equation

$$[(t - t_0)/t_0 \phi]_{\phi \to 0} = 2.5(1 + 3d/r) \qquad (103)$$

Here, too, the dependence of polymer adsorption on bead radius becomes a problem.

ii. Direct force measurements between film-covered surfaces. If a suspension of beads in a dilute polymer solution is sedimented at a low
centrifugal rate, the sedimented volume V is always higher than that of
beads sedimented from a solution without the polymer, V_0. The difference in volume corresponds to the volume of adsorbed polymer layers
which cannot be further compressed by the applied centrifugal force. At
low centrifugal fields the ratio of these volumes approaches $(1 + d/r)$,
from which the film thickness d can be measured [49]. This thickness is
probably smaller than the hydrodynamic thickness but larger than the
average thickness measured by ellipsometry.

The force between two surfaces as they approach each other is
affected by polymer adsorption onto the surfaces, and can be measured
directly. For doing this kind of measurement, the surface must be
smooth enough to permit an unhindered approach at atomic distances.
Cleaved mica surfaces of macroscopic dimensions can be obtained
which are atomically smooth. When bent into half cylinders and arranged with curved surfaces facing one another at a right angle, the two
mica surfaces can approach each other to point contact. The distance
between the two cylinders at the point of closest approach is measured
interferometrically by half silvering the outer surfaces of the mica sheets
and passing light between them [50]. The light is largely destroyed by
interference after multiple reflection within the gap between the mirrors.
Only light of wavelengths which are exact multiples of the distance
between the surfaces can pass through the cavity. If white light is used
for illumination and the emergent light observed through a spectrometer, lines (fringes) of a sharply-defined wavelength can be seen, and
from these lines the separation of the surfaces can be measured to
0.3 nm. The difference between fringes corresponding to an odd and an
even number of half-wavelengths is a measure of the mean refractive
index, i.e. the mean concentration in the surface layer.

In this technique, one of the mica surfaces is mounted on a rigid
moveable support while the other is mounted on a similar support using
a leaf spring. By independently controlling the relative motion between
the two rigid supports it is possible to directly measure the forces acting
between the surfaces at distances ranging between 0 and 1000 nm.
Measurements are first performed using the pure solvent, in which the
surface displacement devices are calibrated and the nearest approach of
the pure surfaces determined. When polymer is adsorbed onto the two
mica surfaces a force between the two adsorbed layers is created and
produces a measureable deflection, from which the force is computed.
Attractive as well as repulsive forces can result, depending upon the
nature of the solvent-polymer interaction. As the surfaces come closer
and closer, the adsorbed macromolecules start overlapping and get
compressed, and the force between them becomes strongly repulsive;
from these repulsive forces film thicknesses can be computed. The film

thickness of proteins adsorbed on glass and on mica were determined by the hydrodynamic method and by force-distance measurements, respectively. In the case of collagen, which has three polypeptide chains wound around each other to form a triple helix, there is a good agreement between the thickness of surface layers inferred from the increase in flow time in a viscometer and that inferred from the distance between two mica plates at which a steep increase in the repulsive force occurs [51]. The thickness of a single layer is about 300 nm and is composed of rodlike triple helices attached head-on to the surface. Presumably the triple helical conformation is unstable at the ends, and attachment to solid surfaces is through unwound single strands; there is no sidewise adsorption of the collagen triple helix. This attachment is very strong and practically irreversible, for no desorption of intact adsorbed collagen molecules is observed within a reasonable time. The distance between two mica surfaces covered by a collagen monolayer at which a steep increase in repulsive forces occurred was 600 nm.

Studies on the adsorption of serum albumin from concentrated solutions onto solid surfaces give nonsensical results; the thickness of the surface layer increases almost linearly with protein concentration to values tens and hundreds of times greater than the diameter of the protein molecule. In this case we are not dealing with ordinary adsorption phenomena, but rather with surface-induced crystallization, nucleation, or some related surface reaction(s). Therefore, any theoretical thermodynamic or statistical mechanical approach to adsorption in this system is completely useless.

4.3. Determination of adsorbed layer concentration and structure on electrode surfaces

4.3.1. Monomeric substances on a polarizable mercury electrode: Adsorption of monomeric substances onto electrode surfaces can be treated thermodynamically with a reasonable degree of accuracy. Thus, the surface pressure and the surface concentration of decylamine were determined from measured capacitance values (Fig. 22a–c) at different temperatures using different concentrations of salt and decylamine [52]. This cannot be done readily on a solid electrode where the interfacial tension is immeasurable and the measured differential capacitance values are thermodynamically ill defined.

Adsorbed organic monolayers on either electrode form a barrier to transport of electrons and of ions. If the adsorbed monolayer is sufficiently thin, and particularly if it consists of aromatic residues or of conjugated double bonds along the hydrocarbon chains, electrons can tunnel across the monolayer. This kind of electron tunnelling was observed across monolayers of quinoline and its derivatives [53]. Thus

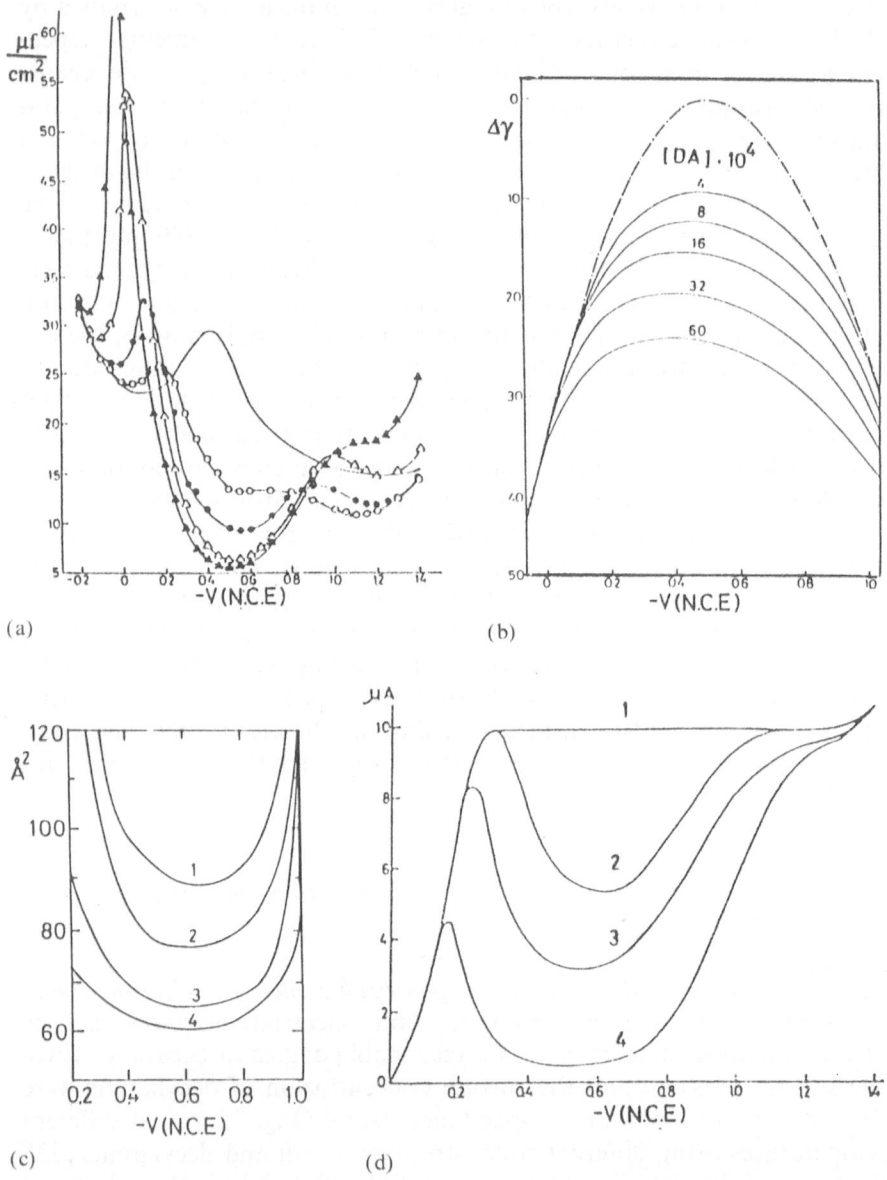

Fig. 22. Physical properties of decylaminium monolayer adsorbed onto the polarized mercury electrode surface from different solution concentrations [52]. (a). Differential capacity against polarization relative to N calomel electrode in the presence of 0.01 N HNO₃ as supporting electrolyte. Concentration of decylamine × 10⁻⁴ M: —, 0; ○, 2; ●, 8; △, 32; ▲, 60. (b). Electrocapillary curves calculated from measured differential capacity values. Supporting electrolyte 0.01 N HNO₄. Decylamine concentrations indicated on the curves; dotted line, no added DA. (c). Areas for adsorbed DA molecules as a function of polarization calculated by Eqn. (1). Supporting electrolyte 0.01 N HNO₃ + NaNO₃, as given in the table. (d). Instantaneous polarographic current of Cu(II) against potential at a drop age of 9 sec in the presence of 0.04 N HNO₃ and of the following decylamine (DA) concentration: 1, no DA; 2, 6 × 10⁻⁴ M DA; 3, 1.5 × 10⁻³ M DA; 4.4 × 10⁻³ M DA.

the adsorbed layer affects the rate constants of the electrode process, which can be expressed [53] by

$$k = b \prod_{ei} (\varepsilon) |L_{if}|^2 \prod_{s} (\varepsilon) \tag{104}$$

where b is a constant, $\Pi_{ei}(\varepsilon)$ and $\Pi_S(\varepsilon)$ are the distribution functions of the energy levels of the electrode and the solution side of the interface respectively, and $|L_{if}|^2$ is the overlap integral representing the probability of electron transition. The expressions for $\Pi_{ei}(\varepsilon)$, $\Pi_S(\varepsilon)$, and $|L_{if}|^2$ depend on the mechanism of the electrode process, but only $|L_{if}|^2$ is affected by the monolayer barrier and the tunnelling through it. $|L_{if}|^2$ decays exponentially with the distance from the electrode surface and can be expressed as follows [54]:

$$|Lif|^2 \sim \exp - 2x(2m_e \Delta U)^{1/2}/f \tag{105}$$

where f is Planck's constant, x the width of the barrier, and ΔU the average height of the energy barrier through which the electron of mass m_e is tunnelling. ΔU may be augmented by the monolayer barrier even though it has been assumed unaffected by the aromatic quinoline monolayer [53]; it is a function of the overvoltage η, and for the case of a linear drop of the overvoltage across the interface.

$$\Delta U = \Delta U_0 + e\eta/2 \tag{106}$$

The width of the barrier in the absence of the monolayer is x_0 and in the presence of a monolayer of thickness δ it is $x_0 + \delta$. The current resulting from the electron tunnelling across the monolayer can then be treated as an ordinary kinetic current. For a stationary electrode surface the solution of the diffusion equation coupled with an electrode kinetic process yields the boundary condition for the current i_t at time t:

$$i_t/(nF) = k_f C(t)_{x=0} = D\left(\frac{\partial C(t)}{\partial t}\right)_{x=0} \tag{107}$$

where F is the charge of a Faraday and n is the number of Faradays per electroacting molecule. This results in the following relation between the reduced current (i/id), namely the measured current divided by the diffusion current and the electrode kinetic constant [57]:

$$i/id = \pi^{1/2} \exp \lambda^2 \, \text{erfc} \, \lambda$$

where

$$\lambda = \frac{k_f t^{1/2}}{d^{1/2}} \quad \text{and} \quad \text{erfc} \, \lambda = 1 - \frac{2}{\sqrt{\pi}} \int_0^{\lambda} \exp(-u^2) \, du \tag{108}$$

The solution for the expanding surface of the dropping mercury electrode was given by Brdicka and Koutecky [58], and the resulting dependence of $(i/i_d)_t$ on $(k_f t^{1/2}/D^{1/2})$ tabulated. Thus k_f can be derived directly from $(i/i_d)_t$ where i_d is the diffusion controlled instantaneous current.

Electron tunnelling becomes negligible across saturated hydrocarbon layers which are thicker than 12 Å as in the case of a saturated monolayer of decylamine. In this case any electrode process is dependent on penetration of the electroactive redox substance into the monolayer to a critical distance, allowing electron tunneling. In these cases the permeability of the monolayer at different temperatures, surface concentrations, and surface pressures was inferred from the reduction of the polarographic current by the adsorbed monolayer [55–56] as shown in Fig. (22d). Since the monolayer is a potential dependent steric barrier, the monolayer penetration by the electroreactant may be considered to be a reaction preceding the electrode process while the back permeation of the product is a subsequent reaction. The permeability rate constant k_f is then obtained from (i/id) by Eqn. 108 or by the method of Brdicka and Koutecky. In addition, k_f for transport through charged lipid monolayers can be expressed in terms of absolute rate theory. The activation energy comprises the following terms:

(a) The elctrostatic energy $-ze\psi_0$ required for transferring the charged electroreactant (e.g. Cu^{2+}) into the plane of charges of the monolayer, where ψ_0 is the electric double layer potential of the electronic charge and z is the valency of the ion.

(b) The energy required to form a hole of area A_{act} in the monolayer. This energy is contributed by the compression work of the monolayer against the surface pressure $A_{act}\Delta\gamma$ and the work required to form a boundary line around the bare hole, $2\pi^{1/2}A_{act}^{1/2}\Omega$, where Ω is the line tension of this boundary. The line tension can be related to the cohesion forces between the molecules in the monolayer, or to the transient creation of a hydrocarbon-water interface (see section 2).

(c) The activation entropy S_{act}.

From these terms, k_f can be expressed as

$$k_f = \frac{kT}{\hbar} \exp\left(\frac{S_{act}}{R}\right) \exp\left[\frac{ze\psi_0 + A_{act}\Delta\gamma + 2\pi^{1/2}A_{act}^{1/2}\Omega}{\kappa T}\right] \qquad (109)$$

where k is Boltzmann's constant and $\hbar$ is Planck's constant. The validity of Eqn. (109) was checked for transport of Cu^{2+} across a charged decyl-ammonium monolayer. So long as the monolayers were not fully

---

Fig. 23. Log of the permeability constant Kc of adsorbed decylaminium monolayers towards Cu^{2+} plotted vs. 1/T ((a) and (b)) and against $\Delta\gamma$ (c) [56]. Plot of log k_0 against (1/T). a, in the presence of 0.04 N HNO_3 + 0.06 $NaNO_3$; b, in the presence of 0.04 N HNO_3 + 0.96 N $NaNO_3$; DA concentrations and polarizations as indicated. The natural logarithm of the transport rate constant ln k_c against pressure $\Delta\gamma$. Supporting electrolytes: ○, 0.01 N HNO_3; △, 0.04 N HNO_3 + 0.06 N HNO_3; ○, 0.04 N HNO_3 + 0.26 N $NaNO_3$; ▲, 0.04 N HNO_3 + 0.96 N $NaNO_2$.

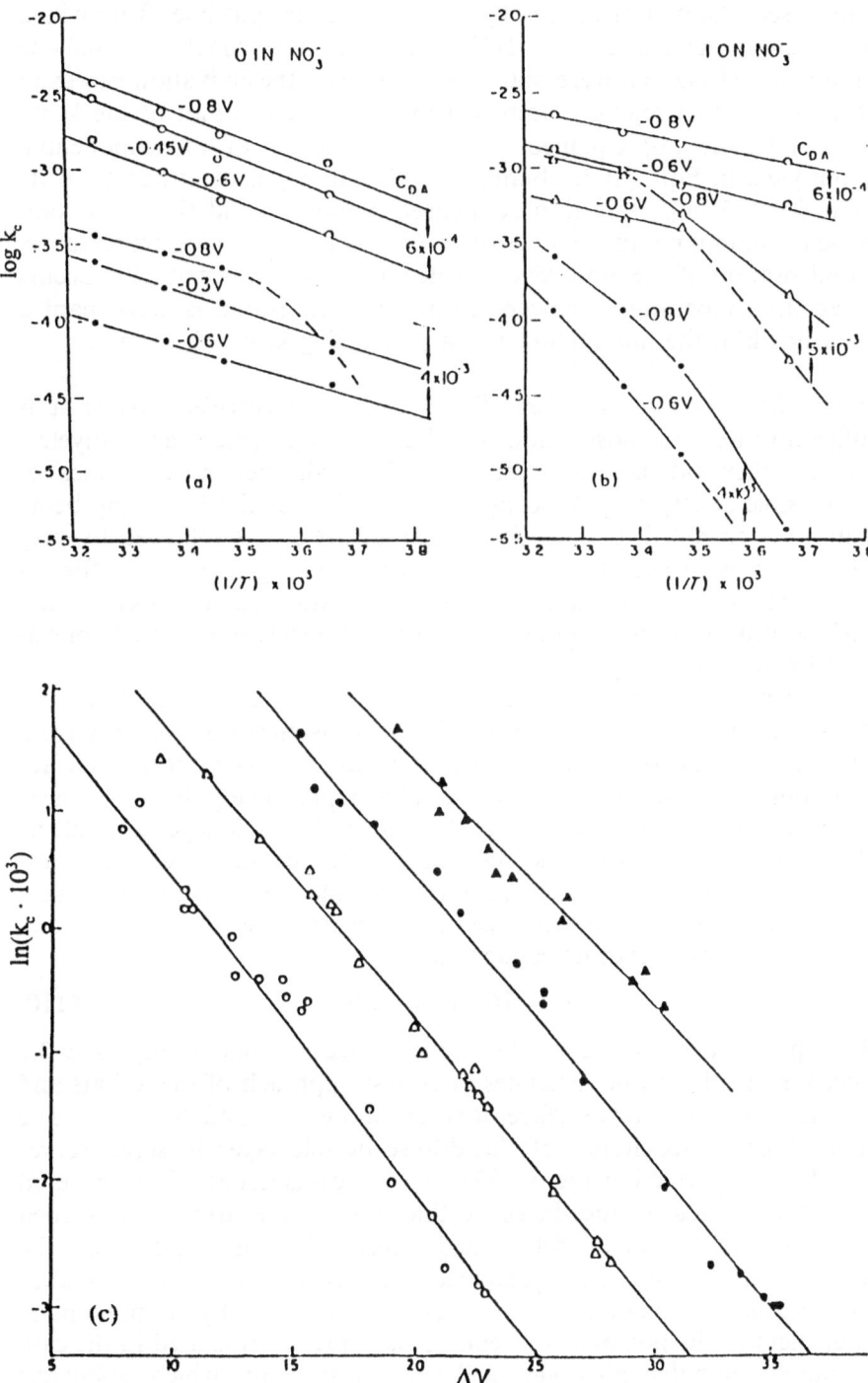

Fig. 23(a)–(c).

condensed, the plot of the log k_f vs. $\Delta\gamma$ gave a straight line, from which an $A_{act} = 80\ \text{Å}^2$ for $0.04\ M$ HNO_3 and an $A_{act} = 65\ \text{Å}^2$ for $0.04\ M$ $HNO_3 + 1\ M$ $NaNO_3$ were obtained. Moreover, the activation energy in this compression region obtained from an Arrhenius plot of log k_f vs. $(1/T)$ (Fig. 23a, b) equalled the first two terms of the exponential component in Eqn. (109) obtained by plotting log k vs. either ψ_0 or $\Delta\gamma$ (Fig. 23) [55, 56]. Only in a condensed monolayer did the third component containing the line tension Ω become of major importance. Condensation of the monolayer formed at a given concentration occurs when the temperature is lowered; this phase transition is accompanied by a break in the line obtained in a plot of log k_f vs. $(1/T)$ [56].

4.3.2. Adsorption of polymers, biopolymers, and polyelectrolytes: It is difficult to treat the adsorption of polymers, biopolymers, and polyelectrolytes thermodynamically, as they show all the characteristics of irreversible adsorption. Adsorption is very fast, at the beginning being diffusion-controlled, but the fast initial adsorption is followed by very slow changes in surface parameters corresponding to slow conformational changes as the monolayer becomes saturated. Desorption into pure solvent is nearly infinitely slow, so that adsorption can be considered irreversible.

As mentioned earlier, the measured capacitance reflects the properties of the Helmholtz layer, which extends to a distance up to 2–3 Å from the surface; thus, only the monomeric residues adhering to the surface can appreciably affect the measured capacitance. The polymer segment adsorbed on the surface displaces the electrical double layer and affects the overall capacitance. The resulting specific capacitance can be assumed to be the sum of the capacitances of the bare fraction of surface $(1 - \theta)$ with a specific capacitance C_0 and of the covered (θ) fractions of the surface with a specific capacitance C_p:

$$C = \theta C_p + (1 - \theta)C_0 \tag{110}$$

The specific capacitance of the bare surface at ionic strengths >0.1 depends mainly on the distances of closest approach of the anions and cations to the electrode surface; it varies between 15 and 40 $\mu f/cm^2$. The much higher capacitance of the diffuse double layer in series serves merely as a correction factor. The specific capacitance of the covered fraction depends on the dielectric thickness of the array of adsorbed residues, and cannot be determined directly. In some cases it can be assumed that the specific capacitance of the covered fraction is equal to the specific capacitance of the area fully covered by a monomeric equivalent of the polymer segments. This approach was used in the case of unionized polymethacrylic acid (at low pH), for which isobutyric acid was considered to be its monomeric equivalent [13]. The degree of coverage θ (Fig. 24) of the adsorption layer calculated under these

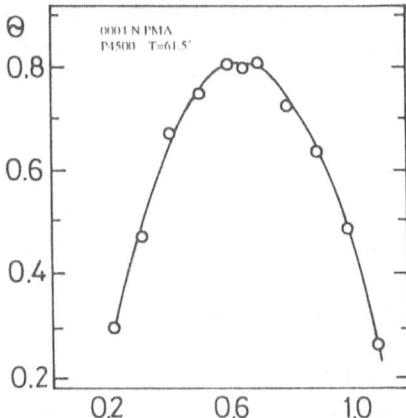

Fig. 24. Degree of surface coverage, θ, as a function of potential [13], with permission.

assumptions reaches a maximal value of 80–85% near the zero charge point and decreases with positive or negative polarization. Inferring the total surface concentration from the adsorption kinetics, assuming that adsorption is diffusion-controlled, one could evaluate what fraction of the adsorbed polymer molecule adheres to the surface and what fraction protrudes as loops from the surface trains into the bulk (see Fig. 25). The results showed that around the zero charge point ca. 25–40% of the molecule (depending on its molecular weight) adheres to the surface,

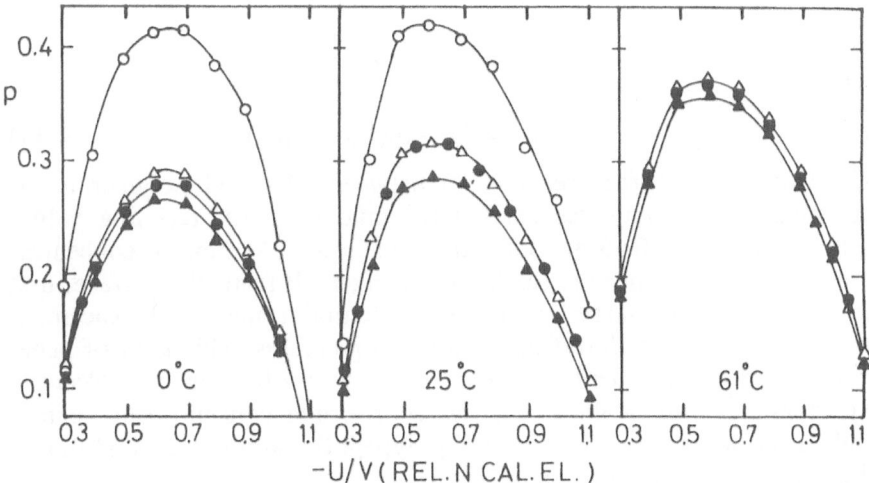

Fig. 25. Anchored fraction of polymethocrylic acid, p plotted against potential relative to N calomel electrode. ○ P = 2900, ● P = 3900, △ P = 5200, ▲ P = 9000 [15], with permission.

while the rest exists as loops. The values of p, just like the values of θ, decrease with electrode polarization, which causes a decrease in the effective standard free energy of adsorption.

The segments in a polymer cannot always be compared with their monomeric equivalents. Large monomeric units as such may be favorably accomodated in the surface but their orientation may be restricted when the monomers are bound in a polymeric chain. Monomeric amino acids have different chemical properties compared to their equivalents in a polypeptide chain. The situation becomes even more complicated for heteropolymers (which include most biopolymers), i.e., polymers composed of chemically-different monomeric units. In all these cases one can use a surface-active probe molecule which tends to adsorb to the surface and fill the loopholes between adsorbed polymeric trains. However, above a certain critical concentration the probe molecules displace the polymer from the surface. This was shown for poly-2-vinylpyridine (P2VP) and poly-4-vinylpyridine P4VP with hexanol as a probe [16]. Hexanol up to a concentration of 0.25% (w/v) lowers the capacitance of PVP-covered electrodes in a wide potential region (see figures 26(a) and (b)). From this decrease in capacitance the free loophole area fraction θ is calculated. The capacitance in the presence of the PVP but in the absence of hexanol, at a certain potential C_p, is

$$C_p = C_{pvp}\theta + C_0(1 - \theta)$$

where C_0 is the capacitance of the water-filled pores with free access of ions and C_{pvp} the capacitance of the surface fraction θ occupied by PVP segments. The capacitance of the hexanol solution in the absence of polymer is a function of its concentration and the electric potential. The capacitance $C_{p,Hex}$ in the presence of PVP and hexanol is

$$C_{p,Hex} = C_p(\theta) + C_{Hex}(1 - \theta)$$

Hence

$$C_p - C_{p,Hex} = (1 - \theta)(C_0 - C_{Hex}) \qquad (111)$$

The value of the free area $(1 - \theta)$ between adsorbed polymer trains obtained in this way decreases with ionic strength, reaching values between 0.18 and 0.15 at ionic strengths >0.2. At hexane concentrations $\geq 0.3\%$, w/v hexane starts displacing PVP from the surface, first around the zero charge point. The potential range of displacement increases as the concentration of hexanol increases. This kind of penetration and displacement can be observed with other probe molecules, e.g. pentanol and heptanol, in the appropriate concentration region, and with other polymers, including polypeptides and proteins, although the results have not been published.

One can also take the ionic permeability of a polymeric monolayer to be an indicator for the layer adhering to the surface, which is the most tightly-packed layer of the polymeric film. The permeability of mono-layers to electroactive molecules can be obtained from the direct polar-ographic current measured across the respective monolayer. It is very instructive to kinetically follow the change in permeability of the poly-meric monolayers during their adsorption [57]. This can be done on a single drop of the dropping mercury electrode, where the surface con-centration of the adsorbing monolayer increases with drop age, as seen in Fig. 27 which shows I-t curves across an adsorbed polylysine mono-layer at different salt concentrations. Above 0.12 M NaNO$_3$ adsorption is diffusion-controlled, this being immediate adsorption of molecules reaching the surface; hence above this salt concentration, surface satura-tion coinciding with the current minimum is reached at the same time for equal concentrations of polysine. The current then increases with continued expansion of the drop; the rate of further diffusion-controlled adsorption, changing with $t^{1/6}$, cannot compensate for the increase in surface area, changing with $t^{2/3}$. The current and hence the permeability of the positively-charged monolayer to Cd^{2+} increases with further increase of ionic strength. At ionic strengths $<0.1\ M$ NaNO$_3$ adsorp-tion of polylysine is slowed down, its equilibrium surface concentration decreases, and a larger fraction of the surface remains free of adsorbed segments and so is available for ion transfer as well as adsorption of

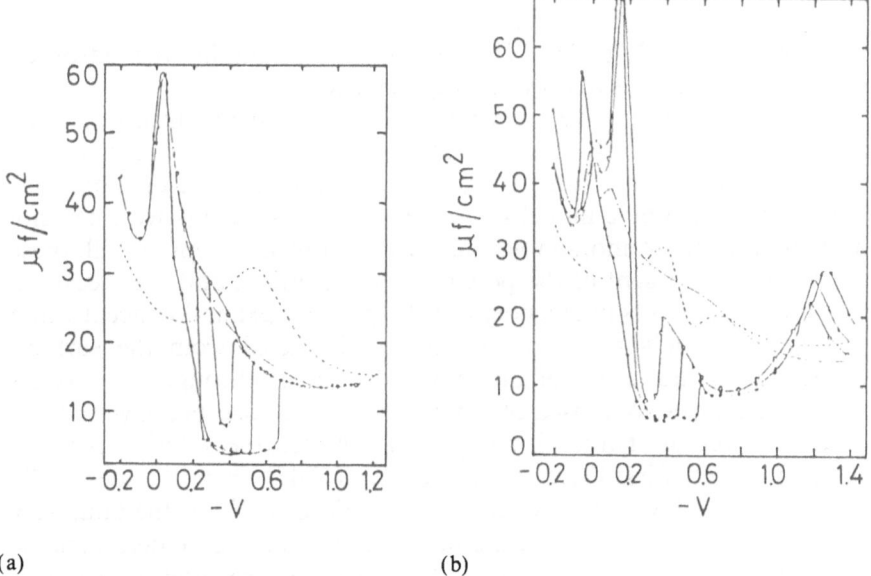

(a)   (b)

Fig. 26(a). Differential capacity curves: ($---$) in the presence of 0.5 N NaNO$_3$ + 0.015 N HNO$_3$; ($-\bullet-$) 0.005 N 4PVP added. Solid lines: various concentrations of hexanol added: ($-\bigcirc-$) 0.26%; ($-\triangle-$) 0.32%; ($-\blacktriangle-$) 0.42%; ($-\bullet-$) 0.55%. (b). Differential capacity curves ($---$) in the presence of 0.015 N HNO$_3$, ($-\cdot-\cdot-$) 0.005 N 4PVP added. Solid lines: various concentrations of hexanol added: ($-\bigcirc-$) 0.3%; ($-\triangle-$) 0.4%; ($-\cdot-$) 0.55% [16].

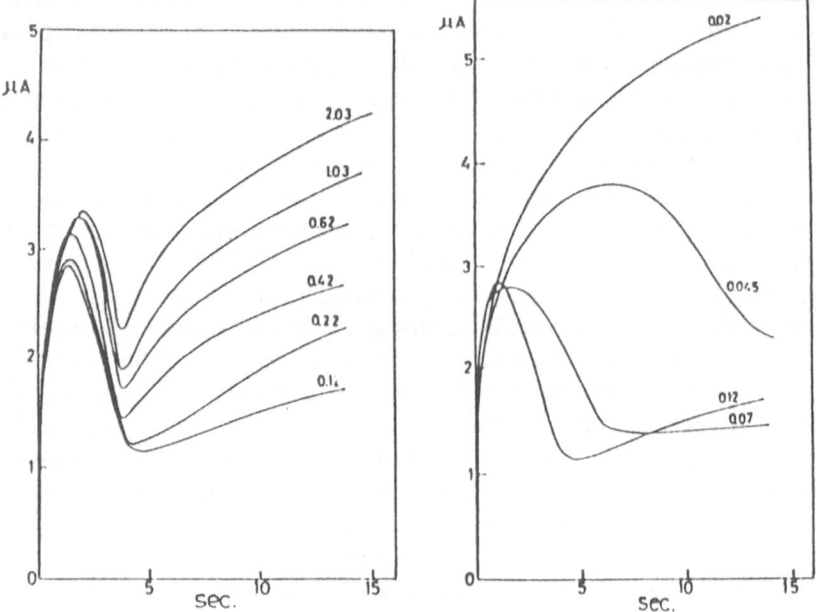

Fig. 27. Current-time (i − t) curves of Cd²⁺ in the presence of $6 \times 10^{-4} N$ polylysine (degree of polymerization, 400) at different concentrations of nitrate [59], with permission.

probe molecules (Fig. 26). The current inhibition by the adsorbed monolayer diminishes and eventually vanishes.

An adsorbed monolayer serving as a barrier to ion transport may allow ion transport through fluctuating pores as in the case of monomeric monolayers (see Eqn. 109), or through a network of loopholes. The constant loophole model fits the experimental results with hexanol or with any other monomolecular probe. Probes seem to fill these loopholes independent of the polymer concentration, and the measured total loophole area is independent of the probe (hexanol) concentration as long as the adsorbed polymer is not displaced from the surface. Modelling the transport of an ionic depolarizer through a polyelectrolyte monolayer composed of a monomolecular adsorbed layer and of a surface layer of charged loops protruding into the solution is not a simple problem. First there is a depletion of Cd²⁺ from the identically charged surface layer. The concentration ratio of Cd²⁺ at the boundary between the surface layer and the bulk solution should be determinable by the Donnan equilibrium. At ≤0.1 M salt electrostatic repulsion is the main factor lowering the rate of access of Cd²⁺ to the electrode. The electrostatic repulsion decreases with increasing salt concentration, and >2 M salt only the barrier of the trains of segments covering the surface remains. Access to the surface then proceeds through the loopholes constituting only a small fraction $(1 - \theta)$ of the surface. However the relative current i/i_d is much larger than $(1 - \theta)$, since radial diffusional fluxes towards the loopholes are much larger than unidimensional diffusional fluxes.

4.3.3. Adsorption on reversible electrode surfaces in the presence of a high current density across the interface: In all the experiments hitherto described the permeant ions reduced on the surface are co-present with a large excess of inert electrolyte. The electrode is polarized and no current besides the relatively small depolarization current crosses the interface. The stability of the monolayer is a function of an electrostatic field emanating from the surface charge, but the currents flowing through the monolayer are always too low to have any appreciable effect on its stability. The situation changes when there is no supporting electrolyte or when the supporting ions themselves participate in the

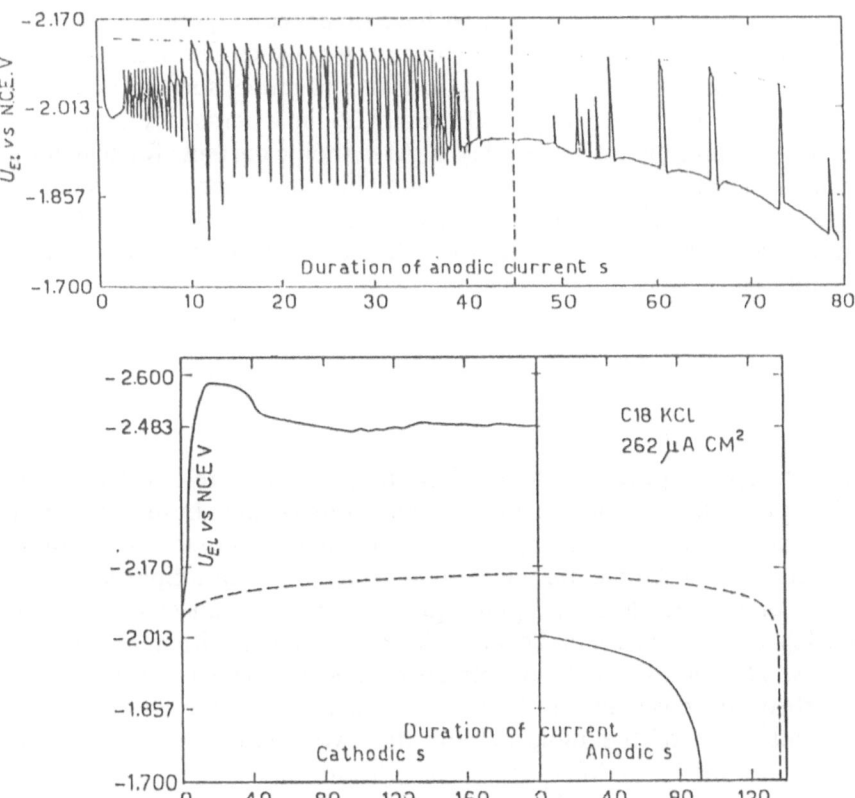

Fig. 28. Electrode potential vs. duration of the cathodic and anodic currents (260 mA/cm²) in the presence of octadecyltrimethylammonium chloride (3.35×10^{-4} M), 0.1 N KCl 25.0°C. Upper square: oscillation starting at higher anodic current density; dashed lines: equilibrium amalgam electrode potentials [60], with permission.

electrode process. For the reduction current of Na^+ to sodium amalgam and its oxidation current back across surface layers of long chain ($\geq C_{18}$) trimethylammonium, potential drops of up to 400 mV are observed across these surface barriers. Similar potential drops are obtained with oxido-reduction currents of halides across long chain fatty acid or ammonium monolayers adsorbed onto the surface of reversible calomel or Ag/AgCl electrodes [60]. At low current densities the potential is proportional to the current giving specific resistances of the order of 10^3–10^4 Ω cm^2. The potential increase then diminishes until a limiting value is reached. Under this potential drop across the surface layer, the monolayer may become unstable and break down with concomitant abolition of the potential across the surface region. This allows reconstruction of the surface layer with subsequent potential build-up, the result of which is a potential oscillation at a constant current (Fig. 28). Surface layers of long-chain quaternary ammonium compounds formed on highly negatively-charged amalgam surfaces have capacitance values of around 0.8 μf/cm^2, corresponding to bilayers. The transient surface layer breakdown which induces potential oscillations at anodic currents when the negative surface charge is relatively low starts at a potential drop of 200 mV across the long-chain quaternary ammonium surface layer. At cathodic currents for which the amalgam electrode potential is considerably more negative, potentials higher than 400 mV do not impair the stability of the positively-charged surface layer. Bilayer adsorption on charged hydrophilic surfaces may possibly be of general biological importance.

4.4. Adsorption on lipid monolayers, lipid bilayers, and biological membranes

The adsorption processes to be described are quite relevant to membrane processes, be they membrane reactions or membrane transport, for in both cases the primary step is access and eventual adsorption of a reactant onto the membrane. The interface between a lipid layer and water can be considered a liquid/liquid interface. A molecule adsorbing at such an interface can be partially immersed in either of the two phases. The relative depth of immersion and the density profile of the adsorbed molecules across the interface depends on the affinity of the different parts of the molecule for the two phases. A phospholipid layer/water interface differs from an oil/water interface by the array of polar groups bounding the oil layer. A molecule adsorbing from the aqueous phase may only interact with the polar boundary, or it may partially penetrate this boundary and also interact with the hydrocarbon chains. Moreover, the adsorbed molecule may perturb the planar structure of the interface, co-immersing some of the lipid polar groups with it. The final configuration of the perturbed phospholipid layer/water interface will be that which minimizes the free energy of the system. The hydrophobic amino acids of a protein molecule will extend across the boundary from the aqueous to the hydrocarbon environment. To lower further the free energy these protein molecules will assume a conformation in which the hydrophilic surfaces of the residues immersed in the phospholipid layer will be shielded, while the hydrophobic surfaces will be exposed to the hydrocarbon medium. An α-helical conformation enables a protein to fulfil this requirement. However, even the surface of an α-helix may not be completely oil-soluble if there is a hydrophilic amino acid within the sequence forming the α-helix or if the hydrophobic residues are not large enough to completely screen a hydrophilic polypeptide chain. To overcome this, additional interactions may be required. Either a polar group from the phospholipid array may bend over to interact with the exposed polar spot on the α-helix, or a group of α-helices could stick together with their dipoles in a parallel or antiparallel orientation normal to the plane of the lipid bilayer. Such an aggregate of helices may undergo polar and hydrophobic interactions in order to render the exterior of the aggregate hydrophobic and oil-soluble, with the possible formation of a hydrophilic pore between the α-helical cylinders (Fig. 29). Such a hydrophilic pore may serve as a channel for the selective transport of ions. Enzyme systems at the mouth of such a channel, together with selective transport, are responsible for biological processes proceeding *across* the biological membrane. These processes usually involve active proton or ion transport, coupled to chemical or photochemical reactions. The resultant proton or ion gradients and their corresponding membrane potentials become the energy source for, e.g., oxidative phosphorylation and macromolecular synthesis. Membrane proteins, with their helical structures spanning the membrane, in fact do not become intercalated into the membrane by adsorption from the aqueous solution; rather they are synthesized near the membrane and are partly incorporated into and partly translocated across it. All cytosolic proteins have to be translocated across the membranes of the organelles in which they have been synthesized; however, these translocation reactions proceed via adsorbed states. (At the same time, mechanisms are involved which are specific to particular biological systems, and are not within the scope of this discussion.) In the following, attention will be focussed on adsorption of proteins and other biomolecules from aqueous solution onto lipid monolayers and bilayers, especially the conformational changes which may be important for subsequent biochemical reactions.

4.4.1. Protein adsorption onto lipid monolayers and their penetration by proteins: Adsorption of proteins and their incorporation into lipid monolayers has attracted the attention of surface chemists since the late

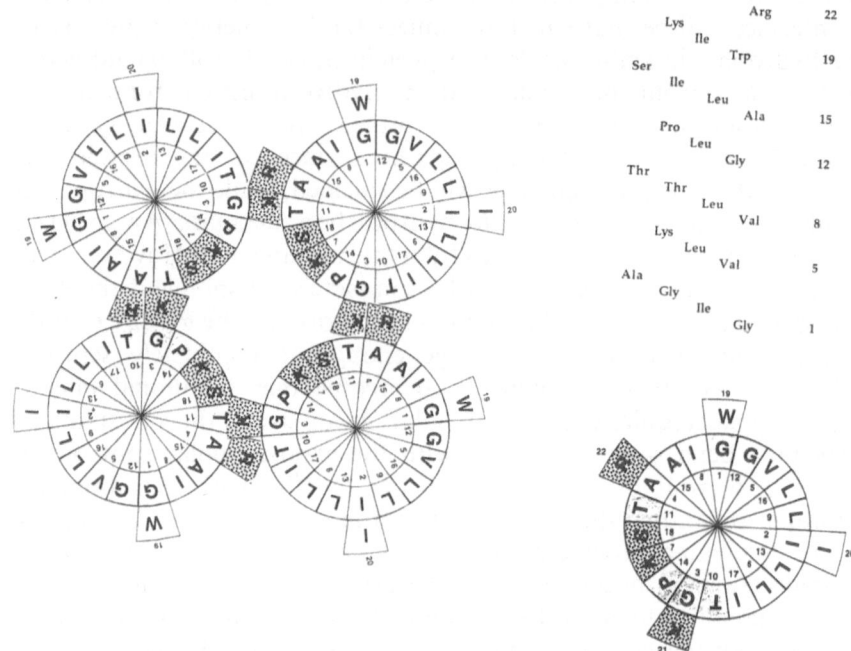

Fig. 29. Schematic illustration of four melittin helices, represented in wheel diagrams, forming a channel. The three different shadings represent the hydrophobicity of the peptides on the α helix surface: white hydrophobic, gray shade intermediate, dotted hydrophilic. The two diagrams on the side show a wheel diagram of a single helix and amino acids along the helix. The amino acids sticking out on the diagram are either in the helix or as in the case of lysine (21) and arginine (22) protruding into the aqueous phase.

1940s when two publications by Schulmann and co-workers appeared [61, 62]. They measured changes in surface pressure and surface potential of lipid monolayers compressed to a given initial surface pressure when different concentrations of proteins were injected into the aqueous solution underneath the monolayer. Today, with clamped surface pressure devices, increases in surface area at constant surface pressures can also be measured. There were attempts to thermodynamically derive the surface concentrations of interacting protein molecules from the changes in surface tension as a function of changes in the chemical potential of the added protein [63]. The results obtained were dubious because: (1) protein adsorption is usually irreversible and thermodynamic equilibria cannot be maintained; and (2) change in the surface concentration of a protein also changes the chemical potential of the lipid monolayer which cannot easily be assessed in this case. Thermodynamic treatment *can* be utilized in the case of interaction between lipid monolayers and oligopeptides such as the neuropeptide substance P [64].

Nonetheless, *qualitative* information can be derived in all of these cases. The increase in surface pressure of a lipid monolayer following injection of a protein to values above those obtained for the same protein monolayer adsorbed at a pure water surface is indicative of a positive interaction between the protein and the lipid monolayer. Similar conclusions can be reached from observations on the increase in surface area of lipid monolayers held at constant surface pressures when proteins are injected underneath them; the increase in area is ascribed to the area of the penetrating protein molecules, and their concentration in the surface is calculated after assuming some specific area per penetrating molecule at the given pressure. These approaches have been used to study interactions between lipids and their apolipoproteins as well as antibodies [63, 64], or with proteins interacting catalytically with lipid surfaces [65, 66]. These types of phospholipid penetration experiments were also done using proteins and their putative signal sequences to try to understand the vectorial transport of proteins through membranes after their synthesis [67].

For more quantitative evaluation of the interaction of proteins with lipid monolayers, the surface concentration of the interacting proteins has to be determined; this cannot be derived by thermodynamic arguments from the change in surface pressure. Even in cases in which thermodynamic equilibria can be achieved, the interaction may involve additional parameters which complicate the treatment. If charged lipid monolayers and protein are involved, the effect on the electrochemical potentials of the counter ions and on their surface excesses has to be considered.

At low ionic strengths addition of a positively-charged biopolyelectrolyte underneath a negatively-charged monolayer brings about a simultaneous increase in the chemical potential of the neutral salt having one ion identical with the counterion of the added biopolyelectrolyte. Adsorption of this polyelectrolyte while simultaneously interacting with the oppositely-charged monolayer induces monolayer counterion desorption, leaving behind in the solution an equivalent amount of polyelectrolyte counterions. Thus, concomittant with adsorption of the biopolymer many counterions will be desorbed, with simultaneous increase in their electrochemical potential which may result in a decrease instead of an increase in the surface pressure. It is clear then, that the direct determination of surface concentration is essential for quantitative evaluation of the interaction between proteins and lipid monolayers. In such cases, the methods used for surface concentration determinations are identical to those described in Section 4.1. Comparing the directly-determined surface concentration [32, 33] of an interacting protein at a given surface pressure of the lipid monolayer with the increase in monolayer area at this surface pressure, one obtains a fair estimate of the degree of penetration of the adsorbed proteins into the lipid layer. However, proteins of interest, such as apolipoproteins, tend to solubilize the lipids to form bulk complexes. Such proteins, when interacting with a lipid monolayer, may extract from the monolayer array some lipid molecules and replace them with some of their own segments; in this case the degree of penetration is higher than that inferred from the increase in surface area. It is therefore crucial to determine any depletion of lipid molecules from the surface that may occur during protein adsorption. This can be determined by measuring the decrease in surface radioactivity of labelled lipid molecules in the monolayer. Alternatively, optical methods like fluorescence can be used together with surface radioactivity for simultaneously monitoring protein adsorption onto and lipid desorption from the monolayer. Displacement of lipid from the monolayer by adsorbing protein molecules can occur when the lipid monolayer is at a high surface pressure under constant pressure conditions. Therefore, a change in surface pressure may not be a good indicator for monolayer penetration under these conditions.

Direct measurement of lipid monolayer penetration by polar molecules can be performed at the interface between electrodes and aqueous solutions. The mercury electrode is an example of a polarizable electrode, as it has a high overvoltage and its interface with water can be polarized to high negative potentials without appreciable hydrogen reduction current; moreover, it gives a well-defined liquid/liquid interface which allows determination of all the relevant thermodynamic parameters, including the differential capacitance. As mentioned earlier, the differential capacitance is sensitive to the Helmholtz layers determined by the closest access of ions to the surface. Any adsorbed or spread and deposited monolayer affects ion access to the surface and thus the capacitance. The simplest way to deposit a well-defined lipid monolayer at the mercury/water interface is to transfer it from the air/water interface [68].

Penetration of lipid monolayers by polar molecules can be investigated by probing the dielectric properties of the monolayer. A condensed lipid monolayer displaces the nearest access of small ions from the electrode surface by a distance determined by the thickness of the hydrocarbon layer. For a hydrocarbon layer with a thickness of 1.2 nm and a dielectric constant of 2.2, the calculated capacitance should be ca. $1.5\ \mu f/cm^2$. The minimal capacitances measured for condensed monolayers of different phospholipids varied between 1.4–$2.0\ \mu f/cm^2$, the precise value depending on the chain lengths and on the head groups. Among phospholipid diesters of identical hydrocarbon chain lengths, phosphatidylethanolamine gives the lowest differential capacitance values [69]. The capacitance increases as the monolayers are expanded or penetrated by interacting polar molecules. If a polar molecule displaces lipid molecules from fraction θ of the surface, the resulting capacitance

C_t (assuming parallel combination of the capacitances of the lipid and penetrant-covered fraction) will be:

$$C_t = (1 - \theta)C_m + \theta C_p$$

In the case of penetration of the polar layer, accompanied only by an expansion of its area and a decrease in hydrocarbon layer thickness from δ_0 to δ,

$$C_t = \frac{\delta_0}{\delta} C_m = \frac{A + \Delta A}{A} C_m \cong C_m/(1 - \theta_p) \tag{112}$$

$1 - \theta_p$ is defined as the fraction of the polar layer area occupied by lipid polar groups. It is evident that penetration of the hydrocarbon layer has a much larger effect on capacitance than penetration of the polar layer only, particularly if the penetrating molecule is very polar, causing an increase in the local dielectric constant.

If the penetrating molecule or some of its penetrating residues are electroactive, a faradaic admittance component is introduced. When the electrode reaction induced by this component is reversible, a pseudo-capacitance peak at the reversible redox potential is observed. Cystine is an electroactive component of most protein molecules; it can undergo electrode reactions only when it is in close contact with the electrode surface. Even small adsorbed protein molecules like RNAse may have only half of their cystines reduced, even though the minor axis diameter of this ellipsoidal molecule is only 29 Å and probably smaller when flattened in the adsorbed state. For larger molecules, e.g. prothrombin, even fewer than half of the cystines may be reduced electrodically [70]. It seems that the electron-tunnelling distance required for reduction of a cystine disulfide bond has to be ≤ 10 Å. Consequently, in order to reduce a cystine when adsorbed on a lipid monolayer, that cystine has to cross the hydrocarbon layer. Cystine and cysteine are endogenous electroactive probes on protein molecules, but different electroactive probes can be attached to different residues on a protein molecule and used to map the penetration tendency of different sections of the molecule. Of course, the attached electroactive probes may modify the penetration tendency, and due caution is required when working with attached probes.

As shown in Fig. 30, the pseudocapacitance peak due to the cystine-cysteine redox reaction across a lipid monolayer is accompanied by an overall increase in capacitance [71], indicating that a large fraction of the protein molecule has to penetrate the monolayer to permit the electrode reaction. The exact redox potential of cystine depends on both pH and on its closest neighbors. It may happen, as in the case of phosphokinase C, that the different cystine residues of a protein are reduced at different potentials [72]. The tendency of different cystine residues to penetrate lipid monolayers, as manifested by their pseudoca-pacitance peaks across the monolayer, may differ as well.

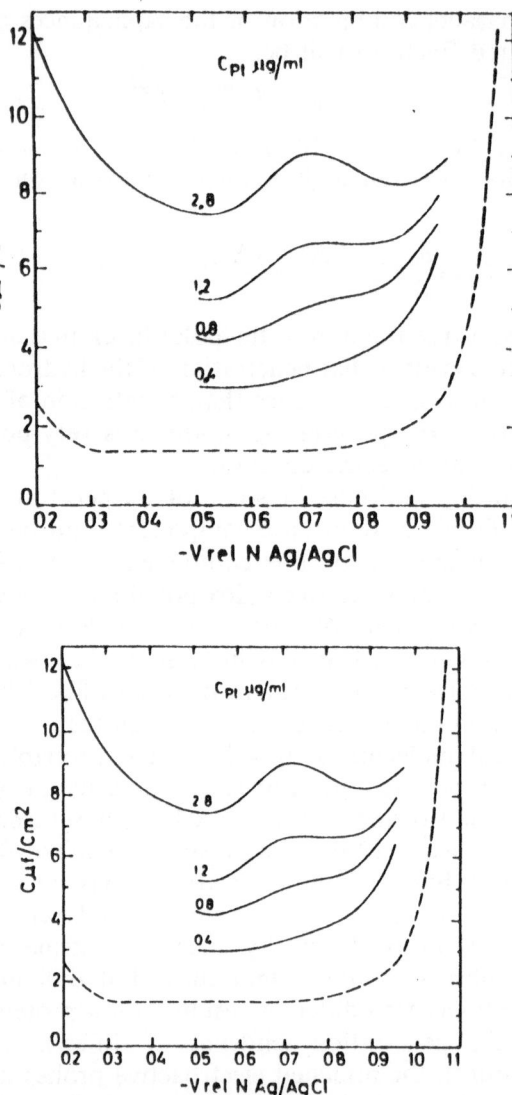

Fig. 30. Differential capacity of a condensed phosphatidylserine monolayer, with different concentrations of prothrombin injected underneath at 0.5 mM Ca^{2+}, as a function of the mercury electrode potential. The dotted line represents the phosphatidylserine monolayer in the presence of Ca^{2+} [71], with permission.

Electrode studies provide information on the penetration and overall effect of proteins on lipid monolayers. To determine the surface concentration of the interacting proteins, the surface layers into which they have penetrated have to be isolated and their protein contents determined independently [68].

Adsorption of proteins on lipid bilayers can be conveniently measured using dispersions of monolayer vesicles employing the usual techniques of separating the dispersed phase from the bathing solution and then determining the protein concentration in one or in the two phases. However, the main interest in studying interactions between proteins and lipid bilayer membranes is the effect on the protein conformation and on the final structure of the protein-containing membrane. To investigate the change in conformation of protein molecules arising from interaction with lipid vesicles of different compositions, circular dichroism, IR, Raman, and NMR measurements can be utilized, all of which provide information on the secondary structure of proteins. In most cases it has been shown that when a polypeptide or a protein section is transferred into a lipid bilayer [73, 74], an α-helical conformation of the type shown in Fig. 29 which shields hydrophilic residues from the hydrocarbon environment, is favored. In many cases, e.g. melittin, aggregates of α-helices exist in aqueous solution, wherein the hydrophobic surfaces are screened by hydrophilic ones which are exposed to the aqueous environment. Upon interaction with a lipid bilayer membrane, the helical conformation is retained; only their stacking geometry is rearranged to expose hydrophobic sites for interaction with the hydrocarbon core of the bilayer [75, 76].

The most important parameters characterizing protein-lipid membrane interactions are related to the functions performed by a protein in the membrane. One such function can be ion transport induced by the protein, whether it be an ion carrier or an ion channel former. A channel formed by different sections of one protein or of several proteins has the general structure depicted schematically in Fig. 29. Induction of ion transport can be easily investigated in bilayer vesicles. The rate of release of small ions or large molecules can be determined by monitoring the increase in their concentration in the outer solution ("extravesicular space"). If the effluxing ions are electroactive their concentration in the outer solution can be determined polarographically or voltametrically, as these methods do not detect ions entrapped in vesicles. If the entrapped molecules are fluorescent and if their fluorescence is quenched at high concentrations, then their release can be monitored by measuring the increase in fluorescence. Only reasonably large channels or pores permit passage of relatively large fluorescent molecules. The release of small ions can be at least qualitatively determined by measuring the short-circuiting of membrane potentials arising from an ion gradient, e.g. K^+ in the presence of valinomycin. The short-circuiting of membrane potentials is monitored by measuring the fluorescence of a potential-sensitive dye [77].

The properties of channels formed from the channel-forming polypeptides or proteins can be investigated in detail on small plasma membranes. In such membranes the rates of opening and closing of single channels and their conductance can be measured.

This chapter on adsorption is an attempt to outline the general principles of adsorption of monomeric and polymeric substances including biopolymers of specific structures; to describe ways of measuring adsorption and the structural changes accompanying adsorption; and to point out some of the biological implications of conformational changes related to adsorption.

Conclusions

Most biological processes take place at interfaces. It is not easy to determine the thickness of an interface which can be correlated with the normal extension of the surface forces [78] and therefore one cannot say what volume fraction of a biological tissue consists of interfaces. The interface concentration in colloid systems and in dispersed adsorbents is defined as area per unit volume. Using this definition the area concentration in an erythrocyte, which may be considered to be a bag filled by haemoglobin, is about $1 \, m^2$ per cm^3 – disregarding the area of the protein molecules, which plays an important role in oxygen binding. In cells which contain different organelles like nuclei with ribosomes, mitochondria, chloroplasts, golgi, etc., the area concentration may reach values of $10 \, m^2/cm^3$ or even higher. The mean distance $\langle x \rangle$ between interfaces across the cytosolution may be of the order of 10 nm and the diffusional contact frequency $(D/\langle x \rangle^2)$ of different cytosolic molecules with surface boundaries may vary between orders of magnitude of 10^3 and $10^6 \, sec^{-1}$.

Most of the interfaces are membranal and they are also the most important ones as they are the adsorption sites of reagents involved in membrane reactions and intercompartmental transport. There are also surfaces of fibrillar substances and globular particles. Not all the surface phenomena related to biological processes can be investigated in vivo or in situ. However, the basic laws of nature are the same and conclusions from investigations of well defined artificial systems can be transferred to dynamic biological systems which elude the same kind of investigation. The artificial model systems are usually better suited for physico-chemical investigation, not only of interfaces of certain well defined composition but also of some specific receptors which can be isolated and embedded into a suitable surface. This line of research will also continue in the future, probably using more sensitive methods which will allow more accurate determination of interfacial dynamic structures and interfacial reaction kinetics. All this effort will continue to be invested in order to achieve a better understanding of the structure-function interrelation of biological interfaces. This endeavour will ultimately lead or at least contribute to an elucidation of the mechanisms of some of the important life processes.

References

1. FM Fowkes, J. Physical, Chem. 67 (1963) 2541–2583.
2. J Frenkel, *Kinetic Theory of Liquids*, Clarendon Press, Oxford, 1947.
3. I Langmuir, Chem. Rev. 13 (1933) 147–191.
4. HT Tien, J. Phys. Chem. 71 (1967) 3395–3401.
5. WD Harkins and FE Brown, J. Am. Chem. Soc. 41 (1919) 499–524.
6. AW Adamson, *Physical Chemistry of Surfaces*, J. Wiley, 1990.
7. KS Cole, J. Cellular Comp. Physiol. 1 (1932) 1–16.
8. DC Grahame, Chem. Rev. 41 (1947) 441–501.
9. JThG Overbeek, in *Colloid Science*, HR Kruyt (ed.), 1953, Vol. 1, p. 113 (see p. 162).
10. AN Frumkin, Z Physik 35 (1926) 792–802.
11. R Defay, I Prigogine, A Bellemans and DH Everett, *Surface Tension and Adsorption*, Longman, London, 1966, Chap. 13.
12. G Ash, DH Everett and GH Findenberg, Trans. Faraday Soc. 64 (1968) 2639–2644.
13. IR Miller and DC Grahame, J. Am. Chem. Soc. 78 (1956) 3577–3585.
14. IR Miller and DC Grahame, J. Colloid Sci. 16 (1961) 23–40.
15. IR Miller, Trans. Farad, Soc. 57 (1961) 301–311.
16. IR Miller, J. Polymer Sci. C16 (1967) 1433–1444.
17. A Takahashi and A Kawakuchi, Advances in Polymer Science 46 (1982) 1–65.
18. B Vincent and SG Whittington, Surface Colloid Sci. 12 (1982) 1–113.
19. I Medalia, in *Surface Science*, E Matijewic, (ed.), J. Wiley, New York, 1971, Vol. 4, pp. 1–92.
20. HL Frish, R Simha and FR Eirich, J. Chem. Phys. 21 (1953) 365–366.
21. HL Frish, J. Phys. Chem. 59 (1955) 633–636.
22. WI Higuchi, J. Phys. Chem. (1961) 487–491.
23. WC Forsman and RE Hughes, J. Chem. Phys. 38 (1963) 2118–2123.
24. R Roe, Proc. Natl. Acad. Sci. USA 53 (1965) 50–57; 56 (1966) 819–824; J. Chem. Phys. 43 (1951) 1591–1598.
25. A Silberberg, J. Phys. Chem. 66 (1962) 1872–1883, 1884–1907.
26. A Silberberg, J. Chem. Phys. 46 (1967) 1105–1114.
27. A Silberberg, J. Chem. Phys. 48 (1968) 2835–2851.
28. G. de Gennes, Macromolecules 14 (1981) 1637–1644.
29. PG de Gennes, *Scaling concepts in polymer physics*. Cornell University Press, Ithaca, New York, 1979.
30. S Lifson and A Katchalsky, J. Poly. Sci. 13 (1954) 43–55; A Katchalsky, Z Alexandrow-icz and O Kedem, in *Chemical Physics of Ionic Solutions*, BE Conway and RG Barradas (eds), J. Wiley & Sons, 1966, p. 295.
31. GS Manning, J. Chem. Phys. 51 (1969) 924–933, 934–938.
32. M. Muramatsu, in *Radioactive Tracers in Surface and Colloid Science*, E Matijevic (ed.), J. Wiley & Sons, 1973, Vol. 6, pp. 101–184.
33. IR Miller, M Teuber and Y Accad, Biopolymers 17 (1978) 1513–1521.
34. MA Frommer and IR Miller, J. Coll. Interface Sci. 21 (1966) 245–251.
35. M-F Le Compte, IR Miller, J Elion and R Benarous, Biochemistry 19 (1980) 3434–3439.
36. G Kortüm, *Reflectance Spectroscopy*. Springer Verlag; Heidelberg 1969.
37. S. Gottesfeld, in *Electrochemistry. I Electroanalytical Chemistry – A Series of Advances*. A Bard (ed.), Marcel Dekker Inc, 1969, Vol. 15, pp. 143–265.
38. RH Muller and JC Farmer, Rev. Sci. Instrum. 55 (1984) 371–374.
39. CT Chen and BD Cahan, J. Electrochem. Soc. 129 (1982) 17–26.
40. MR Deakin and DA Buttry, Anal. Chem. 61 (1989) 1147–1154.
41. GZ Sauerbrey, Z. Phys. 155 (1959) 206–222.
42. BJ Fontana and JR Thomas, J. Phys. Chem. 65 (1961) 480–487.
43. E Killman, J. Eisenlaner and M Korn, J. Polymer Sci. 61 (1977) 413–430.
44. JC Day and ID Robb, Polymer 21 (1980) 408–412.
45. KK Fox, CD Robb and R Smith, J. Chem. Soc. Faraday Trans I 70 (1974) 1186.
46. OE Ohrn, J. Polym. Sci. 19 (1956) 199–200.
47. Z. Priel and A. Silberberg, J. Polym. Sci., Polymer Physics Edition, 1978, Vol. 16, pp. 1917–1928.
48. R Varoqui and D Dejardin, J. Chem. Phys. 66 (1977) 4395–4399.

49. MJ Garvey, TF Tadros and B Vincent, J. Coll. Interface Sci. 55 (1976) 440–453.
50. JN Israelachvili and GE Adams, J. Chem. Soc. Faraday Trans. I. 74 (1978) 975–1001.
51. I Klein, Y. Almoganol and P Luckham, ACS Symp. Series 240 (1984) 227–244.
52. M Blank and IR Miller, J. Coll. Interface Sci. 26 (1968) 26–33.
53. L Lipovski, J, CI Buess-Herman, JP Lambert and L Gierst, J. Electroanal. Chem. 202 (1986) 169.
54. KJ Vetter and JW Schultze, Ber. Bunsenges. Phys. Chem. 77 (1973) 945.
55. IR Miller and M Blank, J. Coll. Interface Sci. 26 (1968) 34–40.
56. IR Miller and H Great, Electrochim. Acta 15 (1970) 1143–1154.
57. P Delahay, *New Instrumental Methods in Electrochemistry*. Interscience, 1954.
58. R Brdicka and J Koutecky, J. Am. Chem. Sci. 76 (1954) 907–908.
59. IR Miller, Y Frei and D Bach, J. Polymer Sci. C16 (1969) 4483–4491.
60. B Masters and IR Miller, Bioelectrochem. Bioenerg. 1 (1974) 466–477.
61. P Doty and JH Schulmann, Disc. Faraday Soc. 6 (1949) 21–26.
62. R Matalon and JH Schulmann, Disc. Faraday Soc. 6 (1949) 27–39.
63. G Camejo, G Colacico and MM Rapport, J. Lipid Res. 9 (1961) 562.
64. A Seelig and PM Macdonald, Biochemistry 28 (1989) 2490.
65. MC Philips and KE Krebs, Meth. Enzymol. 128 (1986) 387–403.
66. MD Bazzi and GI Nelsestuen, Biochemistry 37 (1988) 6776–6783.
67. GD Fidelio, BM Austen, D Chapman and JA Lucy, Biochem. J. 244 (1987) 295.
68. RE Pagano and IR Miller, J. Cell Interface Sci. 45 (1973) 126–137.
69. IR Miller, J. Membr. Biol. 101 (1988) 113–118.
70. MF Lecompte, J Clavilier, C Dode, J Elion and IR Miller, J. Electroanal. Chem. 163 (1984) 345–362.
71. MF Lecompte and IR Miller, Biochemistry 19 (1980) 3439–3446.
72. DS Lester, L Doll, V Brumfeld and IR Miller, Biochim. Biophys. Acta 1039 (1990) 33–41.
73. D Bach, K Rosenheck and IR Miller, Eur. J. Biochem. 53 (1973) 265–269.
74. V Rizzo, S Stankowski and G. Schwarz, Biochemistry 26 (1987) 2751–2759.
75. H Vogel and F Jähnig, Biophys. J. 50 (1986) 573–582.
76. B Stanislawski and H Ruterjans, Eur. Biophys. J. 13 (1987) 1–12.
77. Y Shai, D Bach and A Yanovsky, J. Biol. Chem. 265 (1990) 20202–20209.
78. AI Rustanov, in *Surface & Membrane Science*. JF Danielli MD Rosenberg and DA Cadenhead (eds.), Academic Press, 1971, Vol. 4, pp. 57–114.

Subject Index

Bioelectrochemistry of Cells and Tissues

Bioelectrochemistry: Principles and Practice Vol. II

Edited by
D. Walz, *University of Basel, Switzerland*
H. Berg, *Institute of Molecular Technology, Jena, Germany*
G. Milazzo†, *formerly Istituto Superiore di Sanità, Rome, Italy*

1995. Approx. 380 pages. Hardcover.
ISBN-13: 978-3-0348-7320-8

Bioelectrochemistry: Principles and Practice **provides a comprehensive compilation of all the physicochemical aspects of the different biochemical and physiological processes.**

The role of electric and magnetic fields in biological systems forms the focus of this second volume in the *Bioelectrochemistry* series. The most prominent use of electric fields is found in some fish. These species generate fields of different strengths and patterns serving either as weapons, or for the purpose of location and communication. Electrical phenomena involved in signal transduction are discussed by means of two examples, namely excitation-contraction coupling in muscles and light transduction in photoreceptors. Also examined is the role of electrical potential differences in energy metabolism and its control. Temporal and spatial changes of the potential difference across the membranes of nerve cells are carefully evaluated, since they are the basis of the spreading and processing of information in the nervous system. The dielectric properties of cells and their responses to electric fields, such as electrophoresis and electrorotation, are dealt with in detail. Finally, the effects of magnetic fields on living systems and of low-frequency electromagnetic fields on cell metabolism are also considered.

Further volumes will be added to the series, which is intended as a set of source books for graduate and postgraduate students as well as research workers at all levels in bioelectrochemistry.

Birkhäuser Verlag • Basel • Boston • Berlin